Sustainable Development Dimensions and Urban Agglomeration

*Edited by Alessandra Battisti
and Serena Baiani*

Published in London, United Kingdom

IntechOpen

Supporting open minds since 2005

Sustainable Development Dimensions and Urban Agglomeration
http://dx.doi.org/10.5772/intechopen.94779
Edited by Alessandra Battisti and Serena Baiani

Contributors
Chidambara, Donagh Horgan, Rodrigo Duarte Soliani, Tiago Veloso dos Santos, Carol Archer, Anetheo
Jackson, Ghazal Farjami, Maryam Taefnia, Kasimbazi Emmanuel, Haider Jasim Essa Al-Saaidy, Miguel
Ángel Mejías Vera, Víctor Manuel Romeo Jiménez, Alessandra Battisti, Livia Calcagni, Alberto Calenzo,
Sergiy Kostrikov, Denis Seryogin, Nkeiru Hope Ezeadichie, Vincent Aghaegbunam Onodugo, Chioma
Agatha John-Nsa, Serena Baiani, Paola Altamura

Notice
Statements and opinions expressed in the chapters are these of the individual contributors and not
necessarily those of the editors or publisher. No responsibility is accepted for the accuracy of
information contained in the published chapters. The publisher assumes no responsibility for any
damage or injury to persons or property arising out of the use of any materials, instructions, methods
or ideas contained in the book.

First published in London, United Kingdom, 2022 by IntechOpen
IntechOpen is the global imprint of INTECHOPEN LIMITED, registered in England and Wales,
registration number: 11086078, 5 Princes Gate Court, London, SW7 2QJ, United Kingdom
Printed in Croatia

British Library Cataloguing-in-Publication Data
A catalogue record for this book is available from the British Library

Additional hard and PDF copies can be obtained from orders@intechopen.com

Sustainable Development Dimensions and Urban Agglomeration
Edited by Alessandra Battisti and Serena Baiani
p. cm.
Print ISBN 978-1-83969-560-5
Online ISBN 978-1-83969-561-2
eBook (PDF) ISBN 978-1-83969-562-9

We are IntechOpen,
the world's leading publisher of
Open Access books
Built by scientists, for scientists

6,000+
Open access books available

146,000+
International authors and editors

185M+
Downloads

Our authors are among the

156
Countries delivered to

Top 1%
most cited scientists

12.2%
Contributors from top 500 universities

Interested in publishing with us?
Contact book.department@intechopen.com

Numbers displayed above are based on latest data collected.
For more information visit www.intechopen.com

.

Meet the editor

Alessandra Battisti, Ph.D., is an architect and Full Professor of Technology of Architecture and Environmental Design, Department of Planning, Design and Technology of Architecture, Sapienza University of Rome, Italy, where she is also the director of the second-level university master's degree course in "Valorisation and Enhancement of Small Historical Centres" and member of the faculty board and teaching staff for the Ph.D. in Environmental Design and Planning. Since 1993, Dr. Battisti has been carrying out research, teaching, and experimentation activities on technological innovation for environmentally aware architecture, ecological and energetic efficiency of buildings, bioclimatic approach in design and environmental sustainability, urban analysis, and urban regeneration. She has expertise in sustainable architecture, energy building, and renewable energy, and their relationship with regeneration processes of existing urban fabric and architecture. She is an expert consultant for the European Community - DG XXII in the Energy in Building sector as well as for the Roster of Consultants and Permanent Roster of Auditors launched by the MIUR (Italian Ministry of Education, Universities and Research). She is also a member of the Working Group for Energy Efficiency of Cultural Heritage of MiBACT (Italian Ministry of Culture and Tourism). Dr. Battisti has more than 200 scientific publications and 14 books to her credit. She is also the winner of more than thirty international architectural competitions

Serena Baiani, Ph.D., is an architect specializing in environmental industrial design and Associate Professor of Technology of Architecture and Environmental Design, Department of Planning, Design and Technology of Architecture, Sapienza University of Rome, Italy. She is a board member of the Italian Society of Architectural Technology (SITdA) and a member of the faculty board and teaching staff for the Ph.D. in Environmental Design and Planning; for the Specialisation School in Natural and Territorial Heritage_Architecture of parks, gardens, and natural environmental systems. Since 1996, Dr. Baiani has been carrying out research, teaching, and experimentation activities in technological innovation, active conservation, and adaptive reuse (for enhancement of archaeological contexts, resilience of high-sensitivity areas, ecological and energy efficiency of historic buildings) based on a life cycle approach. She is an external expert for the European Cooperation in Science and Technology (COST) and an expert and peer reviewer for MIUR Italian Scientific Evaluation Register (Italian Ministry of Education, Universities and Research). She is also a member of the Historic Building Committee for Green Building Council, the Sapienza Design Research Interdepartmental Centre, and the Sapienza SAPeri&CO (research and service infrastructure of Sapienza) Committee. She has 130 scientific papers and 12 books to her credit.

Contents

Preface

Currently, the housing conditions of large urban agglomerations present the strongest critical elements facing the human habitat. Faced with epochal climatic, social, and epidemiological challenges, urban areas are suffering recurring calamities, risks, and problems that require profound transformations. From a phenomenon of concentration of human activities, large conurbations, an image of wise use of environmental, territorial and landscape resources and of interaction with human capital capable of imagining, realizing and building an artificial "second nature", maximum and complex abstraction and expression of human ingenuity, during the pandemic, they experienced a negative trend that saw them empty and then repopulate after the positive results of the vaccination campaigns. However, metropolises around the world are still the areas where the most relevant social and environmental challenges are concentrated, and although they cover less than 3% of the earth's surface [1], they are responsible for 71% of global carbon emissions related to energy [2], and in particular, climate change [3]. Long-term shifts in temperatures and weather patterns, which are both global and local issues, are increasing the frequency and intensity of extreme temperature and precipitation events, against which large urban areas play a crucial role as key players in the transition to a sustainable and low-carbon economy. As architects, engineers, and scholars of the sector we are called to provide timely and adequate responses to the phenomenon, investigating approaches and methodologies capable of improving this type of urban form by following and pursuing mitigation of extreme phenomena through policies, studies, circular economy, and resilience design strategies. Furthermore, the pandemic has posed further challenges, especially in places where conditions of overcrowding and instability make the response to the global health crisis even more complex [4].

The United Nations 2030 Agenda for Sustainable Development outlines seventeen Sustainable Development Goals (SDGs) for a sustainable future [5]. SDG 3 aims to ensure healthy lives and promote well-being at all ages. SDG 13 addresses climate change and its impacts and SDG 15 addresses the sustainability of forests, combatting desertification, and halting biodiversity loss [6]. SDG 11 aims to make cities inclusive, safe, resilient, and sustainable. In addition, target 11c plans to support Less Developed Countries (LDCs) through financial and technical assistance in the construction of sustainable and resilient buildings [7].

This book discusses the dynamics of the spatial development of agglomerates, taking into account the SDGs, and analyzes the environmental risks to which they are subject due to climate change. It addresses the development of sustainable transport infrastructures and new socioeconomic transformation and regeneration strategies of urban settlements. It is divided into two sections. The first section includes chapters that address the problem of large urban agglomerations and propose new forms of regeneration from the top down.

The second section includes chapters on the conformation of spaces and their agglomerations. The proposed solutions focus on climate-adaptive strategies to affect the mechanisms of urban transformations and their implementation.

The design attention, at different scales, from territory to architecture, focuses above all on those emerging contexts (Brazil, India, Africa) where 95% of the world's demographic development will be concentrated [8], making structural changes necessary. Central to this approach is the "production of local knowledge," understood as a culture of living and acting, contrasted with housing realities characterized by spatial and social inequalities, lack of quality of living spaces, and greater exposure to epidemiological, environmental, and cultural shocks. Some authors propose a new dwelling paradigm, capable of involving the weakest populations through a bottom-up, shared, and not imposed design. The term resilience, in this sense, implies both an environmental instance, connected to the attempt to mitigate the effects of climate change also through the promotion of local empowerment and social connections, integrating categories of the population excluded from the urban design, within design attention addressing the challenges of the future.

Alessandra Battisti and Serena Baiani
Department of Planning, Design, and Technology of Architecture,
Sapienza University of Rome,
Rome, Italy

References

[1] European Environment Agency (2016). Urban adaptation to climate change in Europe 2016 transforming cities in a changing climate.

[2] Center for International Earth Science Information Network (CIESIN), Columbia University (2018), Documentation for the Gridded Population of the World, Version 4 (GPWv4), Revision 11 Data Sets, NASA Socioeconomic Data and Applications Center (SEDAC), Palisades, NY

[3] Florczyk, A., C. Corbane, D. Ehrlich, S. Freire, T. Kemper, L. Maffenini, M. Melchiorri, M. Pesaresi, P. Politis, M. Schiavina, F. Sabo, and L. Zanchetta (2019), GHSL Data Package 2019, JRC 117104, EUR 29788 EN, Publications Office of the European Union, Luxembourg.

[4] UN-Habitat (2020). Guidance on COVID-19 and Public Space. United Nations Human Settlement Programme (UN-Habitat), Nairobi.

[5] UN-Habitat (2017), New Urban Agenda, United Nations Conference on Housing and Sustainable Urban Development (Habitat III), United Nations, General Assembly, A/RES/71/256, New York.

[6] UN-Habitat (2018). SDG Indicator 11.7.1 Training Module: Public Space. United Nations Human Settlement Programme (UN-Habitat), Nairobi.

[7] Dijkstra, L. and H. Poelman (2014), 'A harmonised definition of cities and rural areas: the new degree of urbanisation', Regional Working Paper 2014, WP 01/2014, European Commission Directorate-General for Regional and Urban Policy.

[8] OECD (2012), Redefining "Urban": A New Way to Measure Metropolitan Areas, Organisation for Economic Co-operation and Development, OECD Publishing, Paris.

Section 1

Sustainability and Resilience of Urban Agglomeration

The Concept of Sustainability in the Brazilian Road Freight Transportation Sector

Rodrigo Duarte Soliani

Abstract

The road cargo transportation system has significant representativeness in the Brazilian economic scenario. Companies depend on transportation to receive inputs from their suppliers and take their products to consumers. This modal is also the major consumer of fuel oil products in the transportation sector. Thus, it is necessary to act with a focus on sustainability, considering the economic, social and environmental aspects. From this perspective, this study aims to present aspects of the concept of sustainability in the Brazilian road freight transportation sector, with emphasis on the environmental, social and economic dimensions. A qualitative research approach was used with the literature research technique in order to build the theoretical basis for discussion on the dimensions of sustainability in Brazilian road freight transportation. The results show that the road cargo transportation sector is relevant for the country and seeks to fulfill its social role and be economically viable; requires attention to environmental awareness and eco-efficiency; should make efforts to use alternatives to avoid damage to the environment, applying technologies to reduce noise pollution, GHG emissions; and seek to comply with environmental legislation acting in a sustainable manner.

Keywords: Sustainable Transport, Environmental Dimension, Social Dimension, Economic Dimension, Road Freight Transportation

1. Introduction

In the most diverse organizations, the debate on sustainability has been intensely present in a constant challenge to the potential to generate value for customers, shareholders and society in general. Nowadays, sustainability has become a fundamental theme, with evidence to avoid the negative impacts of activities, also reflecting technological advances, population growth and consumption [1].

This comprehensive view of the importance of sustainability being undertaken in companies also has its validity for the cargo transportation sector, in this specific case, the road modal is the most representative one, being responsible for handling more than 60% of cargo in Brazil [2]. The preference for road transportation compared with other transportation systems is marked by factors such as: greater flexibility and easier access by trucks to the most different locations; ability to transport goods door-to-door; simplification of the sending of documentation in relation to the railway modal; and faster delivery of the product [3].

Freight transportation operations between cities and states, as well as between the producer and the distribution centers and between these and the final market, show relevant economic importance. However, this activity, due to the large number of trucks, in view of the need for loading and unloading operations, causes impacts, both on cities and on roads [4].

This dynamic of road cargo transportation operations has its share of contribution to the country's development process, however, on the other hand, it highlights the intense use of an energy pattern based on fossil fuels that strengthens the relationship with the increase in gas emissions greenhouse effect (GHG) of anthropic origin [5]. However, this is an aspect that raises the question of how to undertake sustainability in road freight transportation. From this perspective, this study aims to discuss the dimensions of the concept of sustainability in the Brazilian road freight transportation sector, identifying the main attributes for a road freight transportation to be characterized as sustainable.

2. Sustainability

The organizational posture focused on sustainability has been increasingly valued by stakeholders (shareholders, employees, customers and the community itself). The concept of sustainability is based on the concept of the Triple Bottom Line (TBL), as presented by John Elkington in 1994, composed of the economic, social and environmental dimensions, forming the tripod of corporate sustainability [6]. The interaction and integration between these three pillars bring benefits to the environment and to society, and contribute in the long term to the good economic performance and competitive differential of companies [7–9].

Sustainability has evolved from two sources: the first, in biology, with an emphasis on ecology, highlighting the potential for the recovery and reproduction of ecosystems (resilience), in the face of anthropic actions; the second, in the economy, as a developmental factor, due to the verification of the growth of production and consumption in the course of the 20th century, and which signals the continuity of this rhythm, considering the population increase [10].

Patti, Silva and Estender [11] consider sustainability as a strategy that makes up development, resulting in improved quality of life. From this perspective, an integrated management is projected covering aspects of social development, economic growth and environmental protection. Sustainability is evident as a debate present in the daily agenda of organizations, agencies and government sectors, in non-governmental organizations (NGOs), as well as in the academic environment and in the media in general. Companies have shown interest in the issue, constituting a legal order and government recommendations, projecting environmental and social repercussions.

As for the economic dimension of sustainability, companies need to make products available for consumption, however, this process should provide financial return on the investment made. With regard to the social dimension, the organization needs to offer good working conditions, job creation, social inclusion to combat inequality. In the environmental aspect, the company must have its actions guided by eco-efficiency, being concerned with the impacts caused by the use of natural resources and by the pollutant emissions [12]. An eco-efficient operation is one that manages to produce more and better, with less resources and less waste generation [13].

2.1 Environmental dimension

Usually, the environmental dimension is the first sustainability dimension cited as it concentrates the assumptions that production actions and consumption behavior are compatible with the material - basis of the economy, as a subsystem of the natural environment. It is constituted in the processes of producing and consuming in order to ensure that ecosystems can evolve in their self-repair or resilience potential [10].

The environmental dimension of sustainability emphasizes the preservation of ecological processes, with special attention to the capacity of physical and biological systems to withstand adversity and to maintain its structure and functions. In the environmental view, sustainability points to the impacts of actions caused by people in relation to the environment. It is a process that establishes policies for the conservation of energy and natural resources, reduces the use of fossil fuels and the emission of polluting substances, substitutes non-renewable products for renewable ones and transforms used products into more efficient ones [14].

The concept of environmental responsibility goes beyond the mere fulfillment of obligations established by legislation. It contemplates citizenship, social commitment, principles, beliefs and values of an organization, their employees and the communities affected [15]. From sustainability centered on the environmental concept, two actions evolve: the ecological balance that ensures living conditions for people, fauna and flora, and the sustainable use of natural resources by organizations. It involves compliance with environmental laws, the elaboration of projects with reduced environmental impacts, the management of liquid and solid waste, the application of clean technologies, recycling and environmental education [16]. Briefly and directly, all these determinations converge to the term eco-efficiency.

2.2 Social dimension

The social dimension represents the concern to provide society with conditions to live properly. A sustainable society is supported by the theory that all its members have the minimum necessary to live with dignity and that no one practices acts that can harm others [10].

In Gomes and Moretti [17], "social responsibility is the ability to give answers or to seek them". The authors emphasize that social responsibility is one of the active elements in the relationship between organizations and stakeholders, thus, the companies involved in this proposal work to offer answers to social needs.

The sociocultural concept highlights sustainability procedures inserted in performance and profit, being attentive to the social and environmental impacts of the actions, with a focus on the quality of life of communities, cultural memory and economic growth. Sustainability under the social view has its activities directed to people, with care for their well-being and quality of life [18].

2.3 Economic dimension

The economic dimension presupposes increasing production and consumption with better use of natural resources, especially fossil sources of energy, water and mineral resources [10]. Sustainability based on the economic aspect supports development and environmental policies in the face of costs and benefits and an economic assessment that supports environmental protection and raises levels of well-being. From this perspective, sustainability includes allocating and managing resources more efficiently and a regular flow of public and private investment. It aims its activities to maximize profit, however, it seeks to sustain competitiveness in the market and seeks to remain aware of environmental and social aspects [19].

Businesses demand the entrepreneur's comprehensive and holistic perspective in order to remain active and with a high level of competitiveness in the segment and in the market. They require the adoption of a management model that has identification with the market in which it operates, seeking to generate profitability for shareholders and respect and commitment to citizenship and environmental preservation. It is, however, an imposition of the market to apply socially responsible and sustainable management, as well as an advantage for obtaining profit, which is the guarantee factor of longevity of the business [15].

3. Overview of road cargo transportation in Brazil

Road transportation is a complementary modality *par excellence* in different situations of cargo transposition, since it is through trucks that different loads leave the production source and reach the railways, airports or ports [20].

The statistics from the National Transportation Confederation [21] shows the road modal with a composition of more than 2.6 million trucks, 600 thousand mechanical horses, 1.3 million trailers and 900 thousand semi-trailers. This structure handled 485 million tons of useful kilometers (TKU), representing 61.1% of the total transportation cargo. It is worth mentioning that the national fleet has more than 63 thousand autonomous drivers. However, the CNT [22] points out that of the 103,259 km of roads analyzed, 58.2% have some type of problem, whether in the conditions of the pavement, signage or road geometry. Regarding the pavement, 48.3% of the stretches evaluated received a regular, poor or very bad classification.

According to the National Land Transportation Agency [23], the average age of the Brazilian fleet is 13 years, with 16.8 years for autonomous vehicles, 9.5 years for companies and 12.6 years for those who belong cooperatives. Vehicle age is an important factor and is reflected in the type of engine and installed emission control technology [24]. The largest concentration of trucks is in the South and Southeast regions due to the greater economic activity, but they operate throughout the national territory on urban and intercity roads of variable quality [25].

The transportation sector consumes about 30% of the country's total energy, with 92% of that consumption taking place in road transportation. Still, transportation consumes 51% of oil products spent in Brazil [26].

Road freight transportation has its typical body type established in Ordinance 96/2015. It is a wide variety of body types defined to suit different goods. The bodies are: open, tipper, chassis, closed, van, board, tank, among others [27].

Cargo transportation is part of an organization's product distribution logistics system and is represented by three players. The first one is the user: units from the industrial, trade and agribusiness sectors. The second are companies that operate in a cooperative manner such as: Road Cargo Transportation (TRC), Logistic Operators (OL), Multimodal Transportation Operators (OTM) and Logistics Service Quarters (4PL). Municipal, state and federal governments form the third player. Their role is to provide infrastructure for roads and terminals suitable for transportation services, in addition to regulation of operation through regulatory agencies. The decisions to optimize the logistics transportation networks, based on improvements in infrastructure, are objects of the political sphere and, therefore, difficult to influence isolated or cooperated by operators and/or users of the system [28].

Thus, road freight transportation is representative in the country's economy and has great historical and current significance in development. It integrates the Brazilian transportation scenario in a relevantly. However, it needs improvements, both in equipment and infrastructure to carry out operations. One way to achieve sustainable development is to put pressure on national transportation sectors to become more sustainable. The transportation sector is an important field for implementing sustainability strategies as it causes many environmental, social and economic problems, such as air pollution, gas emissions, demand for land and infrastructure, among others.

3.1 Sustainable transportation

Sustainable transportation is motivated by three considerations: 1) the concern with cargo transportation and the direction to the operational structure, mainly with regard to roads; 2) recognition for the reduction of vehicles in circulation; and 3) the growth of sustainability awareness. These are connotations that seek to highlight sustainability to ensure business [29].

In Rodrigue [29], sustainable transportation is one that: allows the basic demands of access by society in general to be carried out safely and without damage to human health and ecosystems, and with balance; if it is consolidated as accessible, it acts efficiently, it has options of type of transportation and it sustains a growing economy; limits emissions and waste within the planet's absorption capacity, reducing the consumption of non-renewable resources, and limits the consumption of renewable resources to the levels of sustainable yield, with recycling and reuse of its components.

Companies that work with cargo transportation, following the logic of competitiveness, are increasingly committed to incorporating technological innovations as an alternative to differentiate themselves from competitors and improve customer service. Putting new technologies into practice is a situation endogenous to the capitalist and globalized system, in which consumption determinations and marketing, sales and customer relations are based on consumer opinion and trends [30].

The ability of an organization to grow in a scenario of strong competition is directly associated with the ability to develop some kind of differential. This factor represents, above all, the development and adoption of innovations for the best relationship with the customer, to generate new products or to be able to establish more efficient processes [30]. In this panorama of the performance of road cargo transportation, the highlights are the innovations of a sustainable character, responsible for directing actions for the preservation of the environment, and also sufficient to optimize resources and reduce costs.

4. Materials and methods

The methodology adopted for the development of this study, according to Prodanov and Freitas [31], is: by the qualitative approach of the problem, since it has a dynamic relationship between the real world and the subject, through interpretation without representation numerical; for its exploratory objective, since it intends to familiarize itself with the problem and present hypotheses; and by the technical, bibliographic procedures, when its elaboration uses material already published in the most varied media.

According to Gil [32], literature research is developed from material already prepared, consisting mainly of books and scientific articles. In qualitative research, the natural environment is the direct source for the search for information. The researcher has direct contact with the environment and the object of study in question. The data are observed in their study environment, without the need for manipulation or the application of statistical methods or techniques. The information collected is descriptive, representing the largest possible repertoire of elements existing in the studied reality [31].

For the collection of articles, we used the databases Ebsco, Scopus and Web of Science for article collection of those aligned with the core of this research, the following keywords were adopted: "Transportation" and "Energy Use" and "Environmental Impacts"; "Sustainable Economy" and "Best practices"; "Sustainable Development" and "Business Strategies"; "Sustainable" and "Supply Chain Management"; "Sustainability" and "Road transportation".

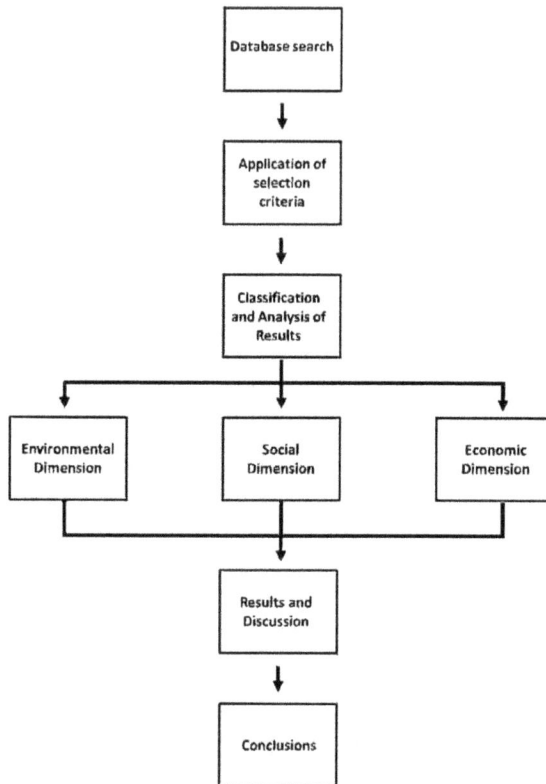

Figure 1.
Methodological Diagram. Source: Author (2021).

These words were used in all defined databases and the filters adopted for the inclusion of articles in the bibliographic portfolio were: the language (English and Portuguese), research area (engineering, administration, logistics), and the articles should have, or in the title, or in the abstract, or throughout the text, any of the keywords used. In this study, only articles available in full text were selected from the database to which they were linked.

Through the pre-selected documents, bibliographic references were searched for other approaches in order to insert them in the research. In total, 58 relevant approaches to the theme were identified, after this survey the research was subdivided into showing the classification, and the relationship with the information available and contributions to the discussion on aspects of sustainability in road freight transportation, as shown in **Figure 1**.

5. Results and discussion

Throughout the researches, a gap was noticed in terms of sustainability in road freight transportation in its most comprehensive definition, that is, from the economic, environmental and social aspects. When it comes to sustainability in transportation, the association with atmospheric emissions is inevitable. However, the other aspects are also of great importance for the well-being of society and many are associated.

The configuration of the transportation system, the performance of its activities, as well as the implementation of actions to achieve sustainable efficiency, require a foundation in the three pillars of the TBL: the economic, with emphasis on the efficiency of the performance of activities, costs of interiorization and compatible price; the environmental, with observation for the prevention of pollution, protection and conservation of natural resources and environmental management; and the social, with attention to people's safety, health and quality of life [29].

A company to be within the parameters of sustainability must be much more concerned than with atmospheric emissions. It is important to be concerned with optimizing the load, the better it is done, the fewer trucks will be needed to transportation the goods, although for this, it may be necessary to increase the waiting time of the goods which would cause a trade-off, increase waiting time and have fewer trucks circulating, or make quick deliveries with trucks circulating below capacity [33].

By reducing the number of trucks, the company reduces the gases emitted, the noise on the roads, the congestion and the number of accidents. By having better infrastructure, land use and truck access, smaller trucks can be placed to circulate in urban areas at times with less vehicle circulation with suitable locations for loading and unloading. This would generate considerable social well-being by decreasing congestion, noise and emissions within urban centers [34].

The number of road accidents can also be reduced through improvements in infrastructure and in the adjustment of the drivers' working hours so that they do not drive under the influence of substances or in conditions of extreme tiredness. Accordingly, in Brazil, Law/2015 was enacted in 2015, which defines rules for road transportation drivers, whether cargo or passengers. It is known as the Driver's Law. It came about in order to ensure the rights and establish the duties of the driver. Therefore, there was a reformulation of some rules, required safety measures, in which the main ones were the readjustment in the workday, waiting and rest periods, toxicological medical exams, fees and fines [35].

The challenge for logistics professionals when embracing sustainability is how to link and balance environmental performance and good business practices. That is, how to identify appropriate approaches or solutions that balance environmental and economic aspects. One of the main objectives of logistics is to increase the efficiency and economic performance of companies. However, the implementation of actions that contribute to changes in other social aspects, such as environmental and social responsibility, depends on the simultaneous fulfillment of short-term economic goals. However, research shows that several benefits generated for companies by adopting environmental policies, such as reducing costs due to resource savings, increasing sales by improving the company's image and adapting to future government regulations [36–40].

5.1 Environmental dimension in cargo transportation

As for the environmental aspect of road cargo transportation, two phenomena stand out: emission of greenhouse gases and noise pollution. Road freight transportation is a major cause of air pollution. The emission of gases has a negative effect on air quality and the health of people and animals. The level of toxicity of the gas emission depends on the fuel used. Although alternative fuels such as biodiesel already exist, the main fuel in use in Brazil to transportation goods is still diesel. This energy source is unsustainable, as it is a finite resource, with potential to damage public health and the environment, through particulate emissions, carbon monoxide (CO), nitrogen oxide (NOx), organic compounds volatiles (VOCs) and greenhouse gases [41]. Noise pollution comes from the operationality of road traffic, resulting from the speed undertaken, which emits noises through propulsion (engine), pneumatic and aerodynamic [42].

5.1.1 Energy consumption

For the National Transportation Confederation [25], the infrastructure used for transportation in Brazil is inadequate. Due to the large territorial area of the country, the use of the road modal causes inefficiency, since this mode is not the most suitable for cargo with low added value and long distances. Still from an economic perspective, the lack of efficiency in Brazilian transportation logistics reduces the country's competitiveness [43].

In Brazil, in 2010, the transportation sector spent 31% of the total energy, of which 91.7% were consumed by road transportation. The energy source of the Brazilian cargo transportation is constituted in a great majority of fossil fuels, showing the potential of environmental impacts of the sector [44]. However, according to the National Petroleum Agency (ANP), renewable sources, especially biofuels, are increasingly being improved with the proposal to increase supply. Biofuels come from renewable biomass, with the potential to replace fuels from oil and natural gas in combustion engines [45].

Diesel oil represents a high consumption rate of the energy matrix that serves the cargo transportation sector in Brazil. However, its use as a fuel, evidencing combustion and exhaust, constitutes a significant element of emission of particulate material and polluting gases that affect the atmosphere [46].

5.1.2 CO_2 emissions

In view of the energy consumption at a level above the world average, the transportation sector in Brazil stands out as a problem with regard to GHG emissions, mainly of CO_2 gas. Transportation emissions - which mainly involve road, rail, air and sea transportation - accounted for more than 24% of global CO_2 emissions in 2016 [47].

The combustion process of diesel oil releases hundreds of chemical compounds in both liquid and gaseous form. In exhaustion, the main elements released are: carbon dioxide, oxygen, nitrogen, water vapor, carbon monoxide, particulate matter and volatile organic compounds, such as hydrocarbons, the latter of high toxicity, among them the most harmful to health are benzene, toluene, ethylbenzene, xylene and polycyclic aromatic hydrocarbons (PAHs) [48].

In summary, the main pollutants that impact the air quality emitted by the combustion of diesel oil are: carbon dioxide (CO_2), particulate matter, nitrogen oxides (NOX), sulfur oxides and other pollutants [49].

CO_2 emissions have been a target of concern and actions to reduce them, so much so that CONAMA Resolution no. 18/1986 created the Air Pollution Control Program for Motor Vehicles (PROCONVE) with the objectives of: reducing the levels of emission of automotive pollutants; promote national technological development; create inspection and maintenance programs for vehicles; promote public awareness of the issue of air pollution by motor vehicles; establish conditions for evaluating the results achieved; and promote the improvement of the technical characteristics of liquid fuels [50].

Encouraging the use of biofuel for road freight transportation in Brazil has been one of the recommendations for reducing CO_2 emissions. Another observation refers to the readjustment of the fleet and vehicle models in order to increase the efficiency of operations, reduce energy costs and increase the competitiveness of

Brazil [45]. Policies for GHG control, in particular the reduction of carbon emissions, the use of new technologies for more efficient and less polluting engines, have been objectives of manufacturers and vehicles, with significant changes in production patterns. The more restrictive emission limit has contributed to environmental awareness with energy efficiency [51].

5.1.3 Noise pollution

The sound is originated by a mechanical vibration that propagates in the air reaching the ear. Noise is just a type of sound, but a sound is not necessarily a noise, subjectively, noise is an unpleasant and undesirable sound. Noise is the physical vibratory phenomenon (in the case of air) as a function of frequency, that is, for a given frequency, there may be, at random, over time, variations in different pressures [52].

The concern with noise and its effects started at the beginning of the Industrial Revolution due to the appearance of powerful machinery both in factories and in construction, as well as new modes of transportation. The development of industry and the growth of cities has now resulted in an essentially urban world. In Brazil, according to the 2015 population census, about 84% of the population lives in an urban environment [53]. This urban expansion eliminated the silence of much of the planet and, today, noise is one of the most harmful contaminating agents to human health.

Road traffic is the main cause of local environmental noise. The maximum amount of noise that human beings can be exposed to continuously, ensuring acoustic comfort and not harming their health is 65 dB, a value ensured by preventive medicine. Exposure to noise of values above this can cause different impacts on the body, such as, for example, disturbed sleep, decreased work performance, hypertension, interference with cardiovascular diseases, among others [54].

Noise tolerance limits set maximum exposure times for certain levels. However, it is known that there is not a single and perfectly constant level of noise during a journey, including, in the Road Freight Transportation there are great variations, mainly with regard to background noise, such as, for example, the passage through the same via trucks, automobiles, motorcycles and, close to rural areas, even tractors [55]. To quantify these exposures, the dose concept is used, which gathers the acoustic variations according to the exposure time and the maximum time allowed during the journey [56].

The three main sources of truck noise on the roads, according to McKinnon et al. [57], are: (a) propulsion noise (engine), which dominates at low speeds (below 50 km / h); (b) pneumatic (contact noise with the road) which is the main cause of noise at speeds above 50 km/h; (c) aerodynamic noise, which increases when the vehicle accelerates.

The National Environment Council (Conama), in its Resolution 001/90, when disposing about criteria of noise emission standards resulting from any industrial, commercial, social or recreational activities, including those of political propaganda, determines that the values and noise emission limits established in the ABNT NBR 10151 standard, with the objective of ensuring public peace and the health of the population [58].

ABNT NBR 10151: 2019 - Acoustics - Measurement and evaluation of sound pressure levels in inhabited areas - General purpose application, Brazilian Association of Technical Standards [58]. In a table, ABNT NBR 10151: 2019 presents the levels of daytime and nighttime noise allowed, in different types of areas possible to exist in a city, such as, for example, strictly residential, urban, hospitals or schools; the mixed with a commercial vocation; the mixed with recreational vocation; and the predominantly industrial one.

5.1.4 Trucks with alternative technologies

All automakers present in America and Europe, and also in Brazil, have been working on vehicle designs with hybrid technology, 100% electric or gas. There are also several startups around the world that also have projects to develop cleaner commercial vehicles [59].

In addition to sustainable issues, trucks with alternative technologies to diesel have low maintenance costs, emit less noise and, therefore, can travel at times or places of greater restriction and, even with the largest initial investment, in a few years, the trucks are pay. Much because of the lower operating cost than diesel guaranteed by the manufacturers [60].

The electric truck is a response from manufacturers to the increasingly strict rules on pollutant emissions. The need to reduce CO_2 levels is such that it has attracted new companies to the transportation sector [60]. This is the case of the Swedish Volta, with the recently introduced HGT, and the American Tesla, with the Semi.

Several tests and attempts to introduce the electric truck in Brazil were made before BYD arrived. The brand started operations in the country in 2015. Currently, according to information from the Chinese company itself [61], it offers the eT7 11,200 and eT8 21,250 models in the Brazilian market for garbage collection operations, and the eT3 van for the urban transportation. BYD's electric truck has lithium iron phosphate batteries. According to the brand, this solution can last up to 30 years and its autonomy is 200 km [61].

JAC Motors is the second brand to bet on electrification. In September 2020, the brand launched the iEV1200T in Brazil. The model is the second electric truck in the country, but the first focused on urban collection and distribution operations. Unlike BYD eT3, which has a load capacity of 720 kg, the JAC model has a Total Gross Weight (PBT) of 7.5 tonnes. The truck's autonomy is up to 250 km, if the truck runs with 2 t of net load. If traveling with 4 t of net charge, the JAC iEV1200T can travel 180 km between battery recharges [62].

Volkswagen Caminhões e Ônibus (VWCO) started production of the e-Delivery electric truck in October 2020. The model, which was developed in Brazil, is being produced at the Resende plant (RJ). Sales will start in the first half of 2021. The e-Delivery electric truck will hit the market with two Total Gross Weight (PBT) models: 11 and 14 tons [63].

The tests started in 2018, after Cervejaria Ambev announced that it will have more than 1/3 of the fleet composed of at least 1,600 Volkswagen electric trucks by 2023, the largest ad of its kind in the world [63].

In about two years of testing, more than 22 tons of CO_2 are no longer emitted into the atmosphere and, so far, e-Delivery has stopped consuming more than 6,500 liters of diesel. The electric truck is recharged with 100% electric energy from clean sources, and 43% of its energy comes from the vehicle's own regenerative brake system. The e-Delivery electric motor generates up to 260 kW (equivalent to 348 hp) of power and its torque is around 233 mkgf [63].

Like e-Delivery, in addition to being supplied with electricity, some models have a braking system that also produces electricity to be stored in the same batteries that supply power to the engine. All of them are non-polluting, both in relation to the emission of harmful gases and in relation to noise [64].

The main application of these vehicles in the near future is in urban traffic, in short and light cargo logistics operations. With the current battery technology, urban vehicles are the most suitable, as they travel short distances and do not need as much energy to operate [65]. Because of their low autonomy and load capacity, electric trucks depend on an operation that prevents, for example, the vehicle from getting stuck in traffic jams. In addition to autonomy, other issues to be addressed to make electric trucks feasible are cost and battery recharge. As with cars, electric trucks also cost more than conventional models. This is mainly due to battery packs, which make up 50% of the vehicle's value [66].

5.2 Social dimension in cargo transportation

In this study, two important points stand out in the social aspect of road cargo transportation: accidents and congestion. Accidents can cause deaths and injuries to those involved, as well as inconvenience to other drivers on the roads. In general, the number of accidents with the participation of heavy vehicles considering the distance covered is lower when compared to automobiles, however, the probability of a truck being involved in a fatal accident is great [26].

5.2.1 Accidents

According to data from the Institute of Applied Economic Research [67], in 2014 the total cost of traffic accidents on federal, state and municipal highways reached an approximate value of R $ 40 billion, with an average cost of R $ 647 thousand by fatal accident. Santana et al. [68], point out that, although road cargo transportation (TRC) is a strategic sector for Brazil, it presents several structural problems, with high social cost, including high mortality due to Work Accidents (AT) with truck drivers.

Every day, Brazil records 14 deaths and 190 accidents on federal highways. In 2018 alone, there were 69,206 accidents, of which 53,963 were victims. These accidents resulted in 5,269 deaths in the year [69]. In the 12 years analyzed by the CNT [69], Brazil had 1.7 million accidents on federal highways, with 751.7 thousand with victims and 88.7 thousand deaths. The highway with the highest number of accidents in 2018 was BR-101, where a total of 8,896 accidents with victims were recorded. Most of the occurrences on Brazilian federal highways have the presence of automobiles (64.6%), followed by motorcycles (44.4%) and trucks (23.4%) [69].

According to Silva et al. [70], drowsiness, physical and mental tiredness, drug use and payment for production are among the main factors that lead to accidents at work with professional drivers. In spite of this, the cargo transportation sector has been organized in such a way as to favor productivity, generally linking drivers' remuneration to the number and extent of trips, which leads them to make long journeys with little rest, a combination considered as one of main factors that contribute to the occurrence of accidents.

5.2.2 Traffic Jams

Urban congestion is one of the main problems generated by the use of road transportation for goods transportation, since the speed is relatively low, compromising the flow of vehicles through the streets and avenues. This situation significantly interferes with the city's routine [26]. It is, therefore, a great challenge to be solved, ensuring the economic development of cities and reducing the negative impacts of congestion.

The National Confederation of Transportation [25] developed the study "Urban Logistics: Restrictions on Trucks?" which presents the current panorama of restrictions on the circulation of trucks and loading and unloading operations in seven Metropolitan Regions of Brazil: São Paulo (SP), Belo Horizonte (MG), Curitiba (PR), Porto Alegre (RS), Goiânia (GO), Recife (PE) and Manaus (AM). To this end, a survey of the laws governing the circulation of trucks and loading and unloading operations was carried out in these municipalities.

The results show that the accelerated urbanization of Brazil, in the last decades, has brought complexity and challenges to the supply logistics of cities where 84% of the Brazilian population lives and 96.7 million motor vehicles circulate. The study found a variety of rules and restrictions on the circulation of trucks in urban centers, added to problems of infrastructure, signaling and inspection, among other deficiencies that have an impact on the transportation activity [25].

The problems found by CNT [25] make it difficult to plan cargo transportation, increase operating costs and decrease the quality of supply services in cities, namely:

- Increase in the operational costs of road cargo transportation. In some cases, the barriers encountered by carriers have generated extra fees that affect the price of freight. Two examples are the Delivery Difficulty Rate (TDE), negotiated from a floor of 20% on the value of the freight; and the Traffic Restriction Rate (TRT), calculated at 15% of freight.

- Low predictability of delivery of goods. In addition to congestion and traffic restrictions, the carrier's planning is often changed in an unpredictable manner due to the lack of clarity and transparency about the restrictions on cargo transportation.

- Increased emission of pollutants and noise. Poorly planned restrictions can lead to congestion, discharge queues, increased number of trips, longer and inadequate routes and other disorders that increase noise produced by traffic and the emission of polluting gases into the atmosphere.

- Risk of accidents. Poor signage or even lack of signage, night time windows and other restrictions are factors that increase the risk of accidents.

The main solutions identified by the study [25] are: to improve public policies and planning, including cargo transportation in urban planning and traffic policies, integrating all municipalities in metropolitan regions; carry out democratic management and expand social control of all interested sectors: transportationers, shippers, buyers, manufacturers, distributors, transportationers, logistics operators, wholesalers, retailers and final consumers; improve traffic signs and inspection, giving more clarity and visibility to restrictions on cargo transportation, publicizing alternative routes and expanding inspection, especially in the areas of loading and unloading; expand the supply of loading and unloading spaces and hourly windows for deliveries and collections; increase security, expanding the offer of rest and rest places associated with goods distribution centers; and expand investment in infrastructure, carrying out maintenance and expansion works on urban infrastructure, especially in highway rings.

The complexity of urban distribution stems mainly from the great variety of demands for goods (at different locations and times), the reduced capacity to expand the road infrastructure and the insufficient offer of routes and alternative modes. In addition, the increase in the total vehicle fleet, congestion, restrictions on the circulation of trucks, the inadequate supply of loading and unloading spaces and the reduced hourly windows are some of the factors that condition the performance of the freight transportation activity in the middle of the country, urban transportation, increasing the costs of road transportation and reducing the predictability of goods delivery.

5.3 Economic dimension in cargo transportation

The economic aspect of road cargo transportation, in addition to focusing on obtaining profits, highlights some significant factors: the configuration of infrastructure for operationalization, access opportunities, cargo optimization and adequate land use. These determinants aim at better organization and distribution of cargo, mainly in the movement in urban centers, with definitions of areas for loading and unloading.

When it comes to infrastructure, access, cargo consolidation and land use, the best organization and distribution of cargo in urban centers is sought. An example of measures related to land use is to reserve areas in urban centers for loading and unloading. When talking about access, it refers to spatial and temporal restrictions and, in the case of public infrastructure, the use of transfer points to improve the load factor of vehicles, that is, the load consolidation centers [71].

According to Novaes [72], in Brazil the occupancy rate in road cargo transportation is only 43%, which results in an excess of trucks on the roads. This is because there are customer demands on specific routes, but the demand does not complete the capacity of a cargo vehicle, nor can it be deactivated due to low demand, as it is necessary to serve the customer.

In an attempt to achieve these economic attributes, Vidal, Laporte and Matl [73], report the use of Information Technology (IT) to achieve some objectives, such as: promoting the exchange of information between interested parties; vehicle routing and scheduling according to the degree of congestion in the transportation network; allocate loads in the compartments, efficiently, for the loading and unloading process; and increase the vehicle occupancy rate. The use of IT to help aggregate freight is of great importance to avoid trips below capacity. In addition, vehicle routing and scheduling systems for using the loading and unloading zones can result in savings in travel time between 10% and 15%, according to the authors.

Some urban centers and cities that are on the side of the roads have tended to build road loops to divert the flow from city centers [74]. Another way to reduce the impact of trucks on urban centers is access restrictions, which are the most common regulations in Europe. These access restrictions can be according to the size restriction of the truck or the time allowed for traffic [75].

5.3.1 Last mile delivery

In freight transportation logistics, the final step "Last-Mile" refers to the transportation in which the goods leave the distribution center for the final destination, that is, for the customer, both B2B and B2C, who purchased a certain product [76].

In terms of innovation, Last Mile Delivery, in addition to transforming the methodology commonly used by the transportation sector, which prioritizes the quality and efficiency of delivery, started to take into account issues such as sustainability [77].

Because of this, it has become increasingly common among companies to use bicycles and scooters to make deliveries, whether on short or large routes. The alternative is quite feasible, since it provides the improvement of urban mobility and does not pollute the environment [78].

The investment in technologies allows for faster delivery, which will not only make the final consumer more satisfied, but will also help the company to gain more time, streamlining processes [79]. Geolocation enables the optimization and improvement of other tools used in the transportation process, being essential for the integration of the company's system with the Google Maps API (Application Programming Interface).

A geolocation tool contributes to the definition of more viable routes, which helps to save time and fuel, directly impacting the maintenance of the means of transportation used. In addition, geolocation allows control of delivery in real time, taking into account the company's particularities. With this, the entire delivery process is streamlined, optimizing material and human resources [79].

According to Joerss et al. [80] the business model conventionally applied to the last mile should be replaced due to the new technologies that reach the market. For these authors, the traditional model that uses light diesel vehicles will be responsible for only 20% of deliveries in the last mile in urban areas, being progressively replaced by autonomous vehicles and delivery services by bicycles, more energy efficient. There is a potential for new technologies to transform deliveries in the last mile, which can lead to a new transportation infrastructure and delivery models [81].

6. Conclusions

The road cargo transportation sector has its representativeness in the country's economic scenario and, however, in order to obtain sustainable gains, attention must be paid to the negative effects that its performance can cause to the environment and establish a reduction in the levels of CO_2 emissions. This is done through logistical planning and the choice of more eco-efficient modes, that is, less polluting, such as the railway. Biofuels, as well as electric trucks, may also help to minimize the serious problems of road transportation in the country, especially urban ones.

The current transportation matrix has proved to be inadequate, since the high dependence on the road modal intensifies the problems of urban mobility, enhances environmental problems and negatively affects people's quality of life. Thus, it is understood that the improvement in the country's economic and environmental results is directly related to changes in the transportation sector. The current model is contrary to the search for a better quality of life for society and to the increase of Brazilian competitiveness in relation to the foreign market.

Organizations around the world have faced the challenge of making their operations more sustainable. In logistics, the focus for the coming years will be on reducing carbon emissions and reducing production waste. Technology and logistics go together, mainly with regard to the development of solutions that help to optimize processes, make the results more satisfactory and guarantee higher quality for companies that are served by companies in the sector.

In the case of logistics, innovation is considered an extremely strategic factor for the success of a business in the sector. In this sense, among the main trends identified in this study in Brazilian road freight transportation, are: the greater use of ecological fuels, which are less polluting and provide less noise, the popularization of hybrid or fully electric vehicles, the growth in use of small vehicles for last-mile deliveries, in addition to the use of transportation management tools, which allow to select and better manage the partners that offer these differentials.

Society faces a challenging time for economic growth and public welfare. The environmental problems resulting from unrestrained progress are already reaching great proportions, becoming the subject of discussions and mobilizations worldwide. In this reality in which sustainability is no longer just a competitive differential, it is indispensable for the future of the planet, the Brazilian transportation sector adopts socio-environmental responsibility as the basis for its performance. Promoting social and environmental responsibility in the Brazilian cargo transportation sector and, thus, collaborating to the preservation of life and the environment, constitute the main contribution of this chapter to the book Urban Agglomeration.

Author details

Rodrigo Duarte Soliani
Federal Institute of Acre (IFAC), Brazil

*Address all correspondence to: rodrigo.soliani@ifac.edu.br

IntechOpen

References

[1] Bengtsson, M.; Alfredsson, E.; Cohen, M, L.; Schroeder, P. Transforming systems of consumption and production for achieving the sustainable development goals: Moving beyond efficiency. Sustainability Science, 13, 1533-1547, 2018. DOI: https://doi.org/10.1007/s11625-018-0582-1

[2] CNT. Confederação Nacional do Transporte. Anuário CNT do Transporte 2018. Brasília: CNT, 2018a. www.anuariodotransportatione.cnt.org.br

[3] Monios, J.; Bergqvist, R. Intermodal freight transportation and logistics. CRC Press, Taylor & Francis Group, Boca Raton, 2017.

[4] Diana, M.; Pirra, M.; Woodcock, A. Freight distribution in urban areas: a method to select the most important loading and unloading areas and a survey tool to investigate related demand patterns. European Transportation Research Review. 2020. DOI: https://doi.org/10.1186/s12544-020-00430-w

[5] D'agosto, M. A. Transportation, Energy Use and Environmental Impacts. Elsevier Inc., 2019. DOI: https://doi.org/10.1016/C2016-0-04814-3

[6] Elkington, J. Towards the sustainable corporation: win-win-win business strategies for sustainable development. California Management Review, v. 36, n. 2, p. 90-100, 1994.

[7] Carter, C. R.; Rogers, D. S. A framework of sustainable supply chain management: moving toward new theory. International Journal of Physical Distribution and Logistics Management, v. 38, n. 5, p. 360-38, 2008. DOI: https://doi.org/10.1108/09600030810882816

[8] Morioka, S. N.; Bolis, I.; Evans, S.; Carvalho, M. M. Transforming sustainability challenges into competitive advantage: Multiple case studies kaleidoscope converging into sustainable business models. Journal of Cleaner Production, Vol 167, 2017. DOI: https://doi.org/10.1016/j.jclepro.2017.08.118

[9] Dhahri, S.; Omri, A. Entrepreneurship contribution to the three pillars of sustainable development: What does the evidence really say?. World Development, Volume 106, June 2018, Pages 64-77. DOI: https://doi.org/10.1016/j.worlddev.2018.01.008

[10] Nascimento, E. P. Trajetória da sustentabilidade: do ambiental ao social, do social ao econômico. Estudos avançados, São Paulo, v. 26, n. 74, 2012. DOI: http://dx.doi.org/10.1590/S0103-40142012000100005

[11] Patti, F.; Silva, D.; Estender, A. C. A importância da sustentabilidade para a sobrevivência das empresas. Revista Terceiro Setor & Gestão, v. 9, n. 1, 2015. http://revistas.ung.br/index.php/3setor/article/view/1997

[12] Passetti, E.; Tenucci, A. Eco-efficiency measurement and the influence of organizational factors: evidence from large Italian companies. Journal of Cleaner Production. Volume 122, Pages 228-239, 2016. DOI: https://doi.org/10.1016/j.jclepro.2016.02.035

[13] Abreu, M. F.; Alves, A. C.; Moreira, F. Lean-Green models for eco-efficient and sustainable production. Energy. Vol 137, 2017. DOI: https://doi.org/10.1016/j.energy.2017.04.016

[14] Owusu, P.; Asumadu-Sarkodie, S. A review of renewable energy sources, sustainability issues and climate change mitigation. Cogent Engineering. 2016 DOI: https://doi.org/10.1080/23311916.2016.1167990

[15] Mello, M. F.; Mello, A. Z. An analysis of the practices of social

responsibility and sustainability as strategies for industrial companies in the furniture sector: a case study. Gestão & Produção, 25(1), 81-93. Epub October 30, 2017. DOI: https://dx.doi. org/10.1590/0104-530x1625-16

[16] Marcon, A.; De Medeiros, J. F.; Ribeiro, J. L. D. Innovation and environmentally sustainable economy: identifying the best practices developed by multinationals in Brazil. Journal of Cleaner Production, Vol. 160, pp.83-97, 2017. DOI: https://doi.org/10.1016/j. jclepro.2017.02.101

[17] Gomes, A.; Moretti, S. A responsabilidade e o social: uma discussão sobre o papel das empresas. São Paulo: Saraiva, 2007.

[18] Di Fabio, A. The psychology of sustainability and sustainable development for well-being in organizations. Front. Psychol. 2017. DOI: https://doi.org/10.3389/ fpsyg.2017.01534

[19] Oláh, J.; Kitukutha, N.; Haddad, H.; Pakurár, M.; Máté, D.; Popp, J. Achieving Sustainable E-Commerce in Environmental, Social and Economic Dimensions by Taking Possible Trade-Offs. Sustainability, 2019. DOI: https:// doi.org/10.3390/su11010089

[20] Soliani, R. D.; Innocentini, M. D. M.; Carmo, M. C. Collaborative logistics and eco-efficiency indicators: an analysis of soy and fertilizer transportation in the ports of Santos and Paranaguá. Independent Journal of Management & Production. 2020. DOI: https://doi.org/10.14807/ijmp.v11i5.1303

[21] CNT. Confederação Nacional do Transporte. Boletim estatístico – CNT - outubro 2016. Brasília: CNT, 2016a. https://www.cnt.org.br/boletins

[22] CNT. Confederação Nacional do Transporte. Pesquisa CNT de rodovias 2016: relatório gerencial. Brasília: CNT,

SEST, SENAT, 2016b. https:// pesquisarodovias.cnt.org.br/edicoes

[23] ANTT. Agência Nacional de Transportes Terrestres. Idade média dos veículos. 2016. http://appweb2.antt.gov. br/revistaantt/ed4/_asp/ed4-custosExternos.asp

[24] CETESB. Companhia Ambiental do Estado de São Paulo. Emissões Veiculares no Estado de São Paulo 2016. Série Relatórios. São Paulo, 2017. https://cetesb. sp.gov.br/veicular/wp-content/uploads/ sites/6/2017/11/EMISS%C3%95ES-VEICULARES_09_nov.pdf

[25] CNT. Confederação Nacional do Transporte. Logística urbana: restrições aos caminhões? – Brasília: CNT, 2018b. https://cnt.org.br/logistica-urbana-restricoes-caminhoes

[26] Castro, N. Mensuração de externalidades do transporte de carga brasileiro. Journal of Transportation Literature. Manaus, v. 7, n. 1, pp. 163-181, jan. 2013. DOI: https://doi. org/10.1590/S2238-10312013000100010

[27] Brasil. Departamento Nacional de Trânsito. Portaria n. 96, de 28 de julho de 2015. Estabelece a Tabela I – Classificação de Veículos conforme Tipo/Marca/Espécie e a Tabela II – Transformações de Veículos sujeitos a homologação compulsória da Resolução CONTRAN nº 291/2008. Brasília, 2015a. https://www.gov.br/infraestrutura/ pt-br/assuntos/transito/arquivos-denatran/portarias/2015/ portaria0962015.pdf

[28] Schlüter, M. R. Sistemas logísticos de transportes. Curitiba: InterSaberes, 2013.

[29] Rodrigue, J. The Geography of Transportation Systems. 5th edition. New York: Routledge, 2020.

[30] Oliveira, G.; Machado, A. Dynamic of Innovation in Services for Consumers

at the Bottom of the Pyramid. Brazilian Business Review. Vol.14 no.6 Vitória, 2017. DOI: https://doi.org/10.15728/bbr.2017.14.6.4

[31] Prodanov, C. C.; Freitas, E. C. Metodologia do trabalho científico: métodos e técnicas da pesquisa e do trabalho acadêmico. 2. ed. Novo Hamburgo, RS: Feevale, 2013.

[32] Gil, A. C. Métodos e técnicas de pesquisa social. 7. ed. São Paulo: Atlas, 2019.

[33] Diaz, I. S.; Palacios-Argüello, L.; Levandi, A.; Mardberg, J.; Basso, R. A Time-Efficiency Study of Medium-Duty Trucks Delivering in Urban Environments. Sustainability, 2020. DOI: https://doi.org/10.3390/su12010425

[34] Browne, M.; Allen, J.; Nemoto, T.; Patier, D.; Visser, J. Reducing Social and Environmental Impacts of Urban Freight Transportation: A Review of Some Major Cities. Procedia - Social and Behavioral Sciences, 39, 19-33, 2012. DOI: https://doi.org/10.1016/j.sbspro.2012.03.088

[35] Brasil. Secretaria-Geral. Lei N° 13.103, de 2 de março de 2015. Brasília, 2015b. http://www.planalto.gov.br/ccivil_03/_Ato2015-2018/2015/Lei/L13103.htm

[36] OECD. Organization for Economic Co-operation and Development. OECD Green Growth Papers. OECD Publishing, Paris, 2011. DOI: https://doi.org/10.1787/22260935

[37] OECD. Organization for Economic Co-operation and Development. Building Green Global Value Chains: Committed Public-Private Coalitions in Agro-Commodity Markets. OECD Publishing, Paris, 2013. https://doi.org/10.1787/5k483jndzwtj-en

[38] Leigh, M.; Li, X. Industrial ecology, industrial symbiosis and supply chain environmental sustainability: a case study of a large UK distributor. Journal of Cleaner Production. Volume 106, 1 November 2015. DOI: https://doi.org/10.1016/j.jclepro.2014.09.022

[39] Weng, H.; Chen, J.; Chen, P. Effects of Green Innovation on Environmental and Corporate Performance: A Stakeholder Perspective. Sustainability, MDPI, vol. 7(5), pages 1-30, April, 2015. DOI: https://doi.org/10.3390/su7054997

[40] Maia, D. A. C.; Saraiva, L. G. M.; Ferreira, A. M. C.; Oliveira, T. E.; Oliveira, P. L. Contabilidade da gestão ambiental como ferramenta fundamental para certificação e sustentabilidade. Revista Diálogos Acadêmicos, v. 8, n. 2, p. 18-30, 2019. http://revista.fametro.com.br/index.php/RDA/article/viewFile/223/197

[41] Shahraeeni, M.; Ahmed, S.; Malek, K.; Drimmelen, B. V.; Kjeang, E. Life cycle emissions and cost of transportation systems: Case study on diesel and natural gas for light duty trucks in municipal fleet operations. Journal of Natural Gas Science and Engineering. Vol 24, 2015. DOI: https://doi.org/10.1016/j.jngse.2015.03.009

[42] Heutschi, K.; Bühlmann, E.; Oertli, J. Options for reducing noise from roads and railway lines. Transportation Research Part A: Policy and Practice. Vol 94, 2016. DOI: https://doi.org/10.1016/j.tra.2016.09.019

[43] Haddad, E. A.; Perobelli, F. S.; Domingues, E. P.; Aguiar, M. Assessing the ex ante economic impacts of transportation infrastructure policies in Brazil. Journal of Development Effectiveness. Volume 3, 2011. DOI: https://doi.org/10.1080/19439342.2010.545891

[44] Brasil. Ministério de Minas e Energia. Estudo associado ao plano decenal de energia – PDE 2021:

consolidação de bases de dados do setor de transportes: 1970-2010. Nota técnica SDB-Abast No 1/2012. Brasília: MME, 2012a. https://www.epe.gov.br/sites-pt/publicacoes-dados-abertos/publicacoes/PublicacoesArquivos/publicacao-250/topico-301/Consolida%C3%A7%C3%A3o%20de%20Bases%20de%20Dados%20do%20Setor%20Transporte%201970-2010%20-%20PDE%202021[1].pdf

[45] Leal Júnior, I. C.; Valva. D. C.; Guimarães, V. A.; Teodoro, P. Análise da matriz de transporte brasileira: consumo de energia e emissão de CO_2. Revista UNIABEU, Belford Roxo, RJ, v. 8, n. 18, jan.-abr. 2015.

[46] Hime, N. J.; Marks, G. B.; Cowie, C. T. A comparison of the health effects of ambient particulate matter air pollution from five emission sources. International Journal of Environmental Research and Public Health, vol. 15, no. 6, 2018. DOI: https://doi.org/10.3390/ijerph15061206

[47] IEA. International Energy Agency. CO_2 Emissions from Fuel Combustion 2018 Highlights, 2018. https://webstore.iea.org/co2-emissions-from-fuel-combustion-2018-highlights

[48] Santos, H. L.; Fialho, M. L.; Reis, K. P.; Franco, M. V.; Oliveira, R. B. Relação entre poluentes atmosféricos e suas consequências para a saúde. Revista Científica Intr@ciência, v. 17, p. 01-24, 2019. Disponível em: http://uniesp.edu.br/sites/_biblioteca/revistas/20190312105045.pdf

[49] Viscondi, G. F.; Silva, A. F.; Cunha, K. B. Geração termoelétrica e emissões atmosféricas: poluentes e sistemas de controle. São Paulo: IEMA, 2016.

[50] Brasil. Ministério do Meio Ambiente, Conselho Nacional de Meio Ambiente. Resoluções CONAMA: resoluções vigentes publicadas entre setembro de 1984 e janeiro de 2012.

Brasília: MMA, 2012b. http://www.mpsp.mp.br/portal/page/portal/documentacao_e_divulgacao/doc_biblioteca/bibli_servicos_produtos/BibliotecaDigital/BibDigitalLivros/TodosOsLivros/Resolucoes-Conama_1984-2012.pdf

[51] Brasil. Ministério do Meio Ambiente. Avaliação dos impactos econômicos e dos benefícios socioambientais do Proconve. Brasília: Edições Ibama, 2016. https://www.ibama.gov.br/sophia/cnia/livros/LIVROPROCONVEDIGITAL.pdf

[52] Sørensen, M.; Hvidberg, M.; Andersen, Z. J.; Nordsborg, R. B.; Lillelund, K. G.; Jakobsen, J.; Tjonneland, A.; Overvad, K.; Raaschou-Nielsen, O. Road traffic noise and stroke: a prospective cohort study. European Heart Journal. 2011. DOI: https://doi.org/10.1093/eurheartj/ehq466

[53] IBGE. Instituto Brasileiro de Geografia e Estatística. Pesquisa Nacional por Amostra de Domicílios - PNAD. 2016. https://www.ibge.gov.br/estatisticas/sociais/trabalho/9127-pesquisa-nacional-por-amostra-de-domicilios.html?=&t=destaques

[54] Oliveira, R. C.; Santos, J. N.; Rabelo, A. T. V.; Magalhães, M. C. O impacto do ruído em trabalhadores de Unidades de Suporte Móveis. CoDAS, São Paulo, v. 27, n. 3, p. 215-222, 2015. DOI: https://doi.org/10.1590/2317-1782/20152014136

[55] Wunderli, J. M., Pieren, R., Habermacher, M.; Vienneau, D.; Cajochen, C.; Probst-Hensch, N.; Röösli, M.; Brink, M. Intermittency ratio: A metric reflecting short-term temporal variations of transportation noise exposure. Journal of Oxposure Science & Environmental Epidemiology. Vol. 26, 575-585, 2016. DOI: https://doi.org/10.1038/jes.2015.56

[56] Girardi, G.; Sellitto, M. A. Medição e reconhecimento do risco físico ruído emuma empresa da indústria moveleira da serra gaúcha. Estudos Tecnológicos. Vol. 7, n° 1:12-23, 2011. DOI: https://doi.org/10.4013/ete.2011.71.02

[57] Mckinnon A., S. Cullinane, M. Browne, A. Whiteing. Green Logistics: Improving the environmental sustainability of logistics. Kogan Page Limited Press, London, UK. 2010.

[58] ABNT. Associação Brasileira de Normas Técnicas. Norma técnica sobre medição de ruídos tem nova edição. São Paulo, 2019. Disponível em: http://www.abnt.org.br/imprensa/releases/6412-norma-tecnica-sobre-medicao-de-ruidos-tem-nova-edicao

[59] Sarlioglu, B.; Morris, C. T.; Han, D.; Li, S. Driving toward accessibility: a review of technological improvements for electric machines, power electronics, and batteries for electric and hybrid vehicles. IEEE IEEE Industry Applications Magazine, Vol 23, Issue: 1, 2017. DOI: https://doi.org/10.1109/MIAS.2016.2600739

[60] Inkinen, T.; Hämäläinen, E. Reviewing Truck Logistics: Solutions for Achieving Low Emission Road Freight Transportation. Sustainability. 2020. DOI: https://doi.org/10.3390/su12176714

[61] BYD. Caminhões 100% elétricos BYD. 2020. Disponível em: https://www.byd.ind.br/produtos/caminhoes/

[62] JAC. Jac Motors. Veículos elétricos. 2020. https://www.jacmotors.com.br/veiculos/eletricos

[63] VWCO. Caminhão Elétrico VW E-Delivery Supera 30 Mil Quilômetros em Testes em Parceria com a Ambev. Volkswagen Caminhões & Ônibus. São Paulo, 2020. Disponível em: Disponível em: http://www.vwtbpress.com/noticia-interna.php?id=1444

[64] EEA. European Environment Agency. Electric vehicles from life cycle and circular economy perspectives. TERM 2018: Transportation and Environment Reporting Mechanism (TERM) report. EEA Report No 13/2018. Copenhagen, 2018. DOI: https://doi.org/10.2800/77428

[65] Juan, A. A.; Mendez, C. A.; Faulin, J.; De Armas, J.; Grasman, S. E. Electric vehicles in logistics and transportation: a survey on emerging environmental, strategic, and operational challenges. Energies, 9(2), 86, 2016. DOI: https://doi.org/10.3390/en9020086

[66] Lutsey, N.; Nicholas, M. Update on electric vehicle costs in the United States through 2030. International Council on Clean Transportation. ICCT, 2019. https://theicct.org/sites/default/files/publications/EV_cost_2020_2030_20190401.pdf

[67] IPEA. Instituto de Pesquisa Econômica Aplicada. Acidentes de Trânsito nas Rodovias Federais Brasileiras - Caracterização, Tendências e Custos para a Sociedade. Brasil, 2015. https://www.ipea.gov.br/portal/index.php?option=com_content&view=article&id=26277

[68] Santana V.; Moura, M. C. P.; Pedra, F.; Corrêa, H.; Venâncio, J.; Belino, L. Morbimortalidade por acidentes de trabalho em motoristas do transporte de carga, 2006-2012. Boletim Epidemiológico Acidentes de Trabalho. Centro Colaborador da Vigilância aos Agravos à Saúde do Trabalhador (ISC-UFBA/CGSAT-MS), 2013. Http://www.ccvisat.ufba.br/wp-content/uploads/2019/07/MORBIMORTALIDADE-POR-ACIDENTES-DE-TRABALHO-EM-MOTORISTAS-DO-TRANSPORTE-DE-CARGA.pdf

[69] CNT. Confederação Nacional do Transporte. Painel CNT de Acidentes

Rodoviários - Principais dados - 2019. CNT, 2019. https://www.cnt.org.br/painel-acidente

[70] Silva, L. G.; Luz, A. A.; Vasconcelos, S. P.; Marqueze, E. C.; Moreno, C. R. C. Vínculos empregatícios, condições de trabalho e saúde entre motoristas de caminhão. Revista Psicologia: Organizações e Trabalho, 16(2), abr-jun 2016. DOI: http://dx.doi.org/10.17652/rpot/2016.2.675

[71] Russo, F.; Comi, A. Measures for Sustainable Freight Transportation at Urban Scale: Expected Goals and Tested Results in Europe. Journal of Urban Planning and Development. Vol, 137, 2011. DOI: http://dx.doi.org/10.1061/(ASCE)UP.1943-5444.0000052

[72] Novaes, A. Logística e Gerenciamento da Cadeia de Distribuição: estratégia, operação e avaliação. Rio de Janeiro (RJ): Campus Elsevier, 2016.

[73] Vidal, T.; Laporte, G.; Matl, P. A concise guide to existing and emerging vehicle routing problem variants. European Journal of Operational Research. Vol 286, Issue 2, 2020. DOI: https://doi.org/10.1016/j.ejor.2019.10.010

[74] Nugmanova, A.; Arndt, W.; Hossain, M. A.; Kim, J. R. Effectiveness of Ring Roads in Reducing Traffic Congestion in Cities for Long Run: Big Almaty Ring Road Case Study. Sustainability. 2019. DOI: https://doi.org/10.3390/su11184973

[75] Holguín-Veras, J.; Leal, J. A.; Sánchez-Diaz, I.; Browne, M.; Wojtowicz, J. State of the art and practice of urban freight management Part I: Infrastructure, vehicle-related, and traffic operations. Transportation Research Part A: Policy and Practice. Vol 137, July 2020. DOI: https://doi.org/10.1016/j.tra.2018.10.037

[76] Cardenas, I.; Borbon-Galvez, Y.; Verlinden, T.; Van De Voorde, E.; Vanelslander, T.; Dewulf, W. City logistics, urban goods distribution and last mile delivery and collection. Competition and Regulation in Network Industries. 2017. DOI: https://doi.org/10.1177/1783591717736505

[77] Kin, B.; Spoor, J.; Verlinde, S.; Macharis, C.; Van Woensel, T. 2018 Modelling alternative distribution set-ups for fragmented last mile transportation: Towards more efficient a sustainable urban freight transportation. Case Studies on Transportation Policy. Vol 6, Issue 1, 2018. DOI: https://doi.org/10.1016/j.cstp.2017.11.009

[78] Shaheen, S.; Chan, N. Mobility and the sharing economy: Potential to facilitate the first- and last-mile public transit connections. Built Environment, 42(4), 2016. DOI: https://doi.org/10.2148/benv.42.4.573

[79] Sampaio, A.; Savelsbergh, M.; Veelenturf, L.; Woensel, T. Crowd-Based City Logistics. Sustainable Transportation and Smart Logistics. Decision-Making Models and Solutions. Chapter 15. 2019. DOI: https://doi.org/10.1016/B978-0-12-814242-4.00015-6

[80] Joerss, M.; Schröder, J.; Neuhaus, F.; Klink, C.; Mann, F. Parcel delivery - The future of last mile. Travel, Transportation and Logistics. McKinsey & Company, 2016.

[81] Viu-Roig, M.; Alvarez-Palau, E. J. The Impact of E-Commerce-Related Last-Mile Logistics on Cities: A Systematic Literature Review. Sustainability. 2020. DOI: https://doi.org/10.3390/su12166492

Chapter 2

The Degraded Insular Landscape in the Urban-Rural Interface – Application to the Urban Agglomeration of the South of the Island of Tenerife

Miguel Ángel Mejías Vera and Víctor Manuel Romeo Jiménez

Abstract

The urban agglomeration of the south of Tenerife is characterized by its accelerated and explosive conformation since the tourist boom of the 80s of the last century. This speed has caused radical landscape changes that have had environmental, economic, social, and spatial repercussions. We try to extract those landscape patterns that characterize this urban model but also to analyze and quantify the landscape degradation of the urban-rural transition zones existing between the tourist and non-tourist nuclei. Through the cartographic and graphic method, typical of spatial thinking and regional geographical analysis, we combine multiple components that characterize and synthesize the substance of the abiotic, biotic, and cultural elements. As a result, we have a diagnosis where the centrality of the tourist nucleus brings together economic activity, the movement of people and vehicles, but at the same time, allows the development of other former rural-based nuclei, transforming them into residential ones, as well as the explosion of buildings dispersed between them. We propose that planning should be based on the landscape patterns that characterize it, starting from the corridor that links the urban centers of the agglomeration.

Keywords: degraded landscape, urban sprawl, soil sealing, green corridor, eco-corridor, urban agglomeration, compact city

1. Introduction

In a finite space such as an island, the fragmentation of the landscape [1, 2] induces a growing deterioration, mainly due to the abandonment of the agricultural space and the increase of urban dispersion [3]. In Tenerife, this process is more severe, when in the last 30 years hardly any new urban plans adapted to the social, economic, and environmental reality have been drafted. Instead, there have only been adaptations to new regulatory texts applied to old territorial and urban plans. Among many consequences, there is a disruption of biodiversity and natural capital flows, but also a break in the continuity of the structures of the cultural paleo-landscape. This pattern is recurrent in many cases, but "studying cities is a never-ending process there is always more to learn" [4]. If we say that an urban agglomeration is built from a central urban core and a series of smaller peripheral urban centers that are under its influence, perhaps we are not making a big difference, but if its expansion is channeled through its network of road corridors that stretch the built-up space, while at the same time, buildings are constructed between the interstitial spaces indiscriminately, the perceived landscape is not only fragmented but, in many cases, it has deteriorated. The rural-urban interface is not defined. Beyond the large nuclei that make up the urban agglomeration of the south, there are multiple population swarms of all kinds: medium-sized nuclei, scattered micronuclei, many clandestine and self-built, individual scattered buildings, urbanizations, all glued between the industrial agricultural space and the abundant abandoned agricultural space. If we add to this problem that this process has an explosive character, forged in just two decades, 1970–1990 and that we are facing a model of urban agglomeration linked to the implementation of the massive and fordist tourist industry [5], we believe that it is a space that has enough entity to investigate what has happened in its past [6] to explain its present, but above all, and this is where we are, to monitor the changes that will occur in the future.

But can this urban typological model be considered a city? Possibly we cannot yet consider it as a consolidated urban structure, but yes, it is in the process of conformation. If we look at the projection of its planning, we would say that it could be, in the future, the largest in the Canary Islands. The large central urban nucleus, Los Cristianos-Las Américas-La Caleta (Arona-Adeje), can be considered the driving force of the tourist industry in Tenerife, although there are authors who consider that, although it does not have a direct relationship with the phenomenon of industrialization, it did affect all settlements, being a modifying factor of the first magnitude [7]. The numerous studies developed in Spain between the 80s and 90s of the last centuries on this phenomenon are considerable and use typologies such as enclave, nucleus, and even call conurbation to the whole Mediterranean coast. The evolutionary process of these enclaves or nuclei goes from being simple tourist urbanizations to tourist cities creating specific urban spaces destined for recreational consumption [8]. The south of the island of Tenerife could be considered, on an insular scale, a large conurbation, which is related to the rest of the island, but also to other national and international scales. This same idea overlaps with its immediate past. Agricultural activity took the leading role in the change, when the export of crops became the first great socio-spatial modifier of the south, between the 40s and 50s of the twentieth century. The arrival of water for irrigation (Canal del Sur S.A.) and the implementation of thousands of hectares of irrigated crops, generated a large labor supply, causing the movement of the insular and regional working population [6]. But the physical characteristics that made the farms ideal were also ideal for tourism, generating a dialectic for the soil, the water, and the worker [9]. In this relationship, undoubtedly, the weight has shifted to the side of the tourist industry. The price of land, the speed of profit, the large economic margins, etc., as demonstrated by Víctor Martín, turned agrarian income into urban income [10]. Small, medium, and especially large landowners put up for sale thousands of hectares of land, in many cases wasteland and unproductive land, but close to the sea. The property map changed from physical properties (individuals) to different corporate legal figures. The owners of rural land changed legal figures from individuals to corporations. But even today, in its urban perimeter both activities coexist, although, spatially, agricultural production is displaced more towards both extremes NW and SE, freeing land in the perimeter closer to the large and medium urban centers (**Figures 1** and **2**).

The tourist landscape is a product, as in the rest of Spain, of mass tourism linked to the sun and the coast, which began in the 1960s, centered on the Mediterranean arc and the Balearic and Canary Islands. In our case, it is developed from the 60s in a punctual way in traditional coastal settlements: El Médano or the sale of large plots of wasteland along the coast that allow the promotion of Ten-Bel. But it was not until well into the 1980s that the tourist nucleus was formalized, combining tourist infrastructures (hotels, apartments, beaches, ports, shopping centers, leisure infrastructure, industrial estates, etc.) with the original small settlements, to which many other scattered, illegal, and non-formal settlements were added, spreading

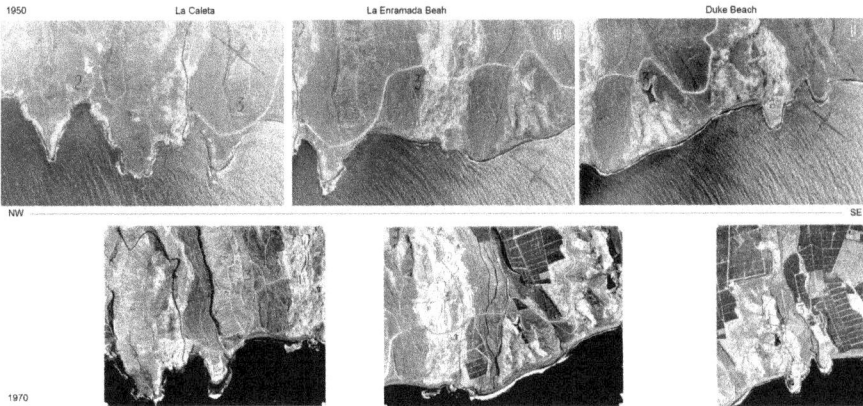

Figure 1.
Landscape of the 1950s and 1970s. Model of transformation from rain-fed agriculture to irrigated agriculture for export. Municipality of Adeje: La Caleta de Adeje (Photo 9), Playa de la Enramada (Photo 10), and Playa del Duque (Photo 11). An area with residual rain-fed agriculture where some small water catchment dams can be observed in the courses of the ravines. Source: 1956 Cadastral Orthophotography, Aerial photograph 1970, and Island Council of Tenerife.

Figure 2.
Urban agglomeration in the Playa del Duque sector (Costa Adeje). 2003. Source: Island Council of Tenerife.

in all directions, following the conception of the processes of the classic ecological system and based on the processes of expansion-aggregation and invasion-succession [11]. The processes of invasion-succession are sustained, when they do not find equivalent resistance, in those cases the type of space occupation is substituted by another. In the south of Tenerife, especially in the coastal areas, the substitution of the agrarian landscape is evident, but also, following H. Gibbard, there is a change in the local population dimension, in the ethnic composition, social stratification, economic activities, residential mobility, the affectation of residential areas, administrative activity, creation of jobs in suburban areas, etc. [11].

Following the ecological method [12] proposed by the Chicago School in the 60s of the last centuries, we could check if the construction of the urban agglomeration in the south of Tenerife followed sustainable patterns adapted to its nature, or on the contrary if it has been built without these logics. That is why we start in this work, from a deterministic point of view, since the urban agglomeration that we analyze is developed from a key environmental justification, the sun, and the sea. Therefore, there are two abiotic and environmental facts of first level, the coast, and the climate. But we also look for cultural components that are the ones that explain the events in a very short timeline. Between them is the whole biotic set that threads them together. This is the transition zone; it is the possible eco-corridor that should be designed.

Our intention is to show a work conceived from spatial thinking.

2. Objectives

This work has two fundamental objectives. On the one hand, to characterize the landscape of the urban agglomeration in the south of the island of Tenerife, placing emphasis on the space occupied by the interface between the different population centers that make it up. It is in these spaces where the different models of growth will be developed, and therefore where it is necessary to intervene on the basis of a landscape policy. On the other hand, it is necessary to construct a method applicable and reproducible to other agglomerations through the combination of multiple landscape components, as well as using different spatial scales, sources, and data models. Therefore, the cartographic and graphical method is substantial.

3. Methodology

Based on spatial thinking, typical of regional geography, we use qualitative and quantitative analyses of spatially based components and variables, following criteria of geographic information processing [13]. To support this method, we use graphics and cartography, supported by geospatial analyses that combine vector data models with raster data models [14]. To do so, it is necessary to proceed with the work by defining the different spatial units of analysis, in this case ranging from a point (location of an activity or the gauging of traffic intensities) to a region (regionalization understood as the sum of municipal entities that share resources and management services). In the middle of this range appear the landscape units [15–17] that are structured from the integrated relationships of abiotic, biotic, and cultural components and that clearly define their identity [18, 19], and functioning. Our study area falls within this pattern and differs from other large island landscape units, such as the metropolitan area or the north. To discover these patterns, it is necessary to follow phases of information processing, that is, at the time of inventory, at the time of processing, and at the time of communication of the results. Clear patterns of graphic semiology [13, 20]. In each of these phases, we proceed to perform qualitative and quantitative analyses of components and variables to characterize their keys [21].

The sources used are multiple. Starting from a general literature review on landscape concepts [15, 22] and their different characterizations [21], spatial distribution measurements [2], and management proposals, catalogs, plans, catalogs, etc. [18, 23]. Strategic concepts of territorial and urban planning, such as the European Territorial Strategy [24] based on the polycentric [24], compact city [25], sustainable [25], and resilient [26, 27] design solutions through ecological and green corridors or green open space [28]. Supported with applied research on the specific region in the geographical field where the hard relations of man in that environment are highlighted but at the same time his adaptation to it. From the slow historical transformation of this space, a vertiginous speed of change took place, first in agriculture, then in tourism, in the second half of the 20th century. In the 1990s, the Geography Department of the University of La Laguna carried out projects, books [9], exhibitions [29], dissertations [6, 30], and theses [10, 31, 32] related to this phenomenon. This space had a great scientific interest, which continues today with new challenges [33]. This work is framed within this line. Cartographic sources: The open data revolution [34] of geographic information allows researchers and analysts of the territory to have a volume of data that were not available until very recently, which is why, from the point of view of the selection processes of sources are directed toward the problems we address, seeking methods of analysis-synthesis that allow us to systematize the multiple relationships that characterize the landscapes. "In the treatment of geographic information, there are three perfectly related and inseparable levels characteristic of any language: data, information, and communication" [20]. The role of spatial data infrastructures in research is marking a new path toward knowledge and this we can implement in our research and results. In this sense, Cartográfica de Canarias (GRAFCAN S.A.), Instituto de Estadística de Canarias (ISTAC), Island Council of Tenerife, Cadastre, and LANDSAT8 support the data that we have converted into the information that we communicate following the following structure:

3.1 Inventory level

From a process of selection and debugging of data sources, we organized the information of this first level. We structure the study area, following landscape science, in thematic spatial components of abiotic, biotic, and cultural character. Each of them is modified in a process of transformation until the objectives of the same are obtained. The analysis of the inventory components will allow us in some cases to characterize the urban agglomeration and check where the conflicts are, in others, to build a space of synthesis were to project the connectors between the nuclei and their natural environments.

3.1.1 Abiotics

Environmental characteristics: From a DEM of the topographic base, we can have a map of insolation. This map, together with the DEM and the location of different meteorological stations, allows us to characterize one of the most important factors in the generation of the current landscape.

3.1.2 Biotic

From the LANDSAT8 satellite image and the combination of 543 NIR bands, we can classify in a supervised way the vegetation space of the bare space that covers the county analysis unit.

3.1.3 Cultural

This space is much more complex. We systematize topographic, cadastral, statistical sources, and raster and vector data models. Each one of them was built for particular purposes of different disciplines.

3.1.3.1 Population structure

We work with 2020 data from the municipal census and select the variables of population, average age, and foreign population. The objective, is to extract the weight of the different settlements and demonstrate which is the driving and attraction core of this urban agglomeration.

3.1.3.2 Structure of the built-up área

We combine topographic data at a scale of 1:5000 for the years 1964, 1987, 1996, and cadastral data of 2015. The objective, is to demonstrate the growth of frequency, dispersion-concentration, and accumulated area and by classes.

3.1.3.3 Road structure

We work on two variables, the main road network and gauging data from different control points of average annual intensities. The objective, is to demonstrate where are the weights of daily mobility of vehicles in the urban agglomeration.

3.1.3.4 Economic structure

Geolocation of economic activity: For punctual implantation, we differentiate the lodging space into different typologies and segregate it from other activities linked to tourism. The objective, is to demonstrate the weight of the economic location of urban centers.

3.1.3.5 Structure of the agricultural space

Using LANDSAT8, extract the cultivated space by means of supervised classification of 654 bands. The objective, is to know the distribution, frequency, and surface of this economic activity. It will serve in the processing phase to combine it in the construction of the geometry of a possible ecological corridor, green or refined eco-corridor.

3.1.3.6 Planning structure

We selected two subjective planning classifications in force, urban planning at the municipal scale of our area of analysis, and the delimitation of protected natural spaces. The objective, is to demonstrate the space projected and committed in the urban planning regulations and its capacity for compactness within the urban agglomeration. The protected natural spaces are those that must be linked through the green corridor with the planned and consolidated urban structures.

3.2 Level of treatment

The Corridor Island [35] as a spatial unit of synthesis. If we reduce the agricultural space and the island is compromised by planning, the result is the eco-corridor island, in short, the unprotected abiotic and biotic space that must thread the urban-rural-natural interface with landscape criteria.

3.3 Level of communication

The whole spatial analytical process leads to a new, more precise unit of analysis: the eco-corridor. On this basis, it will be possible to articulate proposals for landscape integration in future territorial and urban developments.

4. Results

4.1 Inventory level

4.1.1 Abiotics

Environmental characteristics: Without going into the geomorphological and lithological basis that greatly differentiate the landscapes of the unit of analysis, and with a deeper study in this aspect will give us new keys to explain the places of the south, we will focus only on exposing the microclimates as a differentiating factor. Tenerife is a topographically extreme island, but its geographical position and altitude (3718 m) make it differential because it directly affects the generation of microclimates, and that in our study area is very significant. The south of Tenerife, like the rest of the island, from the historical point of view, had a clear settlement pattern, the places with water and fertile soil for agricultural production. For this reason, the humidity factor and the degree of sunshine were very important. Where the environment was less sunny and there was more humidity (**Figure 3**).

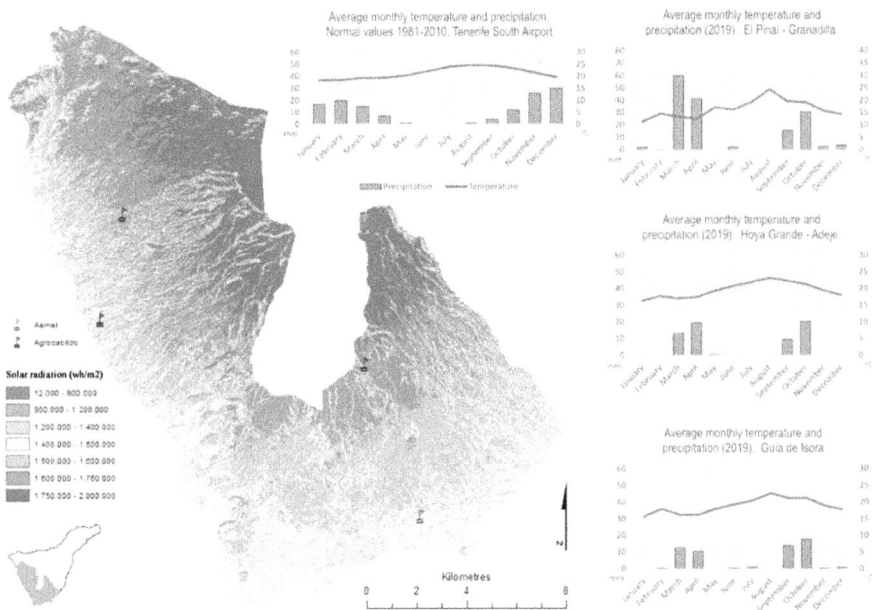

Figure 3.
Representative climograms of coastal and midland areas: (a) Tenerife Sur Reina Sofía Airport. Prolonged series. (b) Hoya Grande, Adeje, SW, elevation 130 m above sea level. (c) Pinal, Granadilla. SE, elevation 850 m. (d) Guía de Isora, SW, 476 m. Sources: AEMET and Agrocabildo. Elaboration: Mejías and Romeo.

These places were not on the coast, but in the middle zones of the island, the so-called "medianías," which range between 400 and 900 m of altitude. As can be seen in the figures, there are enormous differences in temperature and precipitation depending on the altitude, but there are also differences depending on the orientation and position within the island, and this aspect is marked by the wind, the insular space oriented to the SE is swept intensely by the trade winds N-NE to S-SE. On the other hand, the main urban development in the south of Tenerife is protected by the SW. Therefore, coast, light winds, and high insolation are a perfect combination for the development of this industry that formalizes the urban agglomeration of southern Tenerife (**Figure 3**).

4.1.2 Biotic

The vegetation in the area of analysis is irregularly distributed, but clearly has two patterns: the space at high altitudes is occupied by pine forests, broom, and the rest is made up of replacement scrub, often on abandoned cultivation areas, tabaibas, cardones, or balos. The environmental and cultural conditions of intervention in the lowlands make the vegetation very residual and irregular (**Figure 4**).

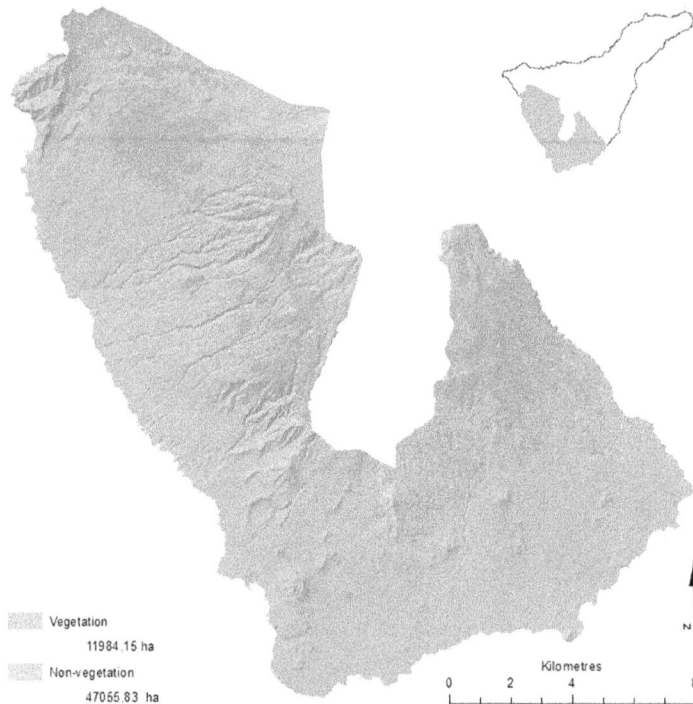

Figure 4.
Biotic synthesis. Source: LANDSAT8. Elaboration: Mejías and Romeo.

4.1.3 Cultural

4.1.3.1 Population structure

We selected the 2020 data. Administratively, the population information of the municipal census of inhabitants is distributed in municipalities, districts, and

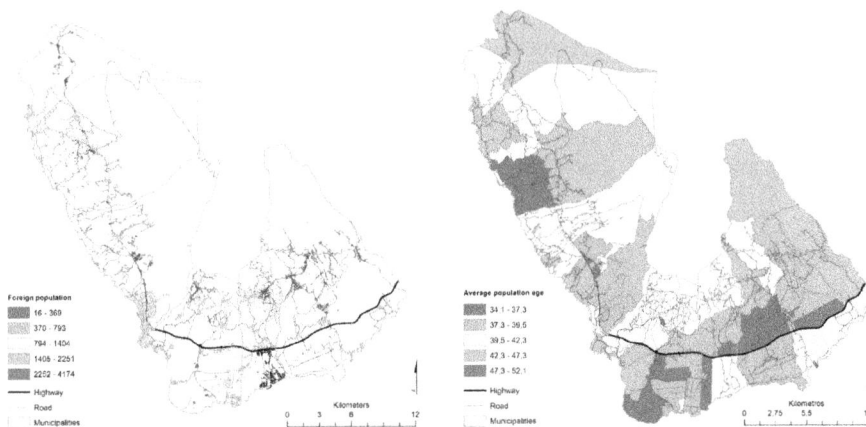

Figure 5.
Distribution of the foreign population by section (a). Population distribution by mean age class and section (b). Source: GRAFCAN S.A. Prepared by: Mejías and Romeo.

MUNICIPALITIES	INHABITANTS
Santiago del Teide	11281
San Miguel	21621
Guía de Isora	21763
Adeje	49030
Granadilla	51233
Arona	82777

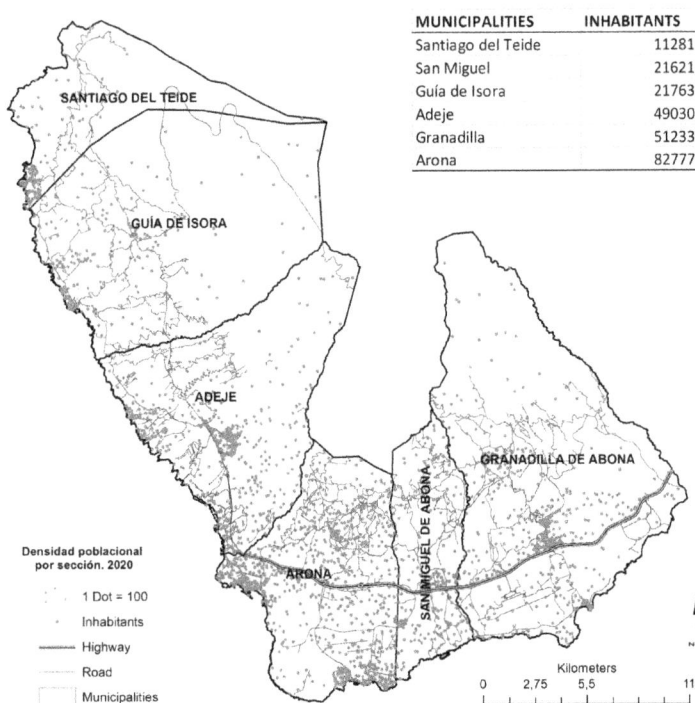

Figure 6.
Distribution of population density by sections. Source: GRAFCAN S.A. Elaboration: Mejías and Romeo.

sections. In order to check the size of the urban agglomeration, we understand that the section offers us more precise information on population movements and we can better check the details of the distribution. The urban agglomeration of the south of Tenerife is inhabited by 244,191 registered persons and an annual transient (tourist) population of 4,601,793 persons. To check the human pressure on the region we must indicate that the total insular total of tourists accommodated in 2019 was 6,071,820, therefore 75% are accommodated in our unit [36]. Undoubtedly this population weight marks the characterization and dynamics of this urban agglomeration. The analysis of this population component currently shows three very significant spatial and statistical patterns: the average age of the area is set at 40 years and its distribution by classes between 32 and 52 years. The gender balance (50.32 men/49.68% women), and the importance of the foreign population (34.85% of the total). Young population: The presence of population between 37 and 41 years of age is the majority group and they are distributed in the sections immediately surrounding the main urban nucleus, while the sections with older average age are distributed spatially between the traditional nuclei of the midlands and the traditional coastal nuclei of Los Cristianos and Las Galletas (Arona). Foreign population: The high percentage of the foreign population has a dispersed spatial distribution, it is distributed throughout the analyzed area, but without a doubt, the highest concentration occurs, first in the sections closest to the coast, secondly, in the central section of the largest urban center and its imme-diate surroundings, thirdly, there is a very significant section where the highest concentration of foreign residents occurs (**Figure 5**). Its location coincides with the proximity of the airport (Granadilla de Abona), the industrial park of Las Chafiras, and Los Abrigos (San Miguel) (**Figure 6**).

4.1.3.2 Structure of the built space

We combined topographic data at scale 1:5000 for the years 1964, 1987, 1996, and cadastral data of 2015. The objective, is to demonstrate the growth of fre-quency, dispersion-concentration, and accumulated area and by classes. The growth dynamics have three very clear phases. The starting point is 1964, the built-up space was in the midlands threaded by the southern general road and the growth to the coast did it by secondary roads to the coast, where were the jetties or coastal ports where all the goods entered and left. For this reason, the development of coastal settlements was minimal. The explosive growth occurred with tourism and the infrastructures created for its formalization: airport, 1978, widening of the TF-1 highway, after the inauguration of the airport, Ferry-Gomera line in 1975 [29]. The structure of the built space grows, but at the same time reduces the average surface area of each polygon, a clear indicator of building dispersion (**Figure 7, Table 1**).

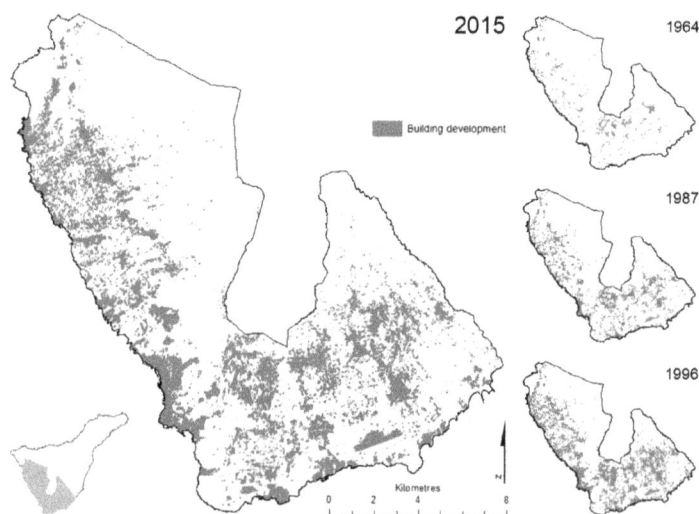

Figure 7.
Evolution of the built-up structure. 1964–2015. Source: GRAFCAN S.A. Island Council of Tenerife, Cadastre. Elaboration: Mejías and Romeo.

Year	Frequency	Area (Ha).	Polygon average (m^2)
1964	5353	123.4	230
1987	13,592	365.3	260
1996	26,568	564.3	210
2015	147,783	1467.6	90

Table 1.
Frequency, cumulative area, and the average size of the polygons of the built-up structure.

4.1.3.3 Road structure

The road structure is divided by hierarchy into a highway linking the metro-politan area of Santa Cruz de Tenerife with the south, which supports the weight of mobility in the region. The southern general highway, parallel to the previous one and with the same purpose, is to connect the island's capital with the south, but link the settlements in the middle of the island. Construction began in the middle of the 19th century and was not completed until the 80s of the 20th century. The structure of roads and secondary roads, many of them rural and unpaved, channel the scattered buildings. To demonstrate the weight of mobility and the centrality factor of the nuclei of the urban agglomeration, we constructed this heat map with the average annual mobility indexes, which indicates the majority weight of the main central nucleus and the axis, more to the E, between the industrial area of Granadilla, the new port, the airport and the urban nuclei of San Isidro and El Médano. This axis is becoming a new and powerful strategic pole of attraction (**Figures 7** and **8**).

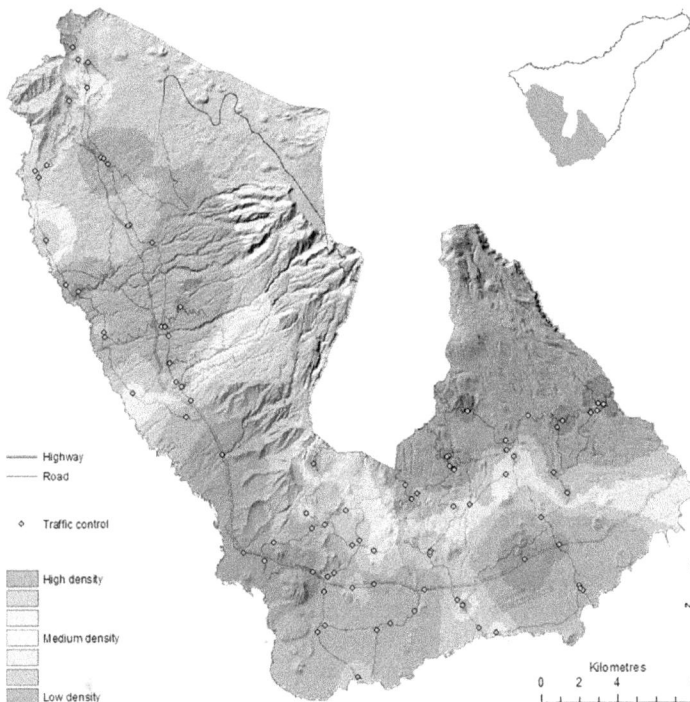

Figure 8.
Level of daily mobility index. Source: Island Council of Tenerife. Elaboration: Mejías and Romeo.

4.1.3.4 Economic structure

Geolocation of economic activity: Undoubtedly, the economic weight generated by the tourist industry is found in the central urban nucleus (Los Cristianos-Las Américas-La Caleta). We have extracted and classified the types of activity separating the lodging activity from the rest of the related activities. The former, in turn, is divided into the traditional hotels and aparthotels, on the one hand, and in the census of vacation homes, on the other, which, in the last decade, have experienced great growth, becoming a variant of the traditional fordist system described above. This has a very significant distribution, is located in the traditional coastal towns (Los Cristianos and La Caleta) and in the urbanizations of the upper coastal zone, exceptionally there are some on the coastline. Therefore, there begins to exist segregation of the lodging activity that we must consider in order to characterize the internal morphology (**Figure 9**).

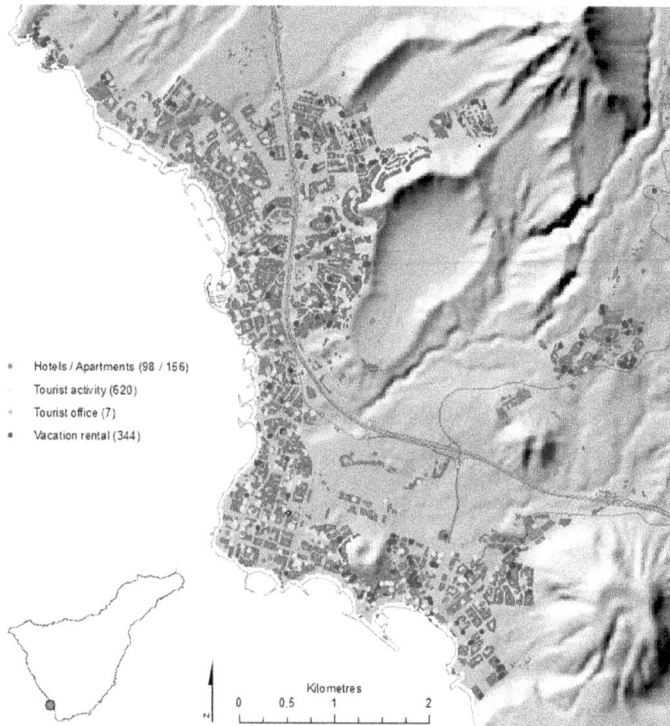

Figure 9.
Distribution of tourism economic activity by type. Detail of the central urban core. Los Cristianos-Las Américas-La Caleta. 2019. Source: Cabildo Insular de Tenerife. Elaboration: Mejías and Romeo.

4.1.3.5 Structure of the agricultural space

The southern landscape was transformed with the arrival of water for irrigation in the 40s and 50s of the twentieth century, the network of canals and secondary water conduction networks allowed the implementation of thousands of hectares of land for export crops (tomato and cotton, in the beginning, banana, mainly at present) [6]. Its current distribution is divided between the coast and the midlands and between the eastern and western sectors. Potatoes are grown in the midland areas of the eastern sector, cultivated on terraces covered with jable

Figure 10.
Synthesis of the agrarian space. Source: LANDSAT8. Elaboration: Mejías and Romeo.

(pumice stone). The coast is mainly reserved for export crops, mainly bananas. The difference between the eastern and western sectors is that the former, which is windier, cultivate under glass, while the western sector, with less wind, cultivates mainly in the open air. This creates totally different landscapes. The construction processes of these agricultural structures are similar to the built structure, land clearing, creation of terraces, soil importation, construction of greenhouses and warehouses, irrigation network. Therefore, we are talk-ing about industrialized agriculture. In the advanced stages of this production system, we have demonstrated processes of the creation of clandestine and self-built micronuclei inside this type of structure. It is a form of conversion of traditional rustic agrarian land to industrial land and then to residential land (**Figure 10**) [32].

4.1.3.6 Planning structure

Based on the current urban planning at the municipal scale, we want to show the urban projection of this urban agglomeration. This synthesis represents the urban-izable corridors. The tendency of the urbanistic model would have been continuity and compactness if, at the end of the 90's, some geomorphologic structures were not declared as protected natural space. This action, at the moment of greatest pres-sure, prevented this continuity, leaving a possible future connection by means of green corridors that link the nuclei with the surrounding agricultural space and the biotic space. This, at least, is our hope, as long as the design linked to the landscape is applied under the criteria of sustainability and resilience. The model we show (**Figure 10**), is the current planning and each of them has approvals from the late twentieth century and adaptations to the new rules of the early twenty-first century [37]. Meanwhile, the growth of urban sprawl at different levels of compaction was explosive (**Figure 11**). The large pockets of developable land give continuity to the

CLASS	FREQUENCY	AVERAGE Has.	AREA Has.
Suspend	1	701,8051	701,81
Land for development	85	34,4867	2931,37
Urban land	142	23,1604	3288,77
Rustic	1014	51,7585	52483,13

Urban planning class
- Suspended
- Developable land
- Urban land

Kilometers
0   3,25   6,5   13

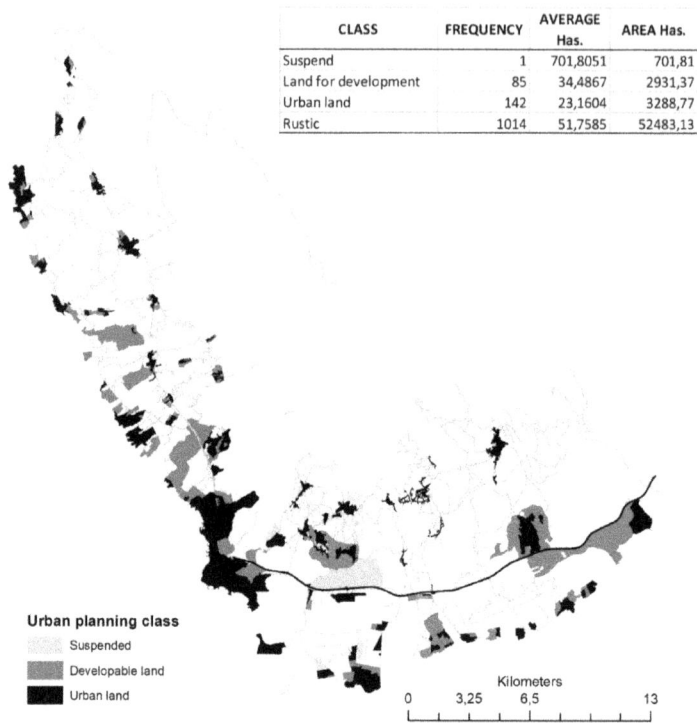

Figure 11.
The projected city. Source: GRAFCAN S.A. Elaboration: Mejías and Romeo.

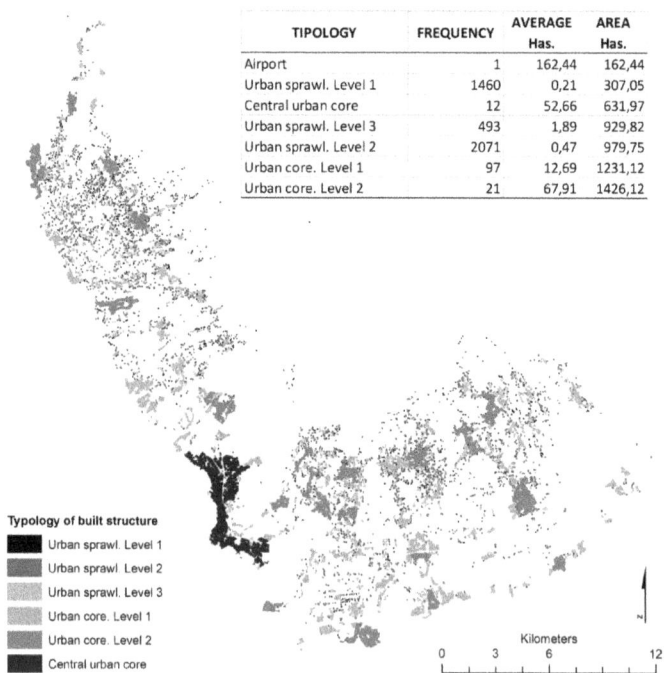

TIPOLOGY	FREQUENCY	AVERAGE Has.	AREA Has.
Airport	1	162,44	162,44
Urban sprawl. Level 1	1460	0,21	307,05
Central urban core	12	52,66	631,97
Urban sprawl. Level 3	493	1,89	929,82
Urban sprawl. Level 2	2071	0,47	979,75
Urban core. Level 1	97	12,69	1231,12
Urban core. Level 2	21	67,91	1426,12

Typology of built structure
- Urban sprawl. Level 1
- Urban sprawl. Level 2
- Urban sprawl. Level 3
- Urban core. Level 1
- Urban core. Level 2
- Central urban core

Kilometers
0   3   6   12

Figure 12.
*Morphology of the built space. Levels of aggregation by area of influence (20 m). Source: Cadastre.
Elaboration: Mejías and Romeo.*

central urban core to the west, the traditional nuclei of medianías are projected in small sectors and the eastern axis of Puerto de Granadilla-San Isidro-Airport projects another large pocket. Arona, and especially the Valle de San Lorenzo, has suspended its urban projection, which was intended to bring together all this dispersed mass (**Figure 12**).

4.2 Level of treatment

Corridor Island [30] as a spatial unit of synthesis. The microanalysis we propose responds to what we have shown in the characterization. We finish as we started, a place as finite as the insular requires processes of spatial microsurgery, so we must refine the units of analysis. The corridor island that we proposed in the previous work, coordinated by Professor Mustafa Ergen [38] proposed a basic corridor island at the island scale. Analyzing larger scale areas allows us to build better models; this is our intention.

4.3 Level of communication

We must project from here the new corridor interface that links the urban agglomeration of the south of Tenerife. The same control that exercised the declaration of natural areas in the 80s will require doing it with new places, more than places, it is necessary to conserve landscapes. The result should help to make planning decisions. The refined eco-corridor will help to do this. We will implement new high-resolution data in the future to continue to monitor these processes (**Figure 13**).

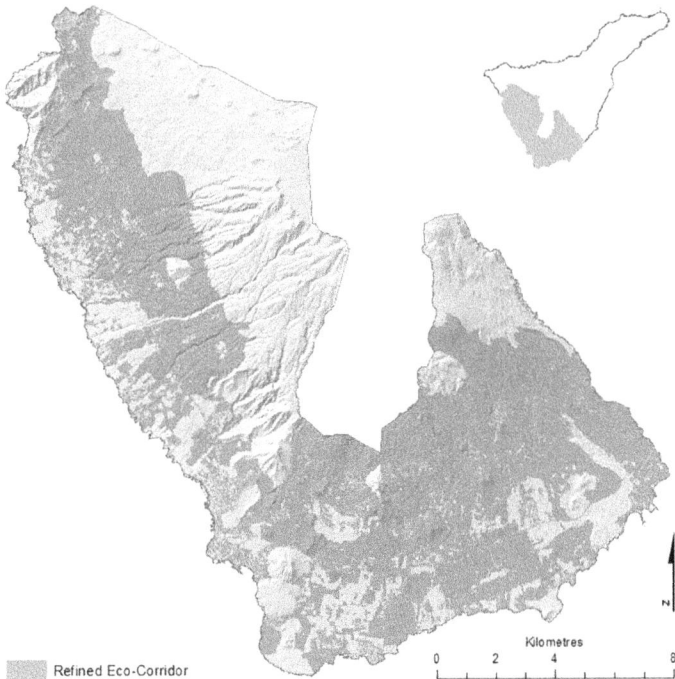

Figure 13.
Refined eco-corridor. Elaboration: Mejías and Romeo.

5. Conclusions

The urban agglomeration that we have tried to show you in this thematic sequence, tells the landscape evolution of an insular region that went in 30 years from oblivion to being a world reference. This perception is reflected and each of the moments of its evolution are impregnated in the landscape of the south. Even today, and despite the passage of time and the indiscriminate and indiscreet intervention of man, the structures of this society remain.

In this current moment of the pandemic by COVID-19, an event has occurred that a few months ago was unimaginable, the tourism machine stopped. The presence of tourists was reduced by 70%, hotels, and aparthotels, travel agencies, stores, schools, transport, airports, ports, were closed. They were prolonged in time. It has been 14 months now. This has shaken the economic structure but above all the social structure. Marginality, hunger, unemployment, and other social consequences denounce if this is the model to be resumed, what has already been called "return to normality." Undoubtedly, we cannot evaluate the effects at this moment, we will do it in the next months or years, but it is necessary to continue investigating the degree of affectation that the population, and therefore, the space of the urban agglomeration.

The analysis shown focuses on the idea of landscape thinking. Planning, managing, educating, and raising awareness in the landscape. It is necessary to interrelate abiotic, biotic, and cultural connections because they are parts of a whole in equilibrium. The imbalances caused by unidirectional decisions provoke crises. The Canary Islands are full of them. We must diversify, invest in what we are powerful in, the renewable energy industry has in the south the most precious source, the sun, and the wind, an industry that also competes spatially, but we have excellent cultural, patrimonial, and natural values. We must not create tourist bubbles. Is it possible to integrate all this, with respect for landscape values without degrading them? We will see.

Conflict of interest

The authors declare no conflict of interest.

Author details

Miguel Ángel Mejías Vera[1*] and Víctor Manuel Romeo Jiménez[2]

1 Faculty of Humanities, Department of Geography and History, University of La Laguna, San Cristóbal de La Laguna, Spain

2 Terrestrial Biodiversity and Island Conservation, Faculty of Science, University of La Laguna, San Cristóbal de La Laguna, Spain

*Address all correspondence to: mmejias@ull.edu.es

IntechOpen

References

[1] Jaeger JA, Mandrinan LF, Soukup T, Schwick C, Kienast F. Landscape Fragmentation in Europe. Joint EEA-FOEN Report. Luxembourg: Publications Office of the European Union; 2011. DOI: 10.2800/78322

[2] Mitchell MGE et al. Reframing landscape fragmentation's effects on ecosystem services. Trends in Ecology & Evolution. 2015;**30**(4):190-198. DOI: 10.1016/j.tree.2015.01.011

[3] Mejías Vera MÁ. ¿Cómo medir el fenómeno urban sprawl a través de indicadores paisajísticos? Aplicación a la isla de Tenerife, Boletín de la Asociación de Geógrafos Españoles. 2013. DOI:10.21138/bage.1569

[4] LeGates R. How to study cities. In: LeGates RT, Stout F, editors. The City Reader. New York: Routledge; 2005. pp. 9-18

[5] Vera JF, López F, Marchena MJ, Antón S. Análisis territorial del turismo. Barcelona, España: Ariel Geografía; 1997

[6] Martín Martín VO. Agua y agricultura en Canarias: el sur de Tenerife. Las Palmas-Santa Cruz de Tenerife: Benchomo; 1991

[7] Precedo A. La red urbana. Madrid, Spain: Editorial Síntesis; 1988

[8] Clavé SA. La urbanización turística. De la conquista del viaje a la reestructuración de la ciudad turística [En línea]. DocumentsAnalisi 1998;0(32). Disponible en: https://www.raco.cat/index.php/DocumentsAnalisi/article/view/31647 [Accedido: 18 May 2021]

[9] Rodríguez Brito W. La agricultura de exportación en Canarias: (1940-1980). Canarias: Consejería de Agricultura y Pesca; 1986

[10] Martín Martín VO. Transformaciones espaciales recientes en el sur de Tenerife. La Laguna: Tesis Universidad de La Laguna; 1997

[11] Castells M. Problemas de investigación en sociología urbana. 4th ed. Madrid, España: Siglo XXI de España editores S.A; 1975

[12] McHarg IL. Proyectar con la naturaleza. Barcelona, España: Gustavo Gili; 2000

[13] Bertin J. La gráfica y el tratamiento gráfico de la información. Madrid, España: Taurus Ediciones; 1988

[14] Bishop W, Grubesic TH. Geographic information. Springer International Publishing AG; 2016

[15] Council of Europe. European Landscape Convention [En línea]. 2000. Disponible en: https://www.coe.int/en/web/conventions/full-list/-/conventions/treaty/176

[16] Antrop M. Background concepts for integrated landscape analysis. Agriculture, Ecosystems and Environment. 2000;77:17-28

[17] Zonneveld IS. The land unit—A fundamental concept in landscape ecology, and its applications. Landscape Ecology. 1989;**3**(2):67-86. DOI: 10.1007/BF00131171

[18] Nogué J, Sala P, Grau J. Els catàlegs de paisatge de Catalunya: metodologia. Olot: Observatori del Paisatge de Catalunya; ATLL, Concessionària de la Generalitat, SA; 2016

[19] Sabaté i Bel J, Mark Schuster J, editors. Designing the Llobregat Corridor. Cultural Landscape and Regional Development. Barcelona, Spain: Universitat Politècnica de

Catalunya & Massachusetts Institute of Technology; 2001

[20] Cortizo Álvarez T. El tratamiento geográfico de la información. Oviedo: Universidad de Oviedo; 2009

[21] Tudor C. An approach to landscape character assessment [En línea]. Natural England. 2014. Disponible en: https://assets.publishing.service.gov.uk/government/uploads/system/uploads/attachment_data/file/691184/landscape-character-assessment.pdf

[22] Antrop M. Landscape change and the urbanization process in Europe. Landscape and Urban Planning. 2004;**67**(1-4):9-26. DOI: 10.1016/s0169-2046(03)00026-4

[23] Cabildo Insular de Tenerife. PTEOP Plan Territorial Especial de Ordenación del Paisaje de Tenerife. Available from: https://www.tenerife.es/planes/PTEOPaisaje/PTEOPaisajeindex.htm

[24] ETE. Estrategia Territorial Europea: hacia un desarrollo equilibrado y sostenible del terrritorio de la UE: acordada en la reunión informal de ministros responsables de ordenación del territorio en Potsdam, mayo 1999. Luxemburgo: Oficina de Publicaciones Oficiales de las Comunidades Europeas; 1999

[25] Rogatka K, Rudge RRR. A compact city and its social perception: A case study. Urbani izziv. 2015;**26**(1):121-131. DOI: 10.5379/urbani-izziv-en-2015-26-01-005

[26] Rosemann J. Permacity. In: Rosemann J, editor. Permacity. Barcelona, Spain: TuDelft; 2007

[27] Naveh Z. Landscape ecology and sustainability. Landscape Ecology. 2007;**22**(10):1437-1440. DOI: 10.1007/s10980-007-9171-x

[28] Gomes M, Pires A, Queiroz da Silva A, Bigate I. Urban agglomeration and supporting capacity: The role of open spaces within urban drainage systems as a structuring condition for urban growth [En línea]. In: Ergen M, editor. Urban Agglomeration. Zagreb, Croatia: IntechOpen; 2018. DOI: 10.5772/intechopen.71658. Disponible en: https://www.intechopen.com/books/urban-agglomeration/urban-agglomeration-and-supporting-capacity-the-role-of-open-spaces-within-urban-drainage-systems-as

[29] Colegio de Arquitectos de Canarias. Demarcación de Tenerife. Comisión de Cultura, El sur de Tenerife: estrategias y paisaje. Santa Cruz de Tenerife: Comisión de Cultura, Demarcación de Tenerife, Colegio de Arquitectos de Canarias; 1991

[30] Sabaté Bel F. Burgados, tomates, turistas y espacios protegidos: usos tradicionales y transformaciones de un espacio litoral del sur de Tenerife: Guaza y Rasca (Arona). Santa Cruz de Tenerife: Servicio de Publicaciones de la Caja General de Ahorros de Canarias; 1993

[31] Sabaté Bel F. El país del pargo salado: naturaleza, cultura y territorio en el Sur de Tenerife (1875-1950). La Laguna: Instituto de Estudios Canarios; 2011

[32] Mejías MA. Tensiones espaciales en el suelo rustico entre las actividades agrarias y Otras Actividades Turístico-residenciales [Tesis doctoral]. La Laguna, San Cristóbal de La Laguna: Sector costero oriental del Municipio de Arona; 2003

[33] García Cruz JI. El impacto territorial del tercer boom turístico de Canarias [Recurso electrónico]. La Laguna: Tesis-Universidad de La Laguna; 2014

[34] de Canarias G. Open data. Available from: http://opendata.gobiernodecanarias.org/opendata/datos-abiertos/que-es/ [accedido: 10 May 2018]

[35] Mejías MA. The corridor island: A new space to redesign the landscape of tenerife [En línea]. In: Ergen M, editor. Urban Agglomeration. Zagreb, Croatia: IntechOpen; 2018. pp. 255-277. Disponible en: https://www.intechopen.com/books/urban-agglomeration/the-corridor-island-a-new-space-to-redesign-the-landscape-of-tenerife

[36] de datos B. Cabildo de Tenerife. Available from: https://www.tenerife.es/bancodatos/

[37] Ley 19/2003, de 14 de abril, por la que se aprueban las Directrices de Ordenación General y las Directrices de Ordenación del Turismo de Canarias. Available from: https://www.boe.es/eli/es-cn/l/2003/04/14/19/con

[38] Ergen M. Urban Agglomeration. Zagreb, Croatia: IntechOpen; 2018

Chapter 3

An Urban Fabric Responsive Last Mile Planning

Chidambara

Abstract

The chapter aims to cover an important and often neglected aspect of transit planning—that of last mile connectivity (LMC). Today most transit systems extend beyond the city to conurbations or metropolitan regions. However, most often LMC planning is on the hindsight or follows a "one shoe fits all" approach, without taking into cognisance the importance of the urban fabric context of the stations. Last mile solutions that do not respond to the built environment context can result in unsustainable mode choice for LMC or in reduced transit appeal. The chapter presents last mile trip characteristics for stations located in different urban fabrics in the city of Delhi and its surrounding town Noida. It explores the attributes of the built environment that impact last mile travel behaviour across the metropolitan region. Additionally, the paper discusses the level of integration, with a lens on the current last mile environment, policy and planning practice for Delhi. The chapter further makes a case for treating LMC planning as integral to transit planning and outlines last mile planning principles suitable for different urban fabrics.

Keywords: last mile connectivity, urban fabric, built environment and travel behaviour, transit access, walk share and urban fabric, Delhi metro, urban rail

1. Introduction

A growing number of cities across varying economies of the world today are nested within urban agglomerations or metropolitan regions. The need for or the factors resulting in co-dependence of the city with its conurbations or satellite towns are well-researched and documented in literature. Geddes in his seminal work, *Cities in Evolution*, nomenclated 'city-regions' or 'town agglomerates' as 'conurbations', identifying them as the future model of urban development; at the same time, underscoring the role of transportation in rendering redundant, the administrative boundaries between the various constituents of a city-region. Describing the absorption of the many villages and boroughs in the development of Greater London, he wrote, "...*Instead of the old lines of division we have new lines of union: the very word "lines" nowadays most readily suggesting the railways, which are the throbbing arteries, the roaring pulses of the intensely living whole;...*" [1]. He further emphasised that different forms of transit systems (rail/trams/buses) will be crucial for such urban development to take shape.

The technological advancements in urban transport (both in automobile and public transport) since that period have been tremendous and we find such models of urban development prevalent economy-wide, albeit, with varying degree of penetration and role of transits. In today's context it is common to see transits extending beyond city boundaries to conurbations or the other entities within their metropolitan regions, playing a vital role in providing several thousands of populations social, cultural, and economic opportunities. Urban rail systems (in all their variants) today have assumed greater significance than ever before, especially in Asian cities. As per a report on World Metro Figures, "at the end of 2017, there were metros in 178 cities in 56 countries, carrying on average a total of 168 million passengers per day. 75 new metros have opened since the year 2000 (+70%). This massive growth is to be credited largely to developments in a few countries in Asia" [2]. Given the pace and nature of urbanisation, the metro rails are likely to play a crucial role in the urban mobility landscape globally, owing to their higher speeds, comfort, safety in comparison to other public transport modes.

In an increasingly globalised economy, the need to connect, both in the physical and the virtual spaces cannot be negated. Travel takes a centre-stage in every urban dweller's life. However, the way people and goods move in a city and across it impacts its socio-economic and physical environment and is one of the key measures of a city's sustainability. Noted economist Colin Clark in his paper *Transport—maker and breaker of cities*, observed that transport is one of the "less tangible implements" that is necessary to create the "end-products" of what is commonly classified as man's basic needs [3]. Developing an effective and efficient public transport thus becomes an indispensable pre-requisite for sustainable mobility and subsequently for sustainable cities. Several scholars recognise the role of transit systems in increasing economic development in cities through the creation of dense urban centres with walking and transit urban fabric [4]. Other benefits of rail transit cited in literature include higher per capita transit ridership, lower per capita traffic fatalities, lower per capita consumer transport expenditure, lower per capita motor vehicle mileage, among others [5]. Through the facilitation of easier access to opportunities, transit systems enhance the catchment and work-sheds which is not only crucial for cities to be globally competitive but also for their overall sustainability. Not surprisingly, we also find that travel patterns and urban forms, in turn, are influenced by the dominance (or absence) of transit outreach.

A substantial volume of scholarly works establishes the link between transit ridership and the surrounding built environment [4, 6–13]. Density (both residential and employment) in particular, is a common indicator across several studies that is found to influence transit ridership. A study of 27 residential areas in California, having different residential densities around metro stations, concluded that higher density residential areas have higher share of transit commute trips [14]. Similarly, transit stations located in higher employment density settings are found to have greater transit shares [15–17]. It is argued that sustainable transport is possible when there is "an emphasis on urban form and density; infrastructure priorities especially the relative commitment to public transport compared to cars; and street planning especially the provision for pedestrians and cyclists", highlighting the importance of other factors apart from densities [18]. This is reiterated through other research studies that have observed higher transit shares in transit and pedestrian-oriented neighbourhoods [10, 19].

1.1 Transit and last mile connectivity

The transit systems in their course, from the city centre to the outskirts and the conurbations traverse different built environment. Alongside, their network density and coverage drop significantly. Planning a transit network that is as dense in the peripheries/suburbs as in the city core might be an almost implausible task. Given this limitation, maintaining the attractiveness of transit, and achieving optimal ridership throughout the system is a big challenge for transit authorities too. It is increasingly accepted that in-transit and out-of-transit experience collectively account for a transit's attractiveness. The last mile connectivity (LMC), referred to in this paper as both the first and the last mile, is an important constituent of the out-of-transit experience, and often, also one of the weakest links of the overall transit journey. The term 'mile' is merely representative, and it can vary from less than half a mile in central parts of the city with dense transit network to significantly over a mile in peripheral areas and conurbations with lower transit network density.

The nature of available options for LMC along with its quality can also have an impact on the catchment sheds of stations located in similar settings, and subsequently on ridership as well. It is important to understand that since the transit coverage itself varies in different parts of the city, the approach to addressing the last mile solution cannot be the same everywhere. While in some areas, it may not be necessary to stress on enlarging the catchment sheds, rather on improving the quality of infrastructure; in other areas the focus necessarily should be on enhancement of the catchment sheds, to enable more areas easier access to transit. This is especially vital for transits that serve metropolitan regions or urban agglomerations. Hence, a pragmatic approach that acknowledges and draws upon the potential and limitations of the physical built context is important to maintain transit attractiveness for higher patronage and greater user experience.

The need for a difference in the approach arises principally out of the difference in the locational context of the stations. Newman, Kosonen and Kenworthy [20] in their 'theory of urban fabric' show that cities are a combination and often overlapping of three distinct types of urban fabric - walking urban fabric, transit/public transport urban fabric and automobile/motor car urban fabric. The 'urban fabric' in this theory signifies "a particular set of spatial relationships, typology of buildings and specific land-use patterns that are based on their transport infrastructure priorities". The three fabrics are distinguishable with respect to aspects such as distance from the city centre, densities, mix of land-use, network typology, characteristics, and quality, among others. The authors further contend that "strategic and statutory planning need to do more than land use and transport integration, and they need to have different approaches in each of the three urban fabrics". Their theory is well applicable and relevant for LMC planning at an agglomeration scale, as well.

This paper includes the findings of a study for Delhi Metro rail (which also serves its satellite towns Noida, Gurgaon, Faridabad and Ghaziabad), which in further sections attempts to show that last mile travel characteristics vary with respect to stations located in different urban fabrics. The paper presents a case for the treatment of LMC planning differently in different urban fabrics. For cities where transits serve an entire agglomeration and/or the suburbs or the surrounding smaller satellite towns, respecting the different urban fabrics in LMC planning becomes even more crucial to maintain the attractiveness of the transit systems and subsequently, for higher transit patronage and greater user experience across all the urban fabrics.

2. Built environment and last mile connectivity

There has been far more research examining the relationship between built environment and transit ridership than on built environment and last mile user trip behaviour. However, from the limited body of literature we can somewhat conclude that urban form surrounding a transit stop is an important decisive factor in transit users' choice of walk, cycle, feeder bus or other forms of transport for the last mile commute. A study conducted in Bogotá examines how the built environment influence walking and cycling behaviour [21]. The authors also observe that while in the developed world, there exists substantial literature that suggest built environment are significant predictors of non-motorised travel, not much research on the same has been carried out in the developing world.

In a study of three European countries namely the Netherlands, UK, and Germany the results indicated that suburbs generate higher levels of cycling-transit users than cities [22]. It would thus be interesting to distinguish the last mile access/dispersal behaviour in city versus the satellite towns in a developing world context. The study also observed that improving the access to railway stations by public transport and non-motorised modes can limit car use. In cities in the developing world this is taken care by a variety of intermediate public transport (IPT) both motorised and non-motorised. Relatively shorter travel distances between common origins and destinations in cities as compared to suburban locations, enables higher walk share. In contrast, in transit-rich, compact cities, transit and walking are attractive alternatives to the bicycle [23]. Moreover, relatively higher densities in cities also makes possible a high-quality feeder bus service with short headways, making them more convenient [24]. The *'Transit Choices Report'* for Santa Clara Valley Transportation Authority corroborates that the pattern of urban development largely determines how many people will be near a stop, whether they can walk to it, and whether transit can follow a path that will be useful to many customers [25]. It identifies density, walkability, diversity of land use, among others as the key indicators of built environment that governs transit ridership.

Yet other studies observe the nature of development around stations influencing non-motorised trip access to stations. Walk/bike share and trip rates were observed to be higher in transit neighbourhoods [10] and walk mode also had higher probability to be used for rail station access in a traditional neighbourhood [11]. Another study found the probability of walking to stations higher when retail uses predominate around stations [26].

Street networks are an important constituent of the built environment. Several studies associate travel behaviour, especially 'walk' share with transportation network. The relative association of street design: local qualities of street environ-ment, street network configuration, spatial structure of the urban grid and land use patterns was studied with the distribution of pedestrian flows in 20 areas in Istanbul [27]. Cervero et al. [21] in their study in Bogota found that the variables that impact most are network characteristics while in developed countries diversity (of land-use) and density impact walking behaviour. They found two network characteristics variables—street density and connectivity index entered the model as significant predictors. Erstwhile, other scholars have used connectivity index [28, 29], street density [30, 31], block length [28, 32], block size [32, 33], block density [11, 34, 35] metric and directional reach and pedestrian detour factor (PDF) (also referred to as pedestrian route directness) [32, 36] as network measures. However, not all these studies have been conducted to understand the last mile travel behaviour per se and it would be interesting to explore whether network characteristics significantly influence station access/egress mode as well.

A study which directly explores this relationship is conducted by the Atlanta Regional Commission [37] which explores "how far urban density, mixed land-uses, and street network connectivity are related to transit walk-mode shares to/from stations". It observes that "local conditions around rail stations are significantly related to riders' choice to walk to/from transit". In particular, the study finds street connectivity to be strongly associated with walk-mode shares when controlling for certain other built and socio-economic attributes.

There is not much conclusive evidence of the relationship between built environment and last mile travel behaviour for cities set in the developing world, which have their own set of uniqueness that set them apart from cities in the developed world. Presence of vertical mixing of land-use areas with fine-grained urban fabric, higher urban densities, poor conditions of walking and cycling infrastructure in several parts of the city, lower automobile-ownership and income levels, presence of the ubiquitous motorised and non-motorised forms of IPT available for individual hire as well as on shared basis, increasing penetration of on-demand/ride hail cabs: all of these present a very different last mile landscape in these cities and city-regions, thereby warranting studies conducted in these settings.

3. Last mile travel behaviour of transit users in different urban fabrics: Delhi, India

The chapter focuses on the relationship of built environment and last mile trip characteristics, based out of a more comprehensive study (also covering last mile mode quality, pedestrian environment and users' socio-economic characteristics) carried out by the author. Hence, the analysis pertaining to only built environment characteristics is presented here. The study was carried out for 10 metro stations of Delhi Metro rail network currently having a network length of 348 kms. Delhi Metro covers the National Capital Territory of Delhi (NCTD), and the surrounding towns of Noida, Gurugram, Faridabad and Ghaziabad. The average daily ridership of the Delhi metro, although not phenomenal, has gradually risen from 0.12 million in 2004–2005 to 2.2 million in 2013–2014, and 2.76 million in 2016–2017. The stations selected for the study lie on the two busiest metro lines and one on a relatively new line, representing low to high ridership levels. The stations were selected also to represent different locational contexts with respect to distance from the city centre, land-use, population and employment densities, and last mile supply/quality. Two of these stations are located in the satellite town Noida. **Table 1** gives the profile of each of the ten case stations and their context.

User surveys were conducted at these 10 stations (collecting 1000 transit user samples) using revealed preference method to understand the users' current first/last mile mode choice and other travel characteristics, their socio-economic characteristics, along with rating and ranking of criteria for mode choice decisions.

3.1 Urban fabric around stations

Each station is set in a built context that represents an urban fabric (although, some overlapping of fabrics is also evident, the dominant fabric is used) discussed in the section1. The core CBD areas which are characterised by high density, mixed land-use (primarily, vertical mixing of residential and commercial at building-use level) and narrow, dense street network, qualifies them as having a walking fabric. Transit fabric are predominantly other medium to high density areas and depending on a combination of criteria such as distance from the core, population/employment density, and contiguous development, they were further sub-classified as representing 'inner' or 'outer' transit fabric. For instance, Noida Sec-15 was classified under 'outer' rather than 'inner' since it is not part of NCT of Delhi and falls in the satellite town of Noida. Similarly, Dwarka Mor and Dwarka Sec-10, although located somewhat close to each other, were categorised differently as having 'inner' and 'outer' fabric respectively owing to the much higher densities in Dwarka Mor vis-à-vis Dwarka Sec-10 and also because Dwarka Sec-10 is still not a fully developed area. The stations qualifying under 'automobile'

Station name	Avg. daily ridership & line name	Adjoining land-use	Population density	Employment density	Representative urban fabric
Chawri Bazar (CB)	30,798 (yellow)	Mixed use Commercial	High	High	Walking
Red Fort (RF)	Low ridership (violet)	Commercial Mixed use Heritage	Medium-high	High	Walking
Dwarka Mor (DM)	42,928 (blue)	Residential	Medium-high	Medium	Inner transit
Green Park (GP)	27,900 (yellow)	Residential Institutional Commercial	Medium-high	Medium	Inner transit
Vishwavidyalay (VV)	23,802 (yellow)	Residential Institutional	Medium	Medium	Inner transit
Mayur Vihar-I (MV)	19,413 (blue)	Residential	Medium	Low	Inner transit
Noida Sec-15 (N15)	29,220 (blue)	Residential Industrial Institutional	Medium	Medium-high	Outer transit
Dwarka Sec-10 (D10)	9761 (Blue)	Residential Institutional	Low	Medium-low	Outer transit
Chhatarpur (CP)	36,036 (yellow)	Residential	Low	Medium-low	Automobile
Noida City Centre (NCC)	37,733 (blue)	Residential Commercial (partially developed)	Medium	Medium	Automobile

Table 1.
Case stations and context profile.

fabric are either in low density peripheral areas or terminal stations in the satellite town. **Figure 1** shows the land-use and network pattern for one station representing each fabric typology.

Population and employment densities for each station were measured from population and employment data available for traffic assessment zones (TAZs) of Delhi from a transport demand forecast study [38]. The density map and locations of the case study stations are given in **Figure 2**.

Ranking the stations from low to high density was a challenge since there is no standard definition across globe of what qualifies as low or high densities within cities. For instance, the Master Plan of Noida has two categories of densities: greater than 500 persons per hectare (PPH) as high density and less than 500 PPH as medium density, while Santa Clara, USA considers below 11.6 PPH as low density and greater than 97 PPH as high density. Besides, there is scant literature available that specifies ranges for employment densities from low to high. As such, the study developed its own ranking methodology of low to high densities: five ranges of densities were identified to distinguish clearly the differences in mode share with varying density conditions. The density ranges and corresponding density rank was developed based on the population and employment density values observed in all the 288 TAZs of Delhi. The low, medium-low, medium, medium-high and high ranges correspond to the densities of all TAZs denoting upto 15th percentile, 15th–25th percentile, 25th–50th percentile, 50th–85th percentile and above 85th percentile respectively. Hence, the low to high densities are relative in the context of the city of Delhi.

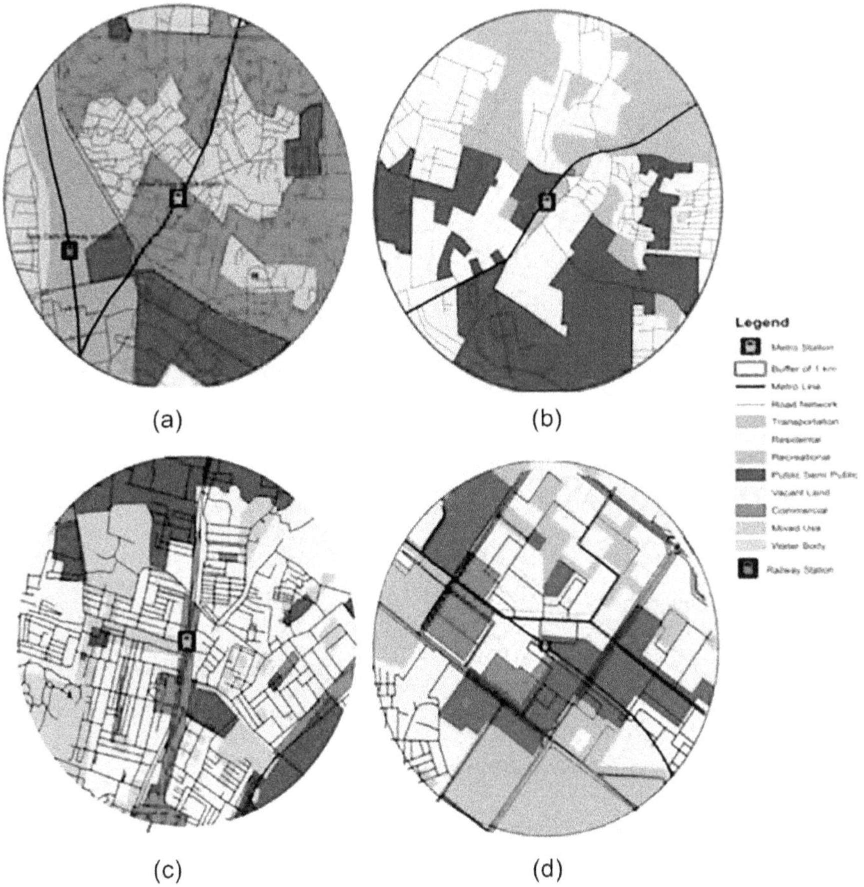

Figure 1.
Land-use and network around stations representing each fabric. a) Chawri Bazar-Walking Fabric. b) Chhatarpur-Automobile Fabric. c) Green Park-Inner Transit Fabric. d) Dwarka Sec-10-Outer Transit Fabric.

3.2 Last mile travel and built environment characteristics of case stations

The last mile trip characteristics including mode share and average trip length (ATL) for the stations are given in **Table 2**. The share of walk trips has a wide variation, the highest being 82.9% while the lowest being 9.4%. The average trip length (ATL) of all modes combined point towards larger catchment-sheds for some stations compared to others. These will be discussed in the context of urban fabric.

Figure 3 gives the distribution of mode share across the ten stations located in different urban fabrics. Several inferences can be drawn from the comparison given in **Table 2** and **Figure 3**. The most evident of these is the decline in the share of walk trips from 'walking' to 'transit' and to 'automobile' fabric and significantly higher share of motorised IPT and private mode trips in transit and automobile fabrics.

High density mixed land-use areas (Chawri Bazar and Red Fort) have higher share of walk trips and shorter overall average trip lengths owing to maximum destinations located within 1 km range. This finding conforms with other studies where it is suggested that people are willing to use slower modes of travel, such as walking, for shorter distances, especially if many trips can be chained [7, 17].

Areas with higher share of institutional use (Vishvavidyalay, Dwarka Sec-10, Green-Park, Noida Sec-15) are observed to have higher share of IPT modes. However, within

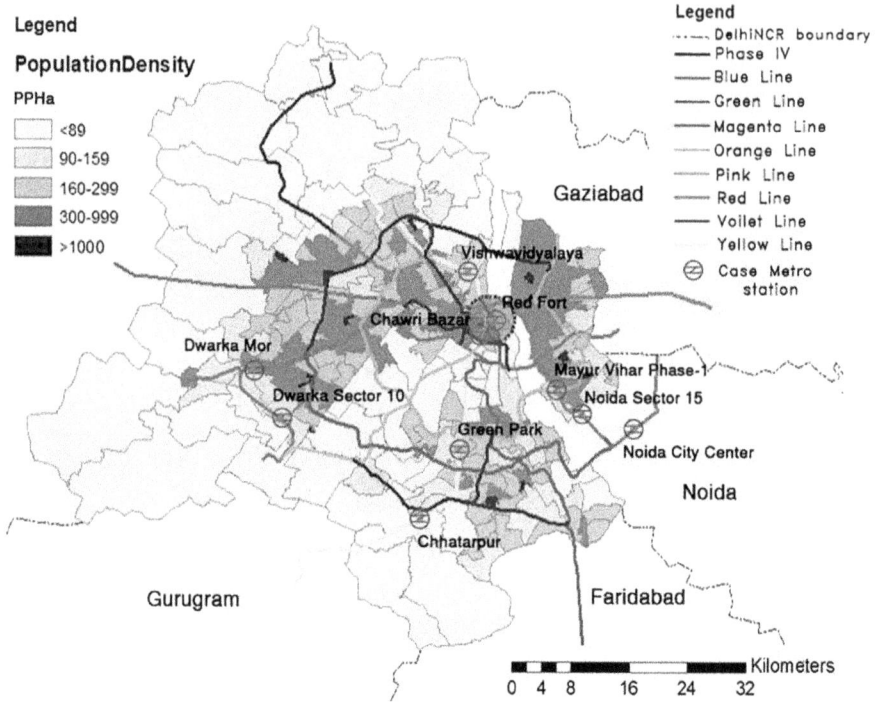

Figure 2.
Population density and station locations on Delhi metro network.

this group, relatively higher activity density areas (Green Park, Noida Sec-15) also have higher share of walk trips. In areas having more than 30% residential land use, it is observed that higher density areas (Dwarka Mor, Green Park, Mayur Vihar) have higher share of walk trips compared to low density residential areas (Chhatrarpur and Dwarka Sec-10). Low to medium density stations located on the peripheries and/or terminal stations (Chhatarpur, Noida City Centre) have the highest overall average trip lengths implying a larger catchment shed. This difference in catchment sheds draws attention to the need for a differential last mile planning approach for stations across a metropolitan region.

The built environment attributes considered for the study were analysed for approximately 1 km buffer around each station. Land-use and network details of areas in 1 km radius around each of the 10 stations were obtained from the *openstreetmaps* and updated through site visits. Network attributes such as network density, node-link ratio (connectivity index), block size and block density were computed on ArcGIS. Another network characteristic of importance is the pedestrian route directness or pedestrian detour factor (PDF) which is the ratio of length of walking distance to the geodesic distance between its start and end points. For the detour analysis all blocks were treated and imported as zones in VISUM. Similarly metro stations were imported as a single zone but with zero area. Two skim matrices were generated: a 'direct distance matrix' and a 'travel distance matrix'. Finally average of all detour factors was taken.

The values of all the network attributes discussed above for the 10 case stations are given in **Table 3**. Most of the stations have network attributes that are within acceptable or recommended levels. However, there is some degree of relative variation, and the models test whether network attributes significantly affect last mile mode shares.

An Urban Fabric Responsive Last Mile Planning
DOI: http://dx.doi.org/10.5772/intechopen.103954

Station name	Mode share (in %)					ATL (in kms)		Predominant land-use share (in%)
	Walk	C.R. + E.R.	A.R. + cab	Bus	Private	Walk	All Modes	
Chawri Bazar	82.9	17.1	0.0	0.0	0.0	0.76	.76	26% mixed 37% commercial
Red Fort	67.3	32.7	0.0	0.0	0.0	0.84	0.89	8% mixed 20% commercial
Dwarka Mor	51.6	25.3	9.9	8.8	4.4	0.70	1.40	70% residential
Green Park	46.0	1.1	51.7	0.0	1.1	0.76	1.59	45% residential 14% institutional
Vishwavidyalaya	41.3	29.1	18.6	5.8	5.2	0.73	1.22	27% residential 30% institutional
Mayur Vihar-I	41.6	21.2	30.1	0.9	6.2	1.18	1.86	50% residential
Noida Sec15	52.4	9.5	36.5		1.6	0.80	1.85	21% residential 30% industrial 20% institutional
Dwarka Sec10	36.1	19.3	28.9	10.8	4.8	0.71	1.75	31% residential 20% institutional
Chhatarpur	9.4	3.5	49.4	9.4	28.2	0.43	2.15	34.8% residential
Noida City Centre	15.4	3.5	74.1	0.5	6.5	1.26	3.15	55% residential 11% commercial

Note: C.R.—cycle-rickshaw; E.R.—E-rickshaw; A.R.—auto-rickshaw.

Table 2.
Mode shares, ATL, and predominant land-use of case stations.

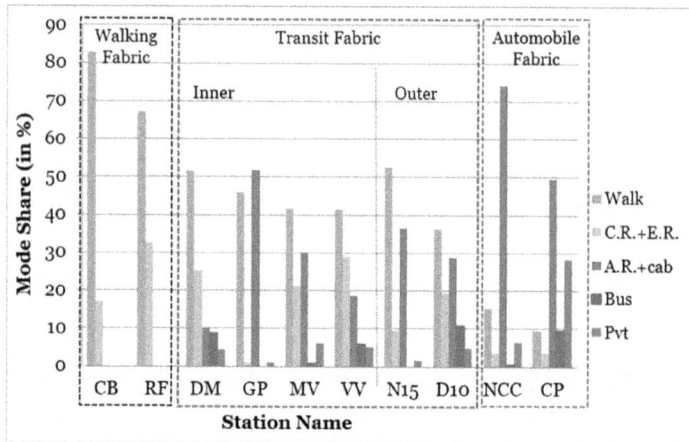

Figure 3.
Mode shares across different urban fabrics.

Station name	Network density (kms/sq.km)	Node-link ratio	Average block size (sq. m)	Block density (no./sq.km.)	Pedestrian detour factor
CB	24.8	1.2	10,705	109	1.49
RF	18.1	1.5	11,886	62	1.41
VV	14.6	1.4	38,952	28	1.17
N15	20.9	1.3	17,502	56	1.71
GP	17.9	1.6	12,287	81	1.59
D10	17.2	1.8	46,234	19	1.67
DM	45.7	1.7	5625	155	1.29
MV	19.9	1.3	16,684	48	1.61
CP	13	1	27,987	29	1.56
NCC	14.9	1.3	19,068	45	1.43

Table 3.
Network characteristics around case stations.

Handy [30] recommends a network density of 26 miles per sq. mile (16.2 km per sq. km) and Mately et al. [31] suggests 18 miles per sq. mile (11.2 km per sq. km) as minimum recommended network density. As can be seen from **Table 3**, almost all case stations have network densities either within these ranges or higher. The station Dwarka Mor has an extremely high network density of 46 km per sq. km. which is due to the presence of exceedingly small block sizes (5625 sqm), which in turn is on account of the area having low-income housing and very small plot sizes.

Further, the recommended and minimum block densities are 160 (62 per sqkm) and 100 per sq. mile (38.6 per sqkm) respectively [11, 34, 35]. Three stations namely Chhatarpur, Dwarka Sector-10 and Vishwavidyalaya have block densities lesser than the minimum figure given above and two stations namely Dwarka Mor and Chawri Bazar have much higher block densities. The remaining five stations have block densities within this range. The connectivity index should be preferably 1.4 or higher and minimum 1.2 [29, 39]. The minimum node-link ratio observed in the 10 case station areas is 1.0 (Chhatarpur). All other stations have connectivity index

An Urban Fabric Responsive Last Mile Planning
DOI: http://dx.doi.org/10.5772/intechopen.103954

higher than 1.2. PDF should preferably be around 1.5 [32] and not more than 1.8 [36]. None of the case stations had a PDF higher than 1.8.

3.3 Impact of built environment attributes on last mile travel

Bivariate regression analysis was carried out between the dependent variables representing last mile travel characteristics and the independent variables representing built environment attributes. Multiple regression was not carried out since the dataset representing the built environment is quite small (just 10—representing 1 for each station), and because some of the network attributes also exhibit multi-collinearity. The dependent variables considered for the models were first/last mile mode share and average trip length (ATL). The specific mode share used in the models was 'walk' since it had the maximum share in almost all case stations except for two; at the same time, it had also wide variations across the stations as reported earlier. Besides, the built environment in 1 km radius around stations is more likely to affect 'walk' shares in comparison to other modes which have much larger catchment area making it unfeasible to study them in detail. The decision to select 'walk' was also guided by the fact that walking is the most common, affordable, and sustainable mode choice (cycling share was quite insignificant across all stations and hence not used) for LMC worldwide, and probably one that is likely to be most affected by the built environment. Hence 'mode share (walk)' and 'ATL (walk)' were selected as the dependent variables.

The independent variables include 'population density', 'employment density', 'network density', 'block size', 'block density' 'link-node ratio' and 'pedestrian detour factor'. Curve Estimation tool under regression module in SPSS was used to check which curve fits best for each of the variable. With the entire dataset, ANOVA value for none of the regression types was observed to be significant. Hence, anomaly (1 data point) in the dataset was identified using 'unusual cases' tool and the models were rerun. The model results (refer **Table 4**) for only the significant variables are shown here.

The regression analysis shows that population density, network density, block density and block size contribute significantly to 'walk' mode share, whereas node-link ratio and network density showed a significant relationship with 'walk' ATL. Population density has the highest and significant relationship with walk share. It

Model	Dependent variable	Independent variable	R^2	ANOVA (P value)	Coefficients	Intercept
1	Mode share (walk)	Population density (in PPH)	0.91	.00025	19.8	−60.8
2	Mode share (walk)	Network density (km/sq.km)	0.72	.004	5.4	−52.7
3	Mode share (walk)	Block density (no./sq.km)	0.55	.023	.60	11.87
4	Mode share (walk)	Block size (sq.m)	0.49	.034	618,438	8.5
5	ATL (walk)	Node Link Ratio Node Link Ratio**2	0.81	.015	4.7 1.6	−2.7
6	ATL (walk)	Network Density Network Density**2	0.52	.05	.06 −.001	−.07

Table 4.
Models results.

has a logarithmic relationship with walk mode share. The finding is substantiated through claims made in other studies where density is thought to shape pedestrian activity by bringing numerous activities closer together, thus increasing their accessibility from trip origins [34, 40]. It is suggested that people are willing to use slower modes of travel, such as walking, for shorter distances, especially if many trips can be chained [7, 17]. However, the Bogota study [21] did not find density and diversity (of land-use) as significant, the reason for which the authors cite could lie on the sample selection of neighbourhoods which consisted of uniformly compact, mixed-use nature. As reported earlier, there is variation in both density and typology of land-use selection in this study and as such contradicts the Bogota study findings.

The models also indicate moderate to high relationship between 'walk' mode share with network density, block density and block size. Among the network attributes, network density has the highest and significant influence on walk mode share. It has a significant linear relationship with walk mode share. There is a significant linear relationship and inverse relationship of block density and block size respectively with 'walk' mode share. There is moderate relationship between 'walk' ATL and network density and link-node ratio (connectivity index). Similar results have been observed elsewhere [21] wherein street density and connectivity index were found to be significant predictors—higher connectivity index and street densities increase the likelihood of walking. The models on 'walk' mode share and link-node ratio (connectivity index) and PDF were not found to be significant. This may be explained by the fact that none of the stations had high PDF values. Also, the relationship between 'walk' ATL versus block Size, block Density, and PDF were not found to be significant and had quite low values of R square.

The study shows that there is distinctive relationship between built environment characteristics and last mile travel behaviour of transit users. Stations located in high activity density mixed land-use areas such as Chawri Bazaar and Red Fort have quite high share of last mile trips made by 'walking'. Within each type of land-use such as those that are predominantly residential, stations located in areas with relatively higher density such as Dwarka Mor have higher 'walk' shares for last mile trips compared to medium to low density areas such as Mayur Vihar-I or Noida City Centre and Chhatarpur. The study also observes that last mile travel behaviour varies across different land-uses, across varying densities within a particular type of land-use and across stations located in peripheries and satellite towns. Unlike the study of European cities cited previously [23] where suburbs had higher share of cycle access to rails, stations located in outer areas in Delhi have higher shares of IPT and private mode usage. Within satellite towns, as densities increase, the share of walk increases. Networks also play a crucial role in influencing walk share for transit access and should be given due consideration in planning of new areas.

4. Last mile mode supply and integration in Delhi

As mentioned in Section 3, Delhi metro provides services in National Capital Territory of Delhi (NCTD) and four surrounding towns of Noida, Ghaziabad, Gurugram and Faridabad, which are part of the National Capital Region (NCR) of Delhi. Although these towns share boundary with NCTD, they lie in different provincial regions (states) of the country, resulting in different administrative jurisdictions. The importance of providing metro connectivity in an integrated manner in the region was acknowledged early on and Delhi Metro Rail Corporation (DMRC) through legislative provisions was given the power for construction, maintenance, and operations of metro rail in these towns. However, the provision of last mile services by DMRC is mostly limited to the NCTD. This undermines the importance of institutional integration at a metropolitan region scale, for the provision of last mile connectivity. It is only recently that some last mile services such as cab aggregator kiosks and authorisation of e-rickshaw services have been initiated at metro stations of the satellite towns as well.

The DMRC's official website has two sections on "passenger information" tab related to last mile connectivity: one for parking and bicycle facilities and another for feeder buses. Interestingly, an important recent addition to the website is pertaining to information on "last mile connectivity" which does not feature on the "passenger information" tab.

Delhi Metro's feeder bus services has a fleet size of just 269 buses [41], with most of them having surpassed their life cycle and with frequency that can be clearly termed as less than satisfactory. There has been no official route rationalisation carried out for the currently operational routes with most of the routes having quite long route lengths and plying primarily on arterial and sub-arterial roads. The website gives a list of feeder buses plying from 32 metro stations [42], giving the names of the location covered under each route. This information is barely useful to commuters since it neither provides a route map of feeder buses nor contains information related to schedule and frequency of service. The physical integration at station is also quite poor. At most of the stations covered in the survey, feeder bus stops were either not clearly demarcated or not integrated with station entry/exit. There was no real time information display of feeder/city bus timings or route guide map, and some stops did not even have a basic bus shelter. While some attempt at fare integration has been recently attempted through introduction of "common mobility card", these are available on an insignificant number of city bus routes, and feeder buses are not covered at all. The DMRC site does not give any information to users on where or how to avail this card.

However, Delhi's metro commuters have the advantage of availability of a wide range of other IPT options for their LMC in the form of cycle-rickshaws, e-rickshaws, auto-rickshaws (for individual hire), shared-auto-rickshaws (plying on semi-fixed routes), jeeps (eg., *gramin sevas*), and mini-buses. These modes are largely demand-driven, and as one can relate from the study findings, also respond to the urban fabric setting. The availability of certain types of modes such as cycle-rickshaws and e-rickshaws at stations are also affected by restrictions on their plying in certain areas. Shared auto-rickshaws and *gramin sevas* plying on fixed or semi-fixed routes are more visible at stations on peripheral areas, thereby enhancing the catchment sheds of these stations. Ride-hailing applications (RHAs) such as Ola, Uber, Meru and other taxi and auto-rickshaw services (individual hire as well as shared) are playing significant role in urban mobility as well as for last mile commute. Besides these IPTs privately and/or company operated chartered bus services also ply to/from some stations. In the last 1 year, Delhi is also seeing some level of penetration of the electric micro-mobility, which could play a significant role in future LMC landscape.

A heartening addition on the DMRC site is the page on "last mile connectivity" which gives some information related to modal integration and/or availability of IPT modes at stations. The referred page gives information on the list of stations where one can avail DMRC-authorised e-rickshaw services, cab aggregator services, e-scooter services and cycle-sharing services. Physical spaces are provided to the operator of these services within the precinct (such as kiosks for cab booking at stations) or outside the station precinct (such as docking facilities for shared cycle services). However, only limited stations have planned spaces available for various IPT modes and the situation is worse for stations in the satellite towns where local agencies are responsible for managing these spaces outside the station precinct; a

Figure 4.
Huge parking spaces for private vehicles but lack of organised space for IPT and buses.

Figure 5.
Chaotic environment outside a station due to lack of physical integration.

few stations have ad-hoc demarcated spaces, primarily located on service roads. The agency has not yet facilitated formal integration of metro system with semi-fixed route shared IPT modes serving the stations and its catchment and subsequently has no information related to the same. The lack of physical integration of the same sometimes results in chaotic and unsafe environment outside station premises (**Figure 5**).

Parking facilities are available at 105 stations with a total area of 32 Ha (for 101 stations) [42] which averages to approximately 3100 sq.m. per station as area under parking. Most stations provide surface parking facilities, and as such this land has not been put to other uses. Various studies have pointed out that the space needed for parking and access of private modes adds significantly to the cost of transit stations and attenuates environmental and traffic benefits of transit service. This negates the very objective of curtailment of automobile usage/dependence in cities like Delhi, which is key to sustainable mobility. While some stations provide huge areas for private vehicle parking, the same cannot be said about planned spaces for IPT modes (Refer **Figure 4**). Operator survey of IPT modes at the 10 case stations revealed 85% citing lack of adequate and designated planned spaces (and subsequently harassment by police/civic agencies) as a key issue.

The National Urban Transport Policy (NUTP), 2014 [43] for the first time explicitly covers "last mile connectivity". The term is mentioned at four places compared to zero in the NUTP, 2006. The new policy document broadens the scope of multi-modal integration to include "private modes of transport i.e., walk, cycle, cars and 2-wheelers and para transit modes i.e., tempos, autos, minibus and cycle rickshaw to the mass rapid transit network" which was limited to "integration of buses with Metro rail" in the previous transport policy [43]. The policy also recognises the significance of improving last mile connectivity to public transport through provision of footpaths and cycle lanes, provision of feeder services, and incorporating design principle to promote safety, accessibility, reliability and affordability, among other measures.

Integration—at levels of physical, operational, fare, information and institutional—of last mile services with the transit service is crucial for enhancing the attractiveness of transit. The integration becomes more critical when transit system extends beyond city border to connect areas that fall in other provincial jurisdictions. While DMRC has taken more pro-active approach towards LMC planning in recent years, the idea of "seamless integration", especially at a metropolitan region level is yet a far cry. There is also a need to develop a strategy or framework to cater to the last mile connectivity that responds to the locational context of the station.

5. Conclusion and last mile planning approach

In view of the present research findings, it is reiterated that users behave differently while using stations located under different urban fabrics in the metropolitan region. A singular approach for addressing the issue of last mile connectivity is thus not appropriate. Last mile strategies and planning need to respond to the urban fabric typology of the station context. The study also reveals that the largest share of last mile commuters walk or use various IPT modes to and from metro stations and policies need to cater to their needs first. Further, the use of clean technology modes in the form of cycle-rickshaw and e-rickshaw which already have a large user base need to be promoted and thus requires appropriate regulatory framework which facilitate their operation rather than adopting a restrictive approach towards them. Policies on last mile should prioritise improving walking and cycling environment around transit stations and facilitating integration of low carbon IPTs, especially in terms of physical integration. It is evident that lack of appropriate last mile planning can result in greater dependence on private modes of transportation to access the transit system, especially in stations lying in the automobile fabric. Automobile usage for access/egress to/from stations generates large number of single occupant vehicular trips at the local level, thereby attenuating the environmental benefits of the transit. The most important policy direction that can be drawn from this study findings, is adopting a multi-pronged planning approach incorporating contextual environment to provisioning of LMC, in place of 'one size fits all' approach. A broad strategy could be focus on enhancing walkability in walking fabric; better physical integration and operational environment for IPT in transit fabric; and high-quality and route-rationalised feeder services and shared IPT services in the automobile fabric.

5.1 Recommended last mile planning approach

Keeping in view the fact that a large share of last mile trips across all case stations are covered by walking and since walking is the most sustainable way of last mile access, it is expected that creation of good walk infrastructure will encourage more people to walk the last mile as well as enhance users walk experience. Replogle [44] developed a transit serviceability index which included components such as 'sidewalk conditions', 'biking conditions', 'land-use mix', 'building setbacks', and transit stop amenities. He observed that zones with high transit serviceability indices not only had higher likelihood of use of transit but also had greater probability of walk access to transits. Provisioning of NMT infrastructure thus also makes economic sense.

Globally there is a lot of stress on improving both pedestrian infrastructure and environment for improving LMC. Provision of extensive network of sheltered and landscaped walkways connecting transit hubs is a pre-requisite for an enabling sustainable last mile ecosystem for Indian cities. This is of utmost significance, given the climatic conditions. However, another part of this study published earlier [45], which examined the effect of walkability on last mile travel behaviour, also suggests that it is not sufficient to merely create sidewalks and cycle lanes; other factors such as safety, aesthetics, etc. that contribute to creating the overall walking and cycling environment also determine how well these facilities are used. Hence, creating vibrant spaces along streets connecting transit hubs should be given due importance. This can be attained through paying attention to the built form design in greenfield areas and 'placemaking' practices in brownfield areas where there is limitation on altering the built form.

The share of bicycles for last mile connectivity as observed in this study was quite low. However, this may be on account of poor cycling environment and supporting infrastructure. Biking as the last mile mode is increasingly being given importance across the globe. The share of cycling for LMC could be enhanced through adequate safe biking and bike parking infrastructure. It is not sufficient to have public bike sharing facility only at the station precinct; there should be a network of deposit and hire facilities at several points in the catchment area (especially in institutional and commercial) for higher usage. Creation of bikeways in low-density peripheral and suburban areas can enhance their catchment sheds. It would also be beneficial in the long run as these areas grow denser in due course of time and transition from an automobile-fabric to transit and walking fabric.

A demand-driven and demand responsive system needs to be in place that caters better to connecting the users to their trip-ends. As the study highlights the vital role that IPTs play in providing LMC, it is important to acknowledge their services by integrating them in transit system planning in a concerted and organised manner. Localised loop or hub-and-spoke systems of e-rickshaws, shared auto-rickshaws, shuttle services can be operated in vicinities ranging from 1 to 5 kms (depending on the location of the station and the mode type). A good feeder service for a wider area can help in increasing the catchment sheds of each station. In this context, high frequency feeder bus services planning in peripheral/terminal stations is especially important given their larger catchment-sheds. Demarcation of planned spaces for all last mile modes at station areas and their adequate integration should be mandatory to avoid chaos and safety hazard to users.

At present there is lack of a set framework for last mile planning in the country. A toolkit containing general guidelines for last mile planning for metro stations should be developed which could guide all cities having or planning for transits. Based on the toolkit more specific area level last mile plans can be prepared for each station. These plans should cater to both station precinct level requirements and catchment area of each station. At station level the focus should be placed on seamless integration of last mile modes with the transit in terms of both spatial and non-spatial integration. Catchment area last mile planning can be more local context specific (responsive to particular urban fabric and socio-economic mix of the population in the area), with focus on making areas safer, active and vibrant for pedestrians and cyclists and facilitating services of modes that are most suited to the locality. However, some facilities should not be compromised upon and kept consistent across all stations, such as, excellent walking and cycling infrastructure and environment.

The planning approach may also be slightly altered for stations located in different urban fabrics in brownfield and greenfield areas. Brownfield stations pertain to those stations that are in areas that are already developed and as such may have limitations in altering of the built elements (except for the redevelopment TOD projects) that are known to encourage walkability. Stations in brownfield areas will be generally located in walking and transit fabrics, and last mile planning should take this into consideration. Greenfield stations pertain to stations that are in the peripheries/fringes in automobile fabric. Although these are generally low-density areas, they offer great opportunity for both station precinct planning and incorporating planning principles that create sustainable built environment and mobility systems. This potential needs to be tapped optimally while planning these areas through incorporating principles of compact development and TOD; mixed use; active frontage; and an efficient road network system that offers connectedness, directness, and permeability. In due course, they can transform to high quality walking and transit fabric.

The study draws our attention to the importance of aligning transit policies with metropolitan region planning as that would enable creating urban fabrics that support sustainable mobility. In the long run it would help in naturally attaining more sustainable last mile behaviour (having higher share of non-motorised trip access to stations) as well as higher transit patronage. Last, but not the least, the role of institutional integration is paramount to providing seamless connectivity, especially for transit systems that serve an entire agglomeration/conurbation/city-region.

Author details

Chidambara
Department of Transport Planning, School of Planning and Architecture,
New Delhi, India

*Address all correspondence to: chidambara17@gmail.com

IntechOpen

References

[1] Geddes P. Cities in Evolution: An Introduction to the Town Planning Movement and to the Study of Civics. London: Williams & Norgate; 1915. p. 27

[2] UITP. World Metro Figures 2018 [Internet]. 2018. Available from: https://www.uitp.org/sites/default/files/cck-focus-papers-files/Statistics%20Brief%20-%20World%20metro%20figures%202018V4_WEB.pdf

[3] Clark C. Transport—Maker and breaker of cities. The Town Planning Review. 1958;**28**(4):237-250

[4] Newman P, Kosonen L, Kenworthy J. Theory of urban fabrics: Planning the walking, transit/public transport and automobile/motor car cities for reduced car dependency. Town Planning Review. 2016;**87**(4):429-458

[5] Litman T. Comprehensive Evaluation of Rail Transit Benefits. Victoria: Victoria Transport Policy Institute; 2004

[6] Sung H, Choi K, Lee S, Cheon SH. Exploring the impacts of land use by service coverage and station-level accessibility on rail transit ridership. Journal of Transport Geography. 2014;**36**:134-140

[7] Marshall N, Grady B. Travel demand modeling for regional visioning scenario analysis. Transportation Research Record: Journal of the Transportation Research Board. 2005;**1921**:44-52

[8] Cervero R. Mixed land uses and commuting: Evidence from the American housing survey. Transportation Research A. 1996;**30**:361-377

[9] Newman P, Kenworthy J. Cities and Automobile Dependence: A Sourcebook. Aldershot, England: Gower Technical; 1989

[10] Cervero R, Gorham R. Commuting in transit versus automobile Neighborhoods. Journal of the American Planning Association. 1995;**61**:210-225

[11] Cervero R, Radisch C. Travel choices in pedestrian versus automobile oriented neighborhoods. Transport Policy. 1996;**3**:127-141

[12] Spillar RJ, Rutherford GS. The effects of population density and income on per capita transit ridership in Western American Cities. In: ITE 1990 Compendium of Technical Papers. Washington, D.C.: ITE; 1990. pp. 327-331

[13] Parsons Brinckerhoff Quade & Douglas Inc. Transit and Urban Form—Commuter and Light Rail Transit Corridors: The Land Use Connection. Transit Cooperative Research Program Report 16. Washington, D.C.: TRB, National Research Council; 1996

[14] Cervero R. Transit-based housing in California: Evidence on ridership impacts. Transportation Policy. 1994;**1**(3):174-183

[15] Cervero R. Rail-oriented office development in California: How successful? Transportation Quarterly. 1994;**48**:33-44

[16] Buch M, Hickman M. The link between land use and transit: Recent experience in Dallas. In: 78th Annual Meeting of the Transportation Research Board; Washington, D.C., 1999

[17] Frank LD, Pivo G. Impacts of mixed use and density on utilization of three modes of travel: Single-occupant vehicle, transit, walking. In: Transportation Research Record 1466. Washington, D.C.: TRB, National Research Council; 1994. pp. 44-52

[18] Newman P, Beatley T, Boyer H. Resilient Cities: Overcoming Fossil-Fuel

Dependence. 2nd ed. Washington: Island Press; 2017

[19] Sasaki Associates Inc. Transit and Pedestrian Oriented Neighborhood. Silver Spring, Md.: Maryland-National Capital Park & Planning Commission; 1993. pp. 47-53

[20] Newman P, Kosonen L, Kenworthy J. Theory of urban fabrics: Planning the walking, transit/public transport and automobile/motor car cities for reduced car dependency. Town Planning Review. 2016;**87**(4):429-458. DOI: 10.3828/tpr.2016.28

[21] Cervero R, Sarmiento OL, Jacoby E, Gomez LF, Neiman A. Influences of built environments on walking and cycling: Lessons from Bogotá. International Journal of Sustainable Transportation. 2009;**3**(4):203-226

[22] Martens K. Bicycle as feedering mode: Experiences from three European countries. Transportation Research Part D: Transport and Environment. 2004;**9**(4):281-294

[23] Keijer M, Rietveld P. How do people get to the railway station? The Dutch experience. Transportation Planning and Technology. 2000;**23**(3):215-235

[24] Krizek KJ, Stonebraker EW, Tribbey S. Bicycling Access and Egress to Transit: Informing the Possibilities. San Jose: Mineta Transportation Institute; 2011

[25] Jarrett Walker Associates. Transit Choices Report. Santa Clara Valley Transportation Authority; 2016

[26] Loutzenheiser DR. Pedestrian access to transit: Model of walk trips and their design and urban form determinants around bay area rapid transit stations. Transportation Research Record. 1997;**1604**

[27] Özbil A, Yeşiltepe D, Argin G. Modeling walkability: The effects of

street design, street-network configuration and land-use on pedestrian movement. ITU A|Z. 2015;**12**(3):189-207

[28] Cervero S, Butler K, Paterson RG. Planning for Street Connectivity— Getting from Here to There. Chicago: American Planning Association; 2003

[29] Ewing R. Best Development Practices: Doing the Right Thing and Making Money at the Same Time. Chicago: American Planning Association; 1996

[30] Handy S. Urban form and pedestrian choices: Study of Austin neighborhoods. Transportation Research Record. 1996;**1552**:135-144

[31] Mately M, Goldman LM, Fineman BJ. Pedestrian travel potential in northern New Jersey. Transportation Research Record. 2001;**1705**:1-8

[32] Hess PM, Moudon AV, Snyder MC, Stanilov K. Site design and pedestrian travel. Transportation Research Record. 1999;**1674**:9-19

[33] Reilly MK. The influence of urban form and land use on mode choice— Evidence from the 1996 Bay Area travel survey. In: Presented at: The Annual Meeting of the Transportation Research Board, Washington, DC. 2002

[34] Cervero R, Kockelman K. Travel demand and the 3Ds: Density, diversity, and design. Transportation Research D. 1997;**2**:199-219

[35] Frank LD, Schmid TL, Sallis JF, Chapman J, Saelens BE. Linking objectively measured physical activity with objectively measured urban form: Findings from SMARTRAQ. American Journal of Preventive Medicine. 2005;**28**(2):117-125. DOI: 10.1016/j. amepre.2004.11.001

[36] Randall TA, Baetz BW. Evaluating pedestrian connectivity for suburban

sustainability. Journal of Urban Planning and Development. 2001;**127**:1-15

[37] Ozbil A, Peponis J. The effects of urban form on walking to transit. In: Greene M, Reyes J, Castro A, editors. Proceedings: Eighth International Space Syntax Symposium. Santiago de Chile: PUC; 2012. pp. 1-15

[38] RITES. Transport demand forecast study and development of an integrated road cum multi-modal public transport for NCT of Delhi. 2010

[39] Handy S, Butler K, Paterson RG. Planning for Street Connectivity— Getting from Here to There. Chicago: American Planning Association; 2003

[40] Kevin JK. Operationalizing neighborhood accessibility for land use-travel behavior research and regional Modeling. Journal of Planning Education and Research. 2003;**22**(3):270-287. DOI: 10.1177/0739456X02250315

[41] Delhi Metro Rail Corporation (DMRC) [Internet]. 2018. Available from: http://delhimetrorail.com/feederbus.aspx

[42] DMRC. [Internet] 2021. Available from: http://www.delhimetrorail.com/feederbus.aspx

[43] Ministry of Urban Development (MOUD). GoI and Institute of Urban Transport India (IUTI). New Delhi, India: National Urban Transport Policy (NUTP); 2014

[44] Replogle M. Computer Transportation Models for Land Use Regulation and Master Planning in Montgomery County, Maryland. Transportation Research Record 1262. Washington, D.C.: TRB, National Research Council; 1990. pp. 91-100

[45] Chidambara. Walking the first/last mile to/from transit: Placemaking a key

determinant. Urban Planning. 2019;**4**(2):183-195. DOI: 10.17645/up.v4i2.2017

Promoting Sustainable Development of Cities Using Urban Legislation in Sub-Saharan Africa

Kasimbazi Emmanuel

Abstract

African countries have been urged to reform their urban policies, practices and laws in order to turn urban areas such as cities and towns into more effective engines of economic growth and play a central role in economic transformation and national development. This chapter examines how urban legislationpromotes sustainable development cities in Africa. Specifically, it discusses the characteristics of cities in sub-Saharan Africa, reviews international legal and policy framework for urban governance and analyses how urban legislation addresses sustainable development aspects in four Africa cities namely: Addis Ababa, Accra, Kampala and Johannesburg.

Keywords: promotion, sustainable development, cities, urban legislation, Africa

1. Introduction

Africa is the most rapidly urbanizing region of the world and has immense urban challenges, such as growing slums and growing poverty and inequality, combined with weak government capacity [1]. Other challenges include land allocation and land use management, provision and management of basic infrastructure/services, such as water, sanitation, and waste management, the movement/accessibility system [2]. The urbanization in Africa need to comply with the goals and principles as developed in the international policies and instruments. The Vancouver Declaration (Habitat I) 1976 recognized the growing impact of urbanization and the need to secure political commitment for sustainable urban. The World Commission on Environment and Development (1987) (Brundtland Report) defined the concept of Sustainable development as "development that meets current needs without jeopardizing future generations' ability to meet their own needs" [3]. Therefore, economic development, social equity, and environmental preservation are the key variables of sustainable urban development. should all be included in sustainable urban development [4]. The 1992 Agenda 21 under Chapter 7 was dedicated to promoting sustainable human settlement development, particularly the urban and rural poor. The Sustainable Development Goal under goal 11 aim to make cities and human settlements inclusive, safe, resilient and sustainable and the World Cities Report 2020 called for well-planned, managed, and financed cities and towns create economic, social, environmental and other unquantifiable value that can vastly improve the quality of life of all [5].

The extraordinary projected rate and scale of urban growth in Africa between now and 2030 underscores the need to urgently develop urban laws and regulations that will create and shape cities that work more efficiently, treat people more fairly and address the urban challenges [6]. New urban infrastructure should be built, new urban growth regions should be established and new city governance and management systems should be implemented [6]. All of this should be done in accordance with laws that provide clarity, ensure that everyone is heard, prepare cities for a climate-change-resilient future, and provide efficient decision-making and administration systems [6]. In doing this, there must be harmonization of the national urban laws with global commitments and calls for urban reforms to enable better urban management [6].

The purpose of this chapter is to analyze how urban legislation promotes sustainable development of cities in Africa. The chapter is containing five sections. Section 1 gives an introduction to the chapter. In Section 2 the characteristics of cities in Sub-Saharan Africa described. These include population, infrastructure development, public utilities services, environmental management, the challenges and challenges of sustainable development of cities in Africa. Section 3 analyses international legal and policy framework for sustainable development of cities in Africa. Section 4 analyses the implementation urban legislation for development for cities selected cities in Africa: Accra in Ghana, Addis Ababa in Ethiopia, Johannesburg in South Africa and Kampala City in Uganda. Section 5 provides concluding remarks and some recommendations.

2. Characteristics of cities in sub-saharan africa

2.1 Population and urbanization in cities in africa

Currently, Sub-Saharan Africa has the lowest proportion of its population living in urban setting and cities, with 472 million people living in urban areas and cities, accounting for roughly 40% of the region's total population [7]. Sub-Saharan Africa, on the other hand, is the world's fastest urbanizing region, with an annual urban population growth rate of 4.1 percent compared to the global rate of 2% [7]. African cities are forecast to urbanize at a rate of 3.65% annually, adding nearly 350 million new city-dwellers by 2030, and a billion more people are expected to be living in African cities by 2063 [8]. Africans are migrating to the cities, and the continent, which already has the world's youngest and fastest-growing population, is urbanizing at a faster rate than any other portion of the globe [9]. By 2050, Africa's 1.1 billion people will have doubled in number, with more than 80% of the growth taking place in cities, particularly slums [9].

This is particularly evident in the continent's spreading urban populations; the top fifteen most populous cities on the continent all have populations above two million [10]. Lagos, Nigeria's capital, is Africa's largest metropolis, with a population of at least nine million people; it is also one of the fastest-growing cities in the world, so the number is sure to increase [11]. This is followed by Kinshasa in the Democratic Republic of Congo, with a population of roughly 7.7 million people [11]. Nigeria and South Africa, two of Africa's most populous countries, have several cities with enormous populations [10]. Nigeria is also home to the cities of Kano and Ibadan, both of which have populations of approximately 3.5 million people, making them large cities in their own right [10]. Cape Town, South Africa, has a population of about 3.5 million people, but Durban, South Africa, is not far behind with 3.1 million [10]. Johannesburg in South Africa has also 2 million residents, as well as Soweto and Pretoria, each with about 1.6 million residents [10].

Southern Sub-Saharan Africa has the biggest proportion of people living in urban areas in Sub-Saharan Africa (more than 70%), followed by West Sub-Saharan Africa, Central Sub-Saharan Africa, and East Sub-Saharan Africa [7]. Demographics across the region show that the urban population is predominantly youthful [7]. Children and youth (0-24 years) accounted for 62.9% of the overall population of Sub-Saharan Africa in 2015, and 19% of the global young population [7]. In Sub-Saharan Africa, the population aged 0-24 years was 628 million in 2017, with an estimated increase to 945 million by 2050, implying that more children and youth will live in metropolitan regions and cities than in rural areas [7]. Notably, the Sub-Saharan Africa region is anticipated to see a positive gain in its child and youth population by 2050, whilst all other regions in the globe are expected to see reductions [7]. Currently, the population pyramid in Sub-Saharan Africa shows a strong child and youth base that anchors the other age groups in the region [7].

2.2 Infrastructure development in cities in africa

Infrastructure shapes cities and its deficiency makes the cities unattractive [12]. On the one hand, cities have physical infrastructure, which includes physical structures such as transportation, electricity grids, drainage systems, sewage systems, and waste disposal systems that are essential for an economy to function. Cities, on the other hand, have social infrastructure, which consists of facilities that support social services and serve as a backbone for communities and societies, such as hospitals, schools, and universities, as well as economic infrastructure (markets) and public facilities such as community housing and prison [7]. Generally, cities in Sub-Saharan Africa have poor infrastructure due to political instability and corruption, complex geographies, cultural barriers, and lack of technology and capital [13]. In addition, a recent World Bank research on infrastructure identified hurdles for continental economic development in this area. It was discovered that deficient infrastructure in Sub-Saharan Africa, specifically electricity, water, roads, and information and communications technology (ICT), lowered annual national economic growth by 2 percentage points and slashed business productivity by as much as 40%.

One of sub-Saharan Africa's top developmental challenges continues to be the shortage of physical infrastructure [14]. Greater economic activity, enhanced efficiency and increased competitiveness are hampered by inadequate transport, communication, water, and power infrastructure. African cities are being held back not only by a lack of urban infrastructure, but also by a lack of city planning, inefficient land use, regulatory barriers, and vested interests [15]. As a result, cities are expansive, fractured, and hyper-informal [7]. African cities are, unsurprisingly, quite expensive to live in. African cities are 29 percent more expensive than non-African cities with equal income levels, according to the World Bank. Locals pay a stunning 100% more for transportation, 55% more for accommodation, 42% more for transportation, and 35% more for food. All of this slows business down, nearly halving firm productivity while drastically raising consumer products input costs. The infrastructure gaps in Africa are not coincidental [7]. One of the main reasons is that municipal governments and city governments are cash-strapped, struggle tax revenue; and often lack the political discretion and financial autonomy to take action [7]. Rapid population growth also places enormous challenges on existing, and often obsolete and poorly maintained infrastructure and resources [14]. In many African countries, infrastructure limitations, notably in power and logistics, inhibit productivity [14].

Sub-Saharan Africa is still a long way from having universal access to the internet [16]. Only 1 in 5 people in Sub-Saharan Africa utilized the internet in 2017, according to the International Telecommunication Union (ITU), which analyzes internet usage internationally and across nation [16]. While internet coverage in Sub-Saharan Africa has increased significantly over the years, it still lags behind the rest of the globe [16]. Sub-Saharan Africa has not achieved the international goal under target 9.C of the Sustainable Development Goals that calls for the achievement of universal and affordable internet access by 2020 [17].

Currently, more than 100 million urban Africans live just beneath a grid but are unable to connect to the grid due to unreasonably expensive connection prices [18]. In Sub-Saharan Africa, 55 percent of urban people live in slum-like conditions, with many of them being without power or connected illegally [18]. Other city inhabitants have electrical connections, but due to frequent outages and voltage changes, they are unable to gain from them [18]. Applications of the Multi-tier Framework in sub-Saharan Africa, for example, show that about 60% of urban households in Ethiopia and 77% in Rwanda experience 4–14 power outages per week [18].

2.3 State of public utilities services in african cities

Majority of the people most African cities live in unplanned urban areas and informal settlements. As a result, the public utilities services in most African cities are well developed. In assessing the performance of Sub- Sharan Africa cities in provision of public utility service due regard must be place on areas that include accessibility to the large portion of the population in the city, safety in terms of water supply, sufficiency in the area of people accessing to at least meet basic health requirements; for reliability in areas of supply of interruptions of limited duration; for affordability and cost-effectiveness dictate the ability of poor households to afford utilities to meet at least basic needs. Water and sanitation are basic human rights, yet in Sub-Saharan Africa, 42 percent of people do not have access to basic water and 72 percent do not have access to basic sanitation [19]. In urban areas, only 56% of the population have access to a piped water supply (down from 67% fifteen years previously), and just 11% to a sewer connection [19, 20]. Simultaneously, the region is quickly urbanizing is urbanizing rapidly—with the urban population anticipated to rise from 345 million in 2014 to 1.3 billion by 2050 [21]. As a result, there is a big and growing demand for services, as well as a growing funding shortfall [22]. To achieve the Sustainable Development Goals for water and sanitation in Sub-Saharan Africa, investments will have to be enhanced by at least threefold [22].

Sewerage systems in most sub-Saharan cities serve few people [23]. They cover only a small fraction of the urban area and even where available, the connection costs are high and unaffordable for poor households [23]. The cost of a sewer connection can be twice as costly as a water connection for individuals living near a sewerage network [24]. Furthermore, once connected, households are subject to a wastewater charge that can account for as much as 50% (and occasionally as much as 90%) of their water bill [24]. The sewerage service network in many cities is limited to better-off, formal, and planned districts, and even here, the rate of connections has been modest because many families already have on-site sanitation facilities [25]. In some circumstances, a mandatory connection policy for residences within a certain radius of the network has been implemented. Even in these locations, however, many homes have yet to connect, and/or utilities have failed to enforce the connection policy because they are unable to provide a consistent water supply to their customers [25]. However, even in these areas, many households have not yet connected and/or utilities have not enforced the connection policy, as they are unable to ensure a regular water supply to their customers [21].

2.4 Status of environmental management in african cities

Sub-Saharan Africa has been (and still is) a region of varying environmental problems which are inherent or human-caused in quest of development [26]. Africa's urban areas are likely to suffer disproportionately from climate change, as the region as a whole is warming up 1.5 times faster than the global average [15]. Africa's fast urbanizing cities are rapidly depleting their natural capital [27]. Unique characteristics of African urbanization, such as considerably lower per capita incomes, a high reliance on biomass fuels, widespread informal settlement with poor service levels, and cities' exposure to natural disasters such as floods, are putting pressure on the natural environment of African cities and eroding the value of environmental assets [27]. As a result, there is a major risk that Africa's cities will be trapped in a "grow dirty now, clean up later" development path that will be costly, irreversible, inefficient, and detrimental to citizens' wellbeing [27].

The impact of urbanization on the natural environment includes a reduction in the amount of the impact of urbanization on the natural environment includes a reduction in the amount of freshwater available and a degradation in its quality; the rate of natural resource consumption is driven by a number of demographic and economic drivers, including population growth, rises in wealth and living standards, and increases in economic productivity [28]. Urbanization has an impact on the city's ecosystems, as well as the volume and value of services generated by these systems. For example, converting wetlands to agricultural or hard surfaces diminishes the value of the water purification services that wetlands frequently provide; the ongoing stress on urban city environments also has an impact on city biodiversity, since species may be destroyed [29]. Slow progress in addressing climate change in many developed countries is wreaking havoc on the world's least developed countries, particularly in Sub-Saharan Africa, where a lack of preparedness for extreme events, as well as socio-economic and environmental resilience, will intensify the negative impacts of climate change and variability [30]. Only a few regulatory regimes in Sub-Saharan Africa explicitly require climate change to be addressed in an environmental impact assessment, indicating that most nations' environmental legislation lags behind the urgent need to tackle climate change [31]. When done right, environmental social and impact assessments can aid in the development and implementation of better projects that address challenging issues such as climate change, biodiversity loss, urban sprawl, conflicts over increasingly scarce resources, inequity, and new technological opportunities [28]. Environmental Social and Impact Assessments can help build a balanced and sustainable future by critically assessing development actions while they are still being conceptualized, as well as molding and improving the society that future generations will live in., which is not operationalized [28].

Water scarcity is a major worry in most cities as a result of urbanization, and adequate, safe water supplies are a key concern [32]. Because majority of Africa is arid or semi-arid, and 41% of African countries are water-stressed, this is posing problems in many towns [32]. Most African cities are characterized by high levels of pollution [33]. The majority of African cities have high levels of air, water, and solid pollution. Pollution is due to the enormous number of households who use wood as a source of energy, and also industrial emissions, fertilizer use in urban and peri-urban farming, and traffic congestion [33]. Insufficient sanitation, as well as industrial discharges and herbicides and pesticides applications, are all linked to water pollution [33]. Low investments in waste collection services are linked to solid waste issues [28]. The majority of these issues are aggravated by inadequate enforcement of the regulations in place [28]. The combination of sewage and poorly managed industrial effluents and agricultural return-flows has led to critical levels of pollution in many urban river systems, to the point of being hazardous to human health [34].

3. Analysis of international legal and policy framework for sustainable development of cities in africa

There are a number of international and regional instruments which provide guidelines and principles for sustainable development of cities in Africa.

3.1 Stockholm declaration 1972

The Stockholm Declaration represented a first taking stock of the global human impact on the environment, an attempt at forging a basic common outlook on how to address the challenge of preserving and enhancing the human environment. Principle 15 requires that planning must be applied to human settlements and urbanization with a view to avoiding adverse effects on the environment and obtaining maximum social, economic and environmental benefits for all. This Principle implies that development of human settlements in cities should consider environmental effects.

3.2 The Vancouver declaration (habitat I) 1976

The Vancouver Declaration of 1976 was the outcome of the first United Nations Conference on Human Settlements held in Vancouver, Canada, 31 May-11 June 1976. The Declaration provides some guidelines for sustainable development of cities in Africa. In its preamble the Declaration recognized the need for socially and environmentally rational human settlements and the dire consequences of "uncontrolled urbanization and consequent conditions of overcrowding, pollution, deterioration and psychological tensions in metropolitan regions," as well as "rural backwardness," especially in the impoverished world.

The Declaration in Principle 1 (b) requires creating more live able, attractive and efficient settlements which recognize human scale, the heritage and culture of people and the special needs of disadvantaged groups. Further, Principle 13 reaffirmed the human right and responsibility of all persons "to participate individually and collectively in the elaboration and implementation of policies and programmers of their human settlements." These two Principles imply that the development of cities should consider the quality of life and human rights.

3.3 Our common future or Brundtland report, 1987

The Brundtland report defined the term sustainable development to mean "development that meets the needs of the present without compromising the ability of future generations to meet their own needs." This term encompasses the three basic variables which are essential for human beings: economic development, social equity and the preservation of the environment.

The report's sixth chapter, "The Urban Challenge," examines the enormous increase in the urban population of developing countries between 1940 and 1980. It also makes predictions about future trends and encourages Third World cities to develop their capacity to generate and manage urban infrastructure and services. Additionally, the Report emphasizes the issues that many cities in both developing and developed nations are facing, and it urges governments to develop and design explicit settlement strategies to manage the urbanization process.

3.4 Agenda 21 and the Rio + declaration 1992

The agenda 21 was one of the documents that were negotiated during the Rio de Janeiro, Brazil, from 3 to 14 June 1992. It is a non-binding action plan of the United Nations with regard to sustainable development whose aim is to achieve global sustainable development.

Agenda 21 provides guidelines that relevant to cities sustainable development in Africa. The goal of Chapter seven of Agenda 21 is to promote sustainable human settlement development, with the objective of improving the social-economic and ecological quality of human settlements as well as the working and living conditions of all people, particularly the rural and urban poor. Paragraph 7.15 requires that to ensure sustainable management of all urban settlements, particularly in developing countries. It emphasizes that improving urban management requires encouraging intermediate city development in order to relieve pressure on large urban agglomerations as well developing and implementing countries policies and strategies towards the development of intermediate cities. Further, Paragraph 7.19 requires all countries to conduct reviews of urbanization processes and policies in order to assess the environmental impacts of growth and apply urban planning and management approaches specifically suited to the needs, resource capabilities and characteristics of their growing intermediate-sized cities.

The other outcome of the Rio Conference was the Rio Declaration on Environment and Development 1992. It also contains guidelines for cities development and management. Principle 4 provides that in order to achieve sustainable development, environmental protection shall constitute an integral part of the development process and cannot be considered in isolation from it while Principle 11 recognizes the importance of enacting effective environmental legislation. Thus, states are required to enact effective environmental legislation. It further requires that environmental standards, management objectives and priorities should reflect the environmental and developmental context to which they apply.

3.5 The Istanbul declaration on human settlements and the habitat II, 1996

The Istanbul Declaration was the outcome of the Habitat II, the second United Nations Conference on Human Settlements that was held in Istanbul, Turkey, 3–14 June, 1996.

The Habitat Agenda's preamble expressly addresses the issue of gradually rising rural-to-urban migration, especially those in developing countries, which has put great strain on already overburdened urban infrastructure and services. Conflicts arising from the expansion of city suburbs have grown as a result of increased rural migration to cities. The haphazard settlement of this land, which is devoid of urban infrastructure, complicates any green space development.

In its Preamble the Declaration recognizes that among the most key problems facing cities and towns, as well as their residents, is the rise of squatter colonies and improper property use. Under paragraph 6, it recognizes the interdependence of rural and urban development, along with the need to focus development, especially in rural areas and small- and medium-sized towns, while minimizing the deprivation causing and resulting from rural-to-urban migration. This Declaration implies that there is need to minimize rural to urban migration in order ensure sustainable development of cities in Africa.

3.6 The Rio declaration 2012 (the Rio + 20) the future we want

The Rio Declaration 2012 was the outcome of the United Nations Conference on Sustainable Development (UNCSD). The Rio + 20 or Earth Summit 2012 was the third international conference on sustainable development aimed at reconciling the economic and environmental goals of the global community. Its objective was to secure renewed political commitment for sustainable development, assess the progress to date and the remaining gaps in the implementation of the outcomes of the major summits on sustainable development, and address new and emerging challenges.

Paragraph 132 notes that transportation and mobility are central to sustainable development because it can enhance economic growth and improve accessibility. It recognizes the importance of environmentally sound, safe, and economical transportation in improving social fairness, health, city resilience, urban-rural links, and rural productivity. We therefore consider road safety as part of our efforts to promote sustainable development in this area.

The establishment of sustainable transportation systems, such as energy-efficient multimodal transport systems, particularly public mass transit systems, clean fuels and cars, and enhanced transportation systems in rural regions, according to paragraph 133. It further recognizes the need to promote an integrated approach to policymaking at the national, regional and local levels for transport services and systems to promote sustainable development.

Paragraph 134 acknowledges that cities can create economically, socially, and environmentally sustainable societies if they are adequately planned and developed, namely through integrated planning and management approaches. It therefore acknowledges the need for a holistic approach to urban development and human settlements that prioritizes slum upgrading and urban regeneration while also providing affordable housing and infrastructure.

3.7 The sustainable development goals 2015

The 2030 Agenda for Sustainable Development, adopted by all United Nations Member States in 2015, provides a shared blueprint for peace and prosperity for people and the planet, now and into the future.

Some of the goals provide some guidelines for urban development and management. Member states are responsible for making cities and human settlements inclusive, safe, resilient, and sustainable, and this according to SDG 11. Target 11.1 is particularly relevant to cities in Africa, as it requires the government to ensure that everyone has access to adequate, safe, and affordable housing and basic services by 2030, as well as upgrade slums, which necessitates making cities and human settlements inclusive, safe, resilient, and sustainable. The targets for achieving this goal include ensuring access for all to adequate, safe and affordable housing and basic services and upgrade slums, providing access to safe, affordable, accessible and sustainable transport systems for all, improving road safety, notably by expanding public transport, notably by expanding public transport, sustainable people settlement management and planning in all countries, minimising the adverse per capita ecological effects of cities, such as through ensuring proper management of air quality and municipal and other waste management, enabling equitable access to green and public areas that are safe, inclusive, and accessible and putting in place integrated policies and programs for inclusion, resource efficiency, climate change mitigation and adaptation, and resilience hazards, and design and implement holistic disaster risk management at all levels, in accordance with the Sendai Framework for Disaster Risk Reduction 2015-2030.

3.8 The paris agreement on climate change, 2016

The Paris Agreement provides a framework for global climate action. In the preamble, the Agreement calls for the inclusion and networking of all levels of government in order to cope with climate change.

The Agreement contains provisions that provide guidelines for cities development and management. The Agreement's Articles 7 and 8 urge for climate change adaptation to be integrated by national adaptation plans that can be utilized to execute policies, programs, and projects. Cities and towns play a vital role in the development of such national plans since they strive to coordinate and integrate efforts to improve the resilience of major infrastructures in the face of climate-related disasters.

3.9 Habitat III: the new urban agenda 2016

The Habitat III or the New Urban Agenda 2016 was the outcome of the United Nations Conference on Housing and Sustainable Urban Development, took place in Quito, Ecuador, 17–20 October 2016.

The New Urban Agenda is a collective vision for a more prosperous and sustainable future. Urbanization, if well-planned and managed, may be a powerful tool for both developing and developed countries to achieve sustainable development. The issues that are covered in the New Urban Agenda include how to plan and manage cities, towns and villages for sustainable development.

The Agenda implored all national, subnational, and local governments, and all relevant stakeholders, to revitalize, enhance, and create partnerships, as well as improve coordination and cooperation, in accordance with national policies and legislation, in order to effectively implement the New Urban Agenda and achieve the shared vision.

The States committed themselves to promote national, sub-national and local housing policies that support the progressive realization of the right to adequate housing for all, equitable and affordable access to sustainable basic physical and social infrastructure for all without discrimination, including affordable serviced land, housing, modern and renewable energy, safe drinking water and sanitation, safe, nutritious and adequate food, waste disposal. The States further committed themselves. The States also agreed to promote appropriate measures in cities and human settlements to ensure that people with disabilities have equal access to the physical environment of cities, including public spaces, public transportation, housing, education and health facilities, public information and communication, and other public facilities and services, in both urban and rural areas, that are safe, inclusive, and accessible open or provided to the public, in both urban and rural areas, safe, inclusive, accessible, green and quality public spaces, including streets, sidewalks and cycling lanes, squares, waterfront areas, gardens and parks.

The states also committed themselves to increasing the supply of a variety of adequate housing options that are safe, affordable, and accessible to people from all walks of life, while also taking into account the socio-economic and cultural integration of marginalized communities, homeless people, and those in vulnerable circumstances, and avoiding segregation.

The New Urban Agenda 2016 sets standards for the development of cities in Africa which include quality urban settlement are dependent upon the set of rules and regulations and its implementation, establishing the adequate provision of common goods, including streets and open spaces, together with an efficient pattern of buildable plots and developing local fiscal systems that redistribute parts of the urban value generated.

3.10 The AU vision 2063

At a regional level Africa adopted Agenda 2063 (The Africa We Want) provides aspirations towards achieving a prosperous Africa based on inclusive growth and sustainable development. The Vision is a Pan-African people-centered vision and action plan that aims to position Africa for growth over the next 50 years. It puts a strong focus on urban development that includes transformational outcomes by 2023 in urban services such as improvements in living standards having access to safe drinking water and sanitation, electricity supply and internet connectivity to be up by 50% and recycling in cities at least 50% of the waste they generate [34]. The Cities are also expected to meet the WHO's Ambient Air Quality Standards (AAQS) by 2025and also make cities and human settlements inclusive, safe, resilient and sustainable [34].

4. Analysis of urban legislation framework for development of selected cities in Africa

Effective urban legislation is an indispensable pillar of sustainable development of cities because it ensures proper planning. The next section analyses the urban legislation development and its implementation in selective African cities. It also describes the adequacy and how urban legislation in supporting sustainable development and transforming cities into more effective engines of economic growth.

Four major cities namely, Kampala, Johannesburg, Accra and Addis Ababa were selected as case studies to provide a comparative analysis of the implementation. Kampala was selected because it is Uganda's biggest city and is reported to be among the fastest-growing cities in Africa. In addition, it is the only city in Africa managed by Kampala Capital City Authority (KCCA) which is the legal entity established by Act of Parliament that is responsible for the operations of the capital city of Kampala in Uganda. Johannesburg was selected because it is South Africa's biggest city and it is recognized as a major world city and the economic capital of both South and sub-Saharan Africa. Accra was selected because it is the capital and largest city of Ghana and was established as the administrative capital of Ghana in 1877 during the British colonial rule. Addis Ababa was selected because it is one of the fastest growing cities in Africa and a capital city of Ethiopia and it also the diplomatic Centre of Africa because it hosts a number of international organizations such as the headquarters of African Union (AU) and the United Nations Economic Commission for Africa (UNECA).

4.1 Kampala city in Uganda

Uganda is in the East African region and situated 800 kilometers inland from the Indian Ocean, across the equator. It is located between the Equator's 10 29′ South and 40 12′ North latitudes, as well as Greenwich's 290 34′ East and 350 0′ East latitudes. Kenya is on the east, South Sudan is on the north, the Democratic Republic of Congo is on the west, Tanzania is on the south, and Rwanda is on the west. It spans a total area of 241,551 square kilometers, with 200,523 square kilometers of land. Kampala is capital city and largest city of Uganda of Uganda. It is the most populous urban centre with 1.5 million persons [35].

Uganda has developed several pieces of legislation that are intended to promote sustainable development of urban areas such as cities. The Constitution of Uganda 1995 sets an objective for the State to promote sustainable development and public awareness of the need to manage land, air and water resources in a balanced and sustainable manner for the present and future generations [36]. The same Constitution mandates Parliament to make laws for the management of the environment for sustainable development [37]. The parent law on environmental management in the country establishes and mandates the National Environment Management Authority to provide for the management of the environment for sustainable development with powers to make statutory instruments that apply to the whole country including the capital city [38].

According to Article 5 (4) of the Constitution of the Republic of Uganda, Kampala is the capital city for Uganda and is administered by the Central Government; Article 5 (6) empowers the Parliament of Uganda to enact a law that provides for the administration and development of Kampala as the capital city [39]. Therefore, Kampala is managed under the Kampala Capital City Authority Act which provides an administrative arrangement for the city of Kampala and provides for its development and physical planning. The same law grants responsibility upon the Minister of Kampala Capital City Authority to coordinate physical planning in the metropolitan area in consultation with the Ministries responsible for urban development and local governments [40].

A specific law on the establishment and regulation of KCCA has been enacted. Section 7 of the KCCA Act specifies functions of KCCA in administration and development that include initiating and formulate policy, enacting legislations in form of ordinances for the proper management of the Capital City, constructing and maintaining major drains, carrying out physical planning and development control among others [41]. Several ordinances have been passed by the City Council while several bye-laws have been passed by division councils to strengthen and support sustainable development and management of the city as an urban area. The deposit or allow any solid waste to be placed or deposited on his or her premises or on private property, on a public street, roadside [42]. The Urban Agriculture Ordinance regulates use of manure, chemicals, disposal of toxic emissions and wastes, disposal of sump oil, prohibits agriculture in certain areas such as green-belts, road reserves, wetlands, an area less than ten feet away from an open drain-age channel, toxic area as well as parks; and also prohibits use of human waste as manure [43]. The Sewage and Fecal Sludge Management Ordinance regulates the disposal of fecal sludge, fecal sludge transportation, and issuance of licenses of providers of environmental sanitation services [44]. KCCA Act alongside the regulations and bye-laws do not express on make the capital city as inclusive, safe, resilient and sustainable just as is stated by Goal 11 of 2030 Agenda for Sustainable Development. This may Uganda to reach Target 11.1 which requires that by 2030, governments must ensure access for all to adequate, safe and affordable housing and basic services, and upgrade slums which requires making cities and human settlements inclusive, safe, resilient and sustainable.

4.2 Johannesburg city in South Africa

South Africa is one of the most geographically varied countries in Africa. It is located at the southern tip of the African continent. The total land is 1,213,090 km^2. It is bordered by Namibia, Botswana, Zimbabwe, Mozambique and Swaziland. In 2019, the population of the city of Johannesburg was estimated to be about 5,635,127 people making it the most populous city in South Africa. In the same year, the population of Johannesburg's urban agglomeration was put at 8,000,000 people.

There are several pieces of legislation that promote and govern sustainable development of urban areas in South Africa. The Constitution of South Africa of 1996 provides a right to every person to have the environment protected, for the benefit of present and future generations, through reasonable legislative and other measures that prevent pollution and ecological degradation; promote conservation and secure ecologically sustainable development and use of natural resources while promoting justifiable economic and social development.

According to Sections 156(2) and (5) of the Constitution a municipality is empowered to make and administer by-laws for the effective administration of matters over which it has jurisdiction, as well as exercise any power over a matter that is reasonably necessary or incidental to the effective performance of its functions [45]. Johannesburg city makes by-laws under Section 13(a) of the Local Government: Municipal Systems Act, 2000 [46]. Some of the by-laws that concern development of a sustainable urban city include: Air Pollution Control By-laws of the City of Johannesburg Metropolitan Municipality which prohibits air pollution by providing for adoption of reasonable measures to prevent and mitigate air pollution [47]. The Waste Management by-laws provide for reduction, re-use, recycling and recovery of waste so as to minimize the environmental harm; to provide for rules on storage, collection, treatment, transportation and disposal of recyclable, industrial and organic waste [48]. The Water Services By-Laws provide for an environmental impact assessment to be carried out before the provision of the water services can be approved or commenced it also prohibits discharge of sewage, industrial effluent or other liquid or substance and installation of pre-treatment facilities [49].

The Municipal Planning By-law seeks to set up and manage land use scheme and municipal spatial development framework through the established of Municipal Planning Tribunals [50]. Public Health By-Laws provides for prohibition on causing public health nuisances and public health hazards, provides for the issuance of Public health permits, demolition orders, provides for compulsory connection to municipal sewage system, prohibits against obstruction of sanitary services, sets requirements in respect of toilet facilities, prohibits pollution of sources of water supply, sets out duties of salon operators overstore or dispose of waste, and also rules on keeping of animals [51].

4.3 Accra City in Ghana

Ghana is a country in West Africa that shares borders with the Ivory Coast in the west, Burkina Faso in the north, and Togo in the east, spanning the Gulf of Guinea and the Atlantic Ocean to the south. Ghana has a total size of 238,535 square kilometers (92,099 sq mi). Accra is Ghana's capital and largest city, with a population of 2.27 million inhabitants. The Greater Accra Metropolitan Area (GAMA) has a population of roughly 4 million people, making it Africa's 11th largest metro area.

Accra City has the Accra Metropolitan Assembly (AMA) which currently derives its legal basis from Local Government Act, 1993, which currently has been amended as the Local Governance Act, 2016, and under Legislative Instrument [52]. It has departments on waste management, works, physical planning among others [53].The functions of the Accra Metropolitan Assembly include provision of a sound sanitary and healthy environment, planning and development control of all infrastructures within Accra and provision of public safety; it has the Metropolitan Planning Committee which has the overall responsibility for the management of the land use plans and physical development activities [53].

The Environmental Protection Bye-law provides a responsibility onto households, industries, waste management operators, corporate bodies, institutions or any other business to take all necessary measures to protect the environment; it creates a duty that any discharge from a factory, industries, commercial mall, market, institutions, office or household must meet the standards set by the Assembly or other regulatory agency [54]. It also provides for promotion of waste treatment systems; sets a requirement of permits for sand winning activities; provides for rules on the protection of wetlands and water bodies, control of tree felling and vegetation; and advocates through permits the replanting of economic trees [54]. Control of Animals Bye-law prohibits keeping of swine, cattle, sheep or goat and other wildlife in any town/community in the area of authority of the Assembly without a permit issued by the Assembly [55]. Sanitation Bye law provides for guidelines on disposal of solid and liquid waste management [56]. The Cleaning Bye-Law prohibits throwing litter, refuse, or other matter into gutters, drains, or unauthorized places which may cause nuisance or block free passage of running water [57]. Building/Physical Development Bye-law, provide for issuance of development and building permits to developers before commences development [58].

4.4 Addis Ababa city

Ethiopia is a landlocked country located in the North Eastern part of the African continent or what is known as the "Horn of Africa." Ethiopia is bounded by Sudan on the west, Eritrea and Djibouti on the northeast, Somalia on the east and southeast, and Kenya on the south. Ethiopia lies between the Equator and Tropic of Cancer, between the 30 N and 150 N Latitude or 330 E and 480 E Longitude. The country occupies an area of approximately 1,127,127 km^2.

According to article 49 (1) of the Constitution of the Federal Democratic Republic of Ethiopia, Addis Ababa is the capital city of the Federal Democratic Republic of Ethiopia [59]. The specific law to manage the administrative affairs of the city is the Addis Ababa City Government Charter Proclamation No. 87/1997 revised by the Charter Proclamation No.361 of 2003 which under Section 9(8) provides one of the objectives of the City Government to make the City a naturally balanced, clean, green and favorable spot through the prevention of environmental pollution [60]. Section 11 (1) empowers the City Government to make laws and exercise judicial powers as well as executive powers and functions over different matters within its jurisdiction, which empowers the city government to legislate on matters that promote sustainable development of the City [60]. The City Government has the powers and functions under Section 11 (2) (g) to administer the land and the natural resources located within the bounds of the City; and under Section 14 (1) (c) to issue the Master plan of the City [60]. Some of these include the Addis Ababa City Government Immovable Property Registration and Information Agency Establishment Proclamation No.22/2010, the Addis Ababa City Government Procurement and Property Administration Proclamation No.17/2009, the Addis Ababa City Master Plan Preparation Issuance and Implementation Proclamation No.17/2004, the Addis Ababa City Government Civil Servants Proclamation No.6/2008, among others.

Section 11 (2) (1) of the Revised Charter Proclamation, provides that the City Government has the powers and functions to: (a) issue and implement policies concerning the development of the City; (b) approve and implement economic and social development plans; (f) identify, ascertain, and organize municipal services to be delivered at the city, sub-city, and Kebele levels; to provide efficient, effective, and fair services by the use of a variety of service delivery options and public participation; and to make sure that a standardized, acceptable system of service delivery is in place. Some of these functions actualize the achievement of the sustainable development goals related to urban development and management. Some of these functions actualize the achievement of the sustainable development goals related to urban development and management.

In addition Section 11 (2) (1) of the Revised Charter Proclamation states that the City Government has the powers and functions to: In addition Section 11 (2) (1) of the Revised Charter Proclamation states that the City Government powers and functions include the following to administer, develop, and sell houses nationalized under the Government Ownership of Urban Lands and Extra Residences Proclamation No. 47/1975 and administered by the City Government, as well as other houses developed or obtained lawfully by the City Government. This is in line with achieving the New Urban Agenda of 2016. The implementation of the New Urban Agenda implies that governments should establish a legal framework for regulating adequate provision of common goods such as streets and open spaces, together with an efficient pattern of buildable plots.

4.5 Challenges of sustainable development of cities in africa

There are several challenges that affect promoting sustainable development of cities in Africa.

First, most cities lack a holistic urban legislation to regulate key issues to sustainable development of cities especially physical planning framework such as physical infrastructure for human settlement or area and public services such as transport, economic activities, recreation and environmental protection. Most cities lack a holistic urban legislation to regulate key issues to sustainable development of cities especially physical planning framework such as physical infrastructure for human settlement or area and public services such as transport, economic activities, recreation and environmental protection.

Second, due to a multiple of pieces legislation that creates conflicts and duplications of mandate there is limited coordination at the various hierarchies of planning by the different authorities which creates a conflict of interest when making decisions that involve consultations from more than one planning authority. For example, in some cases there are conflicts and duplications between Planning authorities and local authorities. The uncoordinated planning and development leads to uncontrolled sprawling of the major towns, growth of slums and informal settlements, lack of public space and weak coverage of basic infrastructure services, notably water, energy, and sanitation, which makes it difficult to improve welfare in either urban or rural environments.

Third, the emerging rapid increase in urban population poses great challenges leading to overcrowding, traffic congestion, growth of slums and informal settlements, dilapidated housing, food security concerns, and poor sanitation.

Fourth, the existing infrastructure and service provision do not correspond to the growing population demands. Investments in urban infrastructure and services have lagged behind the expanding demographic and economic importance of cities, resulting in the expansion of unplanned settlements, urban poverty, insufficient fundamental urban services, and a deteriorating urban environment. Most slum settlements are inaccessible because of poorly planned transport infrastructure and lack access to clean water and waste management systems.

Fifth, there is limited funding of urban development and management as a result in most cities, there are inadequate financial deal with the escalating urban infrastructure challenge.

Sixth, many cities lack the sufficient human resources to develop and implement plans. Developing such capacity within local planning departments, by using other agencies and engaging the community and interest groups, is the key to producing good plans.

Seventh, high level urban poverty has led to the development of slums characterized by the poor housing conditions, high urban crime rates, homelessness, poor medical care, among others.

Eighth, in some cases, there is political interference in urbanization programmes. Political interference sometimes affects enforcement of action plans for example where the development control decision by the planning authorities affects the political position of the government or individuals, such decisions are revoked and authorities instructed to act on the contrary to their directives.

Lastly, corruption of politicians and technocrats leads to poor decision process and increases the cost of and reduces the benefits from development programmers to the society [61]. As a result, the gap between the potential and realized achievements continue to widen thereby further undermining the country's chances of achieving sustainable development [61].

5. Conclusion and recommendations

The analysis in the paper has demonstrated that characteristics of the most cities in the Sub-Saharan Africa include limited regulation of physical planning which has led to widespread illegal and informal development and this has hindered the extension not only of water, electricity, and solid waste collection services but also of adequate sanitation arrangements and road networks. Further, many African cities have not developed legal and regulatory framework that address the realities of urban life. This is in addition to limited ability to implement existing laws and regulations. In most cases, City governments do have adequate legal experts to develop the appropriate regulatory framework that is enforceable. As a result, the citizens are compelled to follow informal methods to conduct land and property transactions, undertake business, acquire means of a livelihood, and even access fundamental services due to the number and rigidity of laws and regulations. Uncoordinated decision making and implementation as well as political interference led to the failure of City Authorities to cope with the challenge of urban growth in Africa. This is in addition to limited financial and human capacity to support sustainable development of cities.

There are recommendations that can be proposed to improve the urban legislation development and implementation in the African cities. First, it is necessary to develop a holistic urban legislation that provides predictability and order in urban development, from a wide range of perspectives, including physical planning framework such as physical infrastructure for human settlement or area and public services such as transport, economic activities, recreation and environmental protection viewpoints. Second, there is need to strengthen the institutional framework on urban development and management so as to stimulate capacity building to sustainably develop cities. This can be done by establishing appropriate urban planning departments with appropriate mandates and governed by the experts. Third, there is need to strengthen capacity of the physical planners, engineers and architects, decision makers and urban dwellers through training and education on issues designed to meet each city's particular circumstances, although most cities and towns share some urban problems whose solutions may be similar. Fourth, there is need to increase funding for activities for cities development and management at various levels in cities. Lastly, development and planning in cities should be participatory at all levels to facilitate more transparent and collaborative decision-making development decisions. This should involve empowerment urban poor communities and provision of pro-poor services.

Author details

Kasimbazi Emmanuel
School of Law, Makerere University, Kampala, Uganda

*Address all correspondence to: ekasimbazi@yahoo.com

IntechOpen

References

[1] UN-Habitat. The state of African Cities 2014: Re-Imagining Sustainable Urban Transitions. Nairobi: UN-Habitat

[2] Smit W. Understanding the complexities of informal settlements: Insights from Cape Town. In: Huchzermeyer M, Karam A, editors. Informal Settlements: A Perpetual Challenge? Cape Town: Juta; 2006. pp. 103-125

[3] Samson K, Alok T. Urban development in Ethiopia: Challenges and policy responses. 2012

[4] City Alliance. Livable Cities: The Benefits of Urban Environmental Planning. Washington DC; 2007. p. 3

[5] UN-Habitat. World Cities Report 2020. The Value of Sustainable Urbanization. United Nations Human Settlements Programme. UN-Habitat; 2020

[6] Berrisford S, McAuslan P. Reforming Urban Laws in Africa: A Practical Guide. New York United Nations Human Settlements Programme (UN-Habitat) and Urban Land Mark: The African Centre for Cities (ACC), Cities Alliance; 2017

[7] Githira D et al. Analysis of Multiple Deprivations in Secondary Cities in Sub-Saharan Africa. UNICEF; 2020. Available from: https://www.unicef.org/esa/media/5561/file/Analysis%20of%20Multiple%20Deprivations%20in%20Secondary%20Cities%20-%20Analysis%20Report.pdf [Accessed: July 4, 2021]

[8] Griffiths P. Bolstering Urbanization Efforts: Africa's Approach to the New Urban Agenda Monday. Africa Growth Initiative, Brookings Institution; 2017. Available from: https://www.brookings.edu/research/bolstering-urbanization-efforts/ [Accessed: July 4, 2021]

[9] Muggah P, Hill K. African cities will double in population by 2050. World Economic Forum; 2018. Available from: https://www.weforum.org/agenda/2018/06/Africa-urbanization-cities-double-population-2050-4%20ways-thrive/ [Accessed: July 4, 2021]

[10] World Population Review. Africa Cities by Population. 2021. Available from: https://worldpopulationreview.com/continents/cities/africa [Accessed: July 4, 2021]

[11] Faria J. Largest Cities in Africa as of 2021, by Number of Inhabitants. 2021. Available from: https://www.statista.com/statistics/1218259/largest-cities-in-africa/ [Accessed: July 4, 2021]

[12] Rahman A. Social and Physical Infrastructure. In: Denial and Deprivation. 2019. pp. 260-279. DOI: 10.4324/9780429058202-11 [Accessed: July 4, 2021]

[13] Oppong RA. Challenges facing Africa's infrastructure development. In: Nkum RK, Nani G, Atepor L, Oppong RA, Awere E, Bamfo-Agyei E, editors. Proceedings of 3rd Rd Applied Research Conference in Africa (ARCA) Conference; 7-9 August 2014; Accra, Ghana. 2014. pp. 13-27. Available from: https://www.academia.edu/8483414/Oppong_R_A_CHALLENGES_FACING_AFRICA_S_INFRASTRUCTURE_DEVELOPMENT?auto=download [Accessed: July 4, 2021]

[14] Deloitte. Addressing Africa's Infrastructure Challenges. Available from: https://www2.deloitte.com/content/dam/Deloitte/global/Documents/Energy-and-Resources/dttl-er-power-addressing-africas-infrastructure-challenges.pdf [Accessed: July 4, 2021]

[15] Muggah R, Hill K. African Cities will Double in Population by 2050.

World Economic Forum; 2018 https://www.weforum.org/agenda/2018/06/Africa-urbanization-cities-double-population-2050-4%20ways-thrive/ [Accessed on July 4 2021]

[16] Mahler D et al. Internet Usage in Sub-Saharan Africa, Poverty and Equity Notes Number 13. World Bank Group; 2019. Available from: https://documents1.worldbank.org/curated/en/518261552658319590/pdf/Internet-Access-in-Sub-Saharan-Africa.pdf [Accessed: July 4, 2021]

[17] International Telecommunication Union, ICTs, LDCs and the SDGs: Achieving universal and affordable Internet in the least developed countries. In: Thematic Report 2018. Available from: https://www.un.org/ohrlls/sites/www.un.org.ohrlls/files/ict-ldcs-and-sdgs.pdf. [Accessed: July 6, 2021]

[18] Odarno L. Closing Sub-Saharan Africa's Electricity Access Gap: Why Cities Must Be Part of the Solution. World Resources Institute; 2019. Available from: https://www.wri.org/insights/closing-sub-saharan-africas-electricity-access-gap-why-cities-must-be-part-solution [Accessed: July 6, 2021]

[19] WHO. Progress on Drinking Water, Sanitation and Hygiene 2017 (Joint Monitoring Programme). WHO/UNICEF; 2017

[20] WHO, Progress on Drinking Water, Sanitation and Hygiene 2017 (Joint Monitoring Programme). WHO/UNICEF. (2017) The percentage access to a sewer connection has also declined

[21] World Urbanisation Prospects 2014. United Nations; 2014

[22] Eberhard R. Access to Water and Sanitation in Sub-Saharan Africa. Deutsche Gesellschaftfür & InternationaleZusammenarbeit (GIZ) GmbH; 2019. Available from: https://www.oecd.org/water/GIZ_2018_Access_Study_Part%20II_Narrative%20Report_Briefing_document.pdf [Accessed: July 6, 2021]

[23] Kariuki M et al. Better Water and Sanitation for The Urban Poor: Good Practice from Sub-Saharan Africa. European Communities and Water Utility Partnership; 2003. Available from: https://www.wsp.org/sites/wsp/files/publications/330200725049_afBetterWaterandSanitationForTheUrbanPoorGoodPracticeFromSSA.pdf [Accessed: July 6, 2021]

[24] Venard JL. Urban Planning and Environment in Sub-Saharan Africa. UNCED Paper no. 5. Washington D.C.: World Bank; 1995. Available from: https://documents1.worldbank.org/curated/en/665641468767651789/pdf/multi-page.pdf [Accessed: July 6, 2021]

[25] World Health Organization. Global Water Supply and Sanitation Assessment Report. Geneva: WHO; 2000

[26] Ogutu ZA. Sustainable Development in Sub-Saharan Africa: What are the Alternatives? Journal of Eastern African Research & Development. 1993;**23**:24-39

[27] White R et al. Greening Africa's Cities: Enhancing the Relationship between Urbanization, Environmental Assets and Ecosystem Services. International Bank for Reconstruction and Development. The World Bank; 2017. Available from: https://openknowledge.worldbank.org/bitstream/handle/10986/26730/P148662%20Greening%20Africa%27s%20Cities_web.pdf [Accessed: July 6, 2021]

[28] UN-Habitat. State of African Cities 2008: A Framework for Addressing Urban Challenges in Africa. Nairobi, Kenya: United Nations Human Settlement Program; 2008

[29] Seto KC, Hutyra LR. Global forecasts of urban expansion to 2030 and direct impacts on biodiversity and carbon pools. Proceedings of the National Academy of Sciences. 2012;**109**(40):16083-16088

[30] Ibeh C, Walmsley B. The Role of Impact Assessment in Achieving the Sustainable Development Goals in Africa. Available from: https://conferences.iaia.org/2021/draft-papers/988_IBEH_The%20role%20of%20impact_Full_Paper.pdf [Accessed: July 6, 2021]

[31] Walmsley B, Husselman S. Handbook on Environmental Assessment Legislation in Selected Countries in Sub-Saharan Africa. 4th ed. Johannesburg, South Africa: Development Bank of Southern Africa; 2019. Available from: https://conferences.iaia.org/2021/draft-papers/988_IBEH_The%20role%20of%20impact_Full_Paper.pdf [Accessed: July 6, 2021]

[32] World Bank. The Cost of Air Pollution: Strengthening the Economic Case for Action. Seattle: The World Bank and Institute for Health Metrics and Evaluation University of Washington; 2016

[33] United Nations Economic Commission for Africa (UN ECA). Economic Report on Africa 2017: Urbanization and Industrialization for Africa's Transformation. Addis Ababa, Ethiopia: United Nations; 2017

[34] Gebre G, Van Rooijen G. Urban water pollution and irrigated vegetable farming in Addis Ababa. Water, sanitation and hygiene: sustainable development and multi-sectoral approaches. In: 34th WEDC International Conference; Addis Ababa, Ethiopia; 2009

[35] Republic of Uganda. National Population and Housing Census 2014 Area Specific Profiles. 2014

[36] Objective XXVII of National Objectives and Directive Principles of State Policy

[37] Article 245 of The Constitution of the Republic of Uganda. 1995

[38] The National Environment Act. 2019

[39] Uganda, Constitution of the Republic of Uganda

[40] Section 21 of Kampala Capital City Authority Act. 2010 (as amended, 2019)

[41] Section 7 of Kampala Capital City Authority Act 2010 (as amended, 2019)

[42] KCCA. The Local Governments (Kampala City Council) (Solid Waste Management) Ordinance, Statutory Instrument. 2006. 243-21. Section. 5

[43] KCCA. Local Governments (Kampala City Council) (Urban Agriculture) Ordinance. 2006

[44] KCCA. (Sewage and Faecal Sludge Management) Ordinance. 2019

[45] The Constitution of the Republic of South Africa No. 108 of 1996

[46] South Africa, Local Government: Municipal Systems Act. 2000 (Act No. 32 0f 2000)

[47] City of Johannesburg, Air Pollution Control By-Laws

[48] City of Johannesburg, Waste Management By-Laws, Local Authority Notice 1012. 2013

[49] City of Johannesburg Metropolitan Municipality, Water Services By-Laws

[50] The City of Johannesburg, Municipal Planning By-Law. 2016

[51] City of Johannesburg Metropolitan Municipality Public Health By-Laws

(Published under Notice No 830 in Gauteng Provincial Gazette Extraordinary No 179 dated 21 May 2004)

[52] Act No.462, ACT No.936, and (L.I) 2034 respectively

[53] Accra Metropolitan Assembly https://ama.gov.gh/theassembly.php [Accessed on July 4 2021]

[54] Accra Metropolitan Assembly (Environmental Protection) Bye-law. 2017

[55] Accra Metropolitan Assembly (Control of Animals) Bye-law. 2017

[56] Accra Metropolitan Assembly (Sanitation) Bye-law. 2017

[57] Accra Metropolitan Assembly (Cleaning) Bye-law. 2017

[58] Accra Metropolitan Assembly (Building/ Physical Development) Bye-law. 2017

[59] Ethiopia, Constitution of the Federal Democratic Republic of Ethiopia. 1995

[60] Ethiopia, Proclamation No. 361/2003: The Addis Ababa City Government Revised Charter Proclamation. 2003

[61] Roy KC, Tisdell CA. Good governance in sustainable development: The impact of institutions. International Journal of Social Economics. 1998. Available from: https://www.emerald.com/insight/content/doi/10.1108/03068299810212775/full/pdf?casa_token=Ol3lfsSlDTAAAAAA:cIYkJ4Mq fqEmAbzT-mTIbNDf_OHxRR1Sga5 DCis0zbn_9jVgEBEeZ2EslSC_ Y6pAB3eJaVwolDSoWUvsq61o PvE6J4YUKKfKgDqdG3QW38 PRn614Rebv6w [Accessed: July 7, 2021]

Chapter 5

Urban Agglomeration and the Geo-Political Status of the Municipality of Portmore, Jamaica

Carol Archer and Anetheo Jackson

Abstract

This chapter will attempt to shed light on the relevant explanations for designating or creating urban agglomeration for the purpose of administration and governance in the case of largest urban space in Jamaica. The main objective of this chapter is to provide an in-depth look at the case of Portmore with specific emphasis on the socio-economic, political, legislative, and relevant policy arrangements that influence the change in sub-national geo-political status from municipality to parish. The researchers explore the literature on local government reform, public financing, urban economics, and urban planning which provided a basis for objectively interrogating the proposed change. From the analysis of existing legislations, policies, and international conventions such as the New Urban Agenda, the proposal to change the geo-political designation of the Municipality of Portmore raises several questions about the economic profile of the area and the capacity to enjoy the benefits of urban agglomeration given its location attributes. Understanding the evolution of the theoretical discourse in urban planning can provide clarity on the relevance of this geo-political designation, the intergovernmental relationship associated with the geo-political designation, particularly as it relates to autonomy, and the allocation of resources for the provision of local government services. This understanding will help to direct the decision makers as to the best designation for Portmore given its current realities and importance of implementing measures to support decentralization and autonomy at the local level for major urban areas. The researchers found that the economic viability of the proposition is at best questionable as the economic base of the municipality is limited and its capacity to generate linkages demand serious considerations. Attention should be given to local government reform in the face of emerging trends and current realities rather than changing status of an urban areas which is the direct result of the principles of urban agglomeration.

Keywords: municipality, local government reform, urban agglomeration, decentralization, political economy, geo-political status

1. Introduction

Recently there has been raging debate regarding changing the geo-political designation of Portmore, Jamaica, from a municipality to a parish. The historical development of Portmore is intrinsically tied to Kingston. Portmore's development began in the early seventeenth century and has developed from being a single community into a vast network of housing schemes. Some of the early communities include Queens Town, now renamed Edgewater and Port Henderson.

Queens Town was established after the 1692 earthquake, which destroyed Port Royal. This community was established as a twin city to Kingston but was destroyed by a storm surge in 1722. Since its inception, Portmore as served as a dormitory of Kingston. In fact, by Jamaican government's definition Portmore is included in the designation of the Kingston Metropolitan Area (KMA) and by extension an agglomeration of Kingston, the capital of Jamaica.

The major justification, for a change in geo-political status, offered by the government representatives, is that parish status would allow for greater autonomy for the elected officials from Portmore. The push for autonomy is influenced by the fact that the population and communities in Portmore increased since 2002 when the area was first designated as a municipality. As reported in the Jamaica Daily Gleaner on December 16, 2020, the communities in Portmore increased from 22 in 1991 to 40 communities in 2001 with the construction of several housing schemes, including that of Greater Portmore.

This chapter will attempt to shed light on the relevant explanations for designating or creating urban agglomeration for the purpose of administration and governance in the case of the largest urban space in Jamaica. Understanding the evolution of the theoretical discourse in urban planning and urban economics can provide a lens through which to examine the efficacy of the proposed geo-political designation, the intergovernmental relationship associated with the geo-political designation, particularly as it relates to autonomy, and the allocation of resources for the provision of local government services. This understanding will help to direct the decision makers as to the most appropriate designation for Portmore given its current realities and importance of implementing measures to support decentralization and autonomy at the local level for major urban areas.

Ebenezer Howard, pioneering urban planner, offers one of the earliest works on designation of urban areas. Howard, in his 1902 seminal writing, *Garden Cities of Tomorrow,* proposed the concept of town clusters [1]. Based on his research, Howard concluded that there was a dynamic relationship between the spatial organization of the cities/urban areas and the semi-urban and rural areas. Furthermore, Howard foresaw urbanized areas comprising several "garden cities around Central Cities. In his vision of the urbanized landscape, the urban form is not only the areas occupied by cities but also an area comprising several peripheral Garden Cities integrated with a Central City. Researchers Fang and Hu ([2], p. 128) argue that Howard's concept eventually evolved into the early forms of the "Garden City" model of urban agglomeration. The growth of urbanized population in Europe, the Americas and Asia, called for further explanation of the relationship between and within these urbanized areas. Geddes [3] coined the term conurbation or continuous urbanization. Unlike Howard that predicted the growth of urban areas contained by greenbelts, Geddes [3] predicted the growth of these urban spaces without the swath of rural lands. Based on Geddes' prediction, the British government recognized the various forms of aggregated local authority to address local land use and service delivery issues.

2. Methodology

An exploratory approach was used to examine the case of Portmore as a Municipality and the efficacy of the proposed change in the geo-political designation from a municipality to a parish. In this regard, the first objective was an in-depth look at the case of Portmore with specific emphasis on the socio-economic, political, legislative, and relevant policy arrangements.

Specifically, detailed information on the proposed change, relevant legislative documents, existing policies, available reports, and prior research were assembled and studied.

The review took account of the existing context of the Municipality of Portmore and its connection to the Kingston Metropolitan Area (KMA), details of the current and proposed geo-political arrangement, the rationale for the proposed changed and the readiness of Portmore for the proposed change from municipality to parish status. In addition to this, economic roles, and functions of the Municipality of Portmore in relation to the neighboring urban centers, the academic literature on local government reform, public financing, urban economics, and urban planning provided a basis for objectively interrogating the proposed change. This also informed the analysis and discussion of the existing fiscal relationships between the local authorities and central government and narrow down key considerations for the government and key stakeholders in determining the way when considering changing sub-national geo-political status. The existing legislative and policy environments and the implications for sustainable development also guided the study.

3. Urban agglomeration/economies and geo-political designation

One of the most recent discourses on geo-political treatment of large, urbanized areas is presented by Fang [4]. Fang suggests that urban agglomerations are very different from the simple clustering of similar administrative units or geographic location. Instead, urban agglomeration is an emerging urban spatial form that is driven by concentrated industries and populations, a highly connected transportation network, an enhanced central city and favorable regional incentive policies. Urban agglomerations are evidently a product of the late stages of metropolitan development. Furthermore, "one can only claim that there is an urban agglomeration when the networks grow in strength and frequency and the socioeconomic ties among the central and peripheral cities become more integrated Fang and Hu ([2], p. 133).

Cities are characterized by urban agglomeration as there exists agglomeration economies. Agglomeration economies are generally either pecuniary or technological in nature [5]. These are resource advantages. The former can result from a large and diverse labor market. Thus, firms in the KMA that rely on Portmore for its labor supply can enjoy greater productivity through lower labor and transportation costs due to proximity to work and stronger competition for jobs. This is of major consideration when other urban areas such as May Pen or Old Harbor are considered (see Map showing designated towns and areas in Jamaica). Portmore is one of the largest residential settlements in Jamaica and the English-speaking Caribbean. It is also one of the fastest growing urban areas in the island. However, as aforementioned, Portmore has been a dormitory for Kingston. Its main pull factors are proximity to Kingston and its affordable housing stock, in comparison to the KMA.

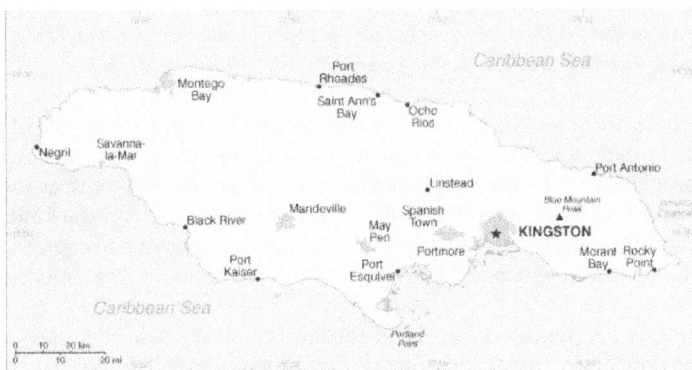

Map showing major urban centers through Jamaica

Technological agglomeration economies result from knowledge spill-over across and within industries. It implies thicker upstream and downstream linkages between economic activities in an urban area. To capitalize on these economies, local governments, with the autonomy to design and attract the right kind of industries will most likely promote the appropriate mix of land uses and production linkages that will trap the benefits of knowledge sharing and the cross-fertilization of ideas, inputs, and outputs, that will maximize returns to firms and by extension to the locality.

Presently, in the Municipality of Portmore, the leading sources of private-sector income are retail and services. There is also a fairly recent trend of a growing number of Business Process Outsourcing (BPO) firms in the Municipality. In addition to this, the steady demand for housing in the area stimulates growth in new outputs of housing and commercial properties. However, the strength of linkages between these economic activities is questionable. For example, there is no evidence of strategic clustering of industries and spatial concentration of suppliers, supply chains, complementary industries, and local institutions to maximize on locational competitiveness [6].

In fact, it is arguable that the informal economic activities in the area perform fairly well in establishing linkages when compared to the formal activities. It must be noted that although the informal sector is a source of employment and income for many residents, it is not a source of revenue for the government. Further, with the exception of the Municipality's fishing industry, these informal economic activities are diverse and very difficult to locate. Strategic clustering in any urban agglomeration is significant as land use patterns are substantial to the maximization of the benefits of agglomeration economies.

A further consideration is the potential intercity dynamics—The economic competition between Kingston and Portmore. If Portmore fails to create and supply the jobs to match its endowment of labor, the region will continue to serve as a dormitory for Kingston's labor input. An alternative arrangement to serving as a dormitory is for Portmore to use its labor supply as a source of competitive advantage to attract the type and quantity of productive activities to the area. According to Reilly's [7] law of gravitation the capacity of two cities to attract retail spending from the catchment area in between them is dependent on the size and the square of the distance from any location between the two cities. However, while large population and labor supply is a potent pull factor for firms, the existing spending power in the locality is also an important consideration. Arguably, higher income cities are more likely to attract higher income residents which makes them more attractive to both large and small businesses. Harvey and Jowsey [8] noted that:

> consideration has to be given to the composition of the population (for example, by age group and working proportion), its earning capacity (for example, whether they are skilled workers), government subsidy policy within the district and the spending habits of different income groups (Harvey and Jowsey [8], p. 97).

When these are taken account of, the pull-factor of a more affordable housing stock in Portmore which makes housing more accessible to residents in lower income groups, may not necessarily translate to making the area more attractive to more businesses. Therefore, Portmore's current ability to attract the kind of economic activities in the quantities that will generate linkages and capitalize on economies, and which will provide job opportunities for its residents is questionable.

As mentioned earlier, Portmore lacks the level of diversity in economic activities and the type of upstream and downstream linkages to fully capitalize on the benefits of agglomeration economies that will likely be needed for the economic viability of the area if it is designated as a parish. However, other considerations include the quality and quantity of its urban infrastructure and services, particularly when compared to neighboring areas. Local governments in Jamaica and throughout the world are preoccupied with identifying the necessary funding to address these issues. It is likely to be more challenging should Portmore be designated parish, to generate the necessary resources to support the demand on its resources which would necessitate increasing investments in some public goods and services and land resources.

The prospects of maximizing scale economies in Portmore will likely depend on intensifying land use in the area. However, given the relatively inelastic supply of land for the various uses in the area, more intensive use will invariably take the form of high-rise developments. High-rise development is the most likely response to increased demand for residential or commercial properties in more urbanized areas where suitable land is relatively scarce. On this point, Portmore's geographical location makes it vulnerable to natural disasters. Consequently, building sound and resilient high-rise structures are likely to be costly. Whereas the land input may be minimized, the capital input would likely be substantial, such that these developments would not be feasible if the cost of capital is not sufficiently low.

Further, if the designation of parish results in altering the land use patterns in Portmore in pursuit of agglomeration economies, through a larger, more diversified economic base, then as mentioned earlier, the demand for services, facilities and urban infrastructure will likely increase. Any foreseeable increase demand on the services and infrastructure must be matched by the revenues generated by economic activities in the area. For this to be feasible, the marginal benefits from additional users in the area should equate the change in costs for the provision of public goods and services by the local government. Public goods are non-rivalry and non-excludable and cannot be efficiently allocated through the market mechanism. These include some levels of school and educational facilities, safety and security services, access to and maintenance of recreational facilities and markets, sewers and sanitation, public health facilities, and roads.

It must be noted, that in countries with decentralized governments, the market could be relied upon to smooth out differences in the marginal costs and benefits of the changes that may accompany a change in the geo-political designation from a municipality to a parish. However, even in a democratic society, as long as central governments have a high level of autonomy over revenues and the sources of income, local governments are not empowered to make the necessary adjustments to manage its provision of public goods and services. This generally results in sub-optimal allocations at the local level. Therefore, the designation of parish may be accompanied by increased fiscal responsibility but not necessarily by the level of autonomy and financial independence of the local government that is needed to avoid welfare loss to its residents.

To evaluate the extent of their financial independence if the designation is changed to parish, Portmore's local government should assess the following:

1. The amount of actual revenue from taxes and fees that will be available to the local authority.

2. The extent and efficiency of its sources of income.

3. The real power and methods by which the local government can influence increase in fees to cover rising costs.

4. The possibility of incurring debt as part of financial autonomy and financial management.

5. The strength and size of the economic base of the locality.

6. Level of diversity in economic activities of the locality.

With the proposed change of the geo-political designation of the area, the arrangements for financing of, for example, public goods and services will necessarily be changed. Thus, the amount of actual revenue from taxes and fees and how these would be allocated to the newly established parish are pertinent to the decision. It must be noted that property taxes and council fees are the primary sources of revenue for local governments in Jamaica. However, property tax rates are set by central government and it is calculated on the unimproved land value, the buildings are not taxed. Consequently, increased tax revenues to the local authority are largely dependent on the demand for land. A noteworthy point on this topic is that the costly re-valuation exercise that is required to capture gains from appreciation in property value, is arguably not undertaken as often as is needed for local governments to capture its share of the gains in property value that is needed to finance investments in public goods and services to support residents and businesses in the area. In fact, disinvestment, and a lack of maintenance of public goods and services by local governments as they rely on this income to cover some of these costs is not far-fetched.

4. Political economy and urban agglomeration

The corollary of the foregoing is that alongside the potential benefits of parish designation touted by government officials, decision-makers are tasked to take an objective look at the Municipality's opportunities to expand and deepen its economic activities to cover both internal and external costs. This objective evaluation should include serious considerations as to whether or not the type, size and diversity of current and potential economic activities in Portmore can sufficiently capture the benefits of agglomeration economies and what changes will be made to empower the local government to manage the fiscal affairs of the proposed parish.

In Jamaica, as in most developed country that have adopted a market driven economy, the bulk of funding for local government services is appropriated by Central government. The assumption behind central government's control of funding for local services has been influenced by the notion of achieving economies of scale and efficiency in the delivery of these services.

For a small island state such as Jamaica, with a parliamentary system almost akin to a unitary form of government, centrally controlled funding for local government services might seem logical at first. However, further examination of income and expenditure at the local level reveals that services are duplicated across agencies and there are no clear lines of accountability towards the allocation of these funds.

Historically there has been a "tug-of war" between the national government and the local authority about who should bear the cost for delivering services. Central to the conflict is the control of powers. The central government wants to maintain control and access to revenues which are generated at the local level. In Jamaica, like any other modern democracy there is the tension between the nation state and local authorities, as the officials at the local level become more concerned with increasing their share of powers and resources visa-vis the claims of other localities and the national state ([9], p. 35).

There are several treatises on local government reform which Jamaica can adopt to carry out the much-needed reform. One line of argument, at the broader macro level, placed the discourse into two camps: neo-Marxist left and the neo-conservative right. The neo-Marxist argues that growth and decline at the local level, particularly in the urban areas is influenced by decisions made at the national level as a function of economic forces [9]. The neo-conservatives believe that decisions at the local level should be left to 'invisible' force of the market and those local areas that can grow as a result will do so. Central to both lines of arguments is that national government can and does influence the growth and decline of local municipalities. This is the main point carried by political economists.

Another theory on local government reform, while arguing from the perspective of market forces, introduces a more complex variable, that of globalism. According to Clarke [10] "global economic change process general transcend scale, yet in some instances are very sensitive to local contextual factors, including state actions" ([10], p. 12). According to Clarke, local officials over the world operate under heightened conditions of economic and political uncertainty. They have new social and economic roles and responsibilities that are unanticipated. Local officials must reconstruct relations between public and private sector at the local level, in the context of the "new globalism" in addition to concentrating on providing the most basic governance issues.

Proponents of the "new global reality of local government" further argue that to the extent that the legitimacy of democratic regimes is tied to economic performance rather than governance based on civic values, the new localized state suggest a subordination of political will to private interest, particularly at the local level. Furthermore, democratization trends at the national level have left unclear the autonomy of local government relative to other ties. Local governments are victims of structural adjustment programmers in developing countries, national government focus on complying with international debt programmer rather than providing financial and administrative resources for the delivery of local services.

Attempts have been made at redistributing local influence. This is evidence by increased participation of the citizenry in the decision-making process. However, full participation is hampered by the fact that those who tend to "benefit" from the new privatization of services is the party 'faithful's' and the often-gained new position of influence in newly privatized local government. Local government's inability to carry out new roles and functions are constrained by "poor capital" and uneven development patterns. Uneven allocation restricts the ability of local governments to compete for new private investment.

New localism generates extensive competition among states and between local areas. This heightened level of competition has been known to exacerbate uneven development. In this era of globalism, local leaders must now move in a decision-making arena where pivotal investors operate at global scales, and at a magnitude and pace that defies local involvement. However, local leaders are asked to perform in this market as entrepreneurs under public constraints. National government also influences local leaders' performance in the global context.

The national government is wary of the fiscal and political consequences of proactive local development efforts. The local decision makers are also limited by the knowledge base and skills to act as public entrepreneurs. In the past local councilors were tied to their local context and the skills there are less likely to transfer when dealing in a global context. Schoburg [11] supports this position. She argues that most local government authority in Jamaica are hampered by an organizational framework built on values that are no longer compatible with contemporary leadership and management technologies or development norms. Furthermore, the operations of local government in the Caribbean and in Jamaica specifically, lack a culture of high performance ([11], p. 19).

Peterson [12] also acknowledged constraints associated with local government. He is of the view that local governments are limited by their ability to implement and redistribute polices or social welfare type policies. These local decision-makers fear that this would encourage the movement of capital. Instead, local decision-makers are likely to seek out developmental policies that serve to increase local capital base. In their view, this represents a maximization of "the public interest." Stone [13], among others, expanded Peterson's argument by pointing out that the political culture of the locality also plays a significant role in the policies developed for that locality. The influence of political culture is most evident in policy making at the local government level here in Jamaica.

Recognizing the limitations associated with resource mobilization, local governments have opted to privatize the provision of certain services. Theories of privatizing local government services have also influenced the discourse on local government reform. Boyne [14] in his analysis of local government reform in Great Britain, points out that since the 1980s there has been efforts to privatize government services. This effort of privatizing is driven by the belief that the market can produce goods and services more efficiently. It is also driven by the belief that resource decisions are seen as more rational when left to individuals who choose alternatives based on their own preferences through a market bidding process. Market like disciplines will prevent consumers of public goods and services from over consuming their services and smooth out inefficiencies in the provision of these goods and services.

The historical overview of local government reform in Great Britain, provided by Boyne [14] mirrored that of Jamaica. Boyne notes that in Great Britain, local authorities have been shaped by central government into service providers whose primary role lies in supplementing the welfare state rather than economic production.

The industrialists shaped municipal policies to give their businesses the competitive edge. As large companies expanded, they began to focus on national and international base of power and gave limited attention to local politics. The national government oftentimes step in and control the provision of infrastructure and social services. In the 1980s, like Great Britain, there was restricted funding for local government in Jamaica. To address funding constraints, the more independent local government in Great Britain developed strategies for economic development. Greater London Council and the more radical councils developed schemes for allocating funds to stimulate local businesses.

In Great Britain, the Local Government Reform Act of 1980 forced local authorities to operate many services as if they were private companies without the capacity to generate a trading loss for a few years or to expand their business through national or international trade. Under this Act, local authorities could not use power to expand entrepreneurial skills to expand into markets other than those directly tied to the needs of the local authority. In other words, the local authorities were limited by Central government.

Hero [15] introduced a departure from the market-centered approach theory to local government reform. Hxe criticized efforts to evaluate local government solely in terms of the empirical distribution of concrete goods and services. People must be placed in the equation. Hero states that the value of ordinary individuals should be measured by the degree to which "outputs" of the system, in the form of security services and material support benefit them ([10]. p. 41). Unfortunately, Hero's suggestions were not taken into consideration until in the 1990s in Great Britain and New Zealand. In the 1990s, under the leadership of Prime Minister Blair, the "Best

Value" system was introduced as a means of ensuring effective and efficient service delivery. This system set targets for local authorities. Included in these targets is a strong customer service component.

In general, existing research on local government reform points to the influence of global economic forces. These forces limit the decision-making capacity of local government, particularly as it relates to economic development. This also impact on service delivery. Although the local authorities are constrained by global forces, the average taxpayers are still concerned about effective and efficient delivery of local services such as fire, sanitation, and provision of social services.

In Jamaica there is an increasing demand for efficient and effective service delivery. With this demand there is a tendency to argue for prioritizing some of these services as a means of achieving these goals. Walker and Davis [16] have conducted extensive research on impact on contracting government services in Great Britain. They argue that services that are easily defined, monitored, and for which appropriate measures for performance can be implemented are the best to be contracted. They also warned that the difficulty in contracting service the recipients (taxpayers) do not often get the best possible service but rather the best service according to the contract.

The promulgation of the various Acts governing local government inclusive of the Parish Council Acts of 1887, the Parochial Rate and Finance Act of 1990, the Kingston and St. Andrew Act of 1925 and Municipal Act of 2003, and most recently the Local Governance Act of 2016, among others, serve to further blur the distinction between the provision of and funding for services at local and central government levels. Increasingly, research on local government recognizes that the powers of national or central government have increased significantly since the 1960s. The central government have assumed more interest in maintaining order and authority at the local level as well as securing public revenues and preserving the interest of public officials in pursuit of the government overall goals relating to growth and development.

In Jamaica, as in other developing states, government spending on public service delivery at the local government level has seen a mercuric rise since the 1980s. Concomitant with this increase in spending, central government has increased its regulatory powers and the decision-making authority have become more centralized. This is evident by the passage of various acts of parliament to centralize funding mechanisms. This is with the aim to achieve economies of scale, eliminate duplication of services and increase service efficiency.

The emphasis on market driven forces to influence the provision of and funding for local government services also served to blur the lines between the responsibilities of local and central government to provide these services. However, Gurr and King [9] argue that "the provision of such public services as water supply, for protection and waste management cannot be left up to the market neither can it be the sole responsibility of one layer of government" Gurr and King ([9], p. 33). According to Gurr and King, "Whatever other interest is pursued by the national government, some minimum level of the public services must be provided if the local authority is to survive at all" Gurr and King, ([9], p. 35).

Ragoonath [17], in his essay, "Challenges for Local Government in the Caribbean" is of the view that Jamaica and other Small Island Developing States of the region is faced with issues of rebuilding credibility to a local government system. The rebuilding is necessary since for many years local government structures of the region has been "'hijacked' and even emasculated by central governments, and at the same time being an arena for corruption and mismanagement, all in the context of self-interest, political expediency and even party paramount." ([17], p. 100). To remedy the situation Ragoonath calls for, reform of the legislative and the administrative structures in order that participation is enhanced, and citizens are empowered. He further argued that such reform must be because of consensus. Consensus will only come after consultation, and when trust is developed or inculcated. Moreover, the opinions of those at the bottom must be considered, alongside all other opinions [17].

5. Urban agglomeration, legislative provisions, policies, and implications for sustainable development

In justifying the proposed geo-political parish status for Portmore, a newly elected government official representing Portmore stated that,

> as a model parish, Portmore has the potential to counter a lot of the ills that ail most parishes: Particularly the blight spawned by unplanned developments and the resulting urban sprawl. The orderly development and zoning of the parish is of tremendous value in making Portmore the parish to live, work and do business. The creation of high-occupancy vehicle lanes at the tolls to encourage carpooling and reduce traffic and pollution; the creation of a model inter- and intra-parish public transportation system; the provision of designated public markets and strict enforcement of vending rules; a modern waste disposal and sewage treatment systems are all potentially practical benefits of parish status. Miller (2020), "Practical Benefits of Portmore As a Parish, Friday, January 29, 2021, Jamaica Gleaner https://jamaica-gleaner.com/article/commentary/20210129/ robert-miller-practical-benefits-portmore-parish.

It is important to note that the existing Town and Country Planning Act (1957) and other legislative provisions make allowances for the Portmore Municipality and the other parishes to prepare its land use plan to address issues of urban blight, transportation, and settlement needs. Unfortunately, in this statement, the elected official failed to recognize these existing legislations and failed to mention the enabling legislative enactments and or policies and programmers that will be implemented for the Jamaican society to achieve the principles of sustainable development as measured by the Sustainable Development Goals (SDG) 11—safe, inclusive, and resilient cities and communities. Elsewhere in Small Island Developing States in the Caribbean and throughout the world, elected officials at the city level, policy makers, and city planners are actively working to create environmentally friendly, safe, and resilient communities, neighborhoods, and cities for their citizens without compromising the needs of the future generation.

The current administration in Jamaica must take care to adhere to the principles of the agreement under the New Urban Agenda which is the Framework for the implementation of SDG 11. By signing the New Urban Agenda, the government agreed to:

1. Support appropriate policies and capacities that enable subnational and local governments to register and expand their potential revenue base, while ensuring that poor households are not disproportionately affected.

2. Promote sound and transparent systems for financial transfers from national governments to subnational and local governments based on the latter's needs, priorities, functions, mandates, and performance-based incentives, as appropriate, to provide them with adequate, timely and predictable resources and enhance their ability to raise revenue and manage expenditures.

3. Support the development of vertical and horizontal models of distribution of financial resources to decrease inequalities across subnational territories, within urban centres and between urban and rural areas, as well as to promote integrated and balanced territorial development.

4. Provide the transparency of data on spending and resource allocation as a tool for assessing progress towards equity and spatial integration.

5. Promote best practices to capture and share the increase in land and property value generated because of urban development processes, infrastructure projects and public investments. Measures such as gains-related fiscal policies could be put in place, as appropriate, to prevent its solely private capture, as well as land and real estate speculation.

6. Support the creation of robust legal and regulatory frameworks for sustainable national and municipal borrowing, based on sustainable debt management, supported by adequate revenues and capacities, by means of local creditworthiness as well as expanded sustainable municipal debt markets when appropriate.

To ensure that the government is not in breach of the New Urban Agenda and is able to achieve the SDGs, it is imperative that the necessary steps are taken to further enact the recommendations of the National Advisory Committee on Local Government Reform. The detailed Report prepared by the National Advisory Committee on Local Government Reform in 2009 for the government of Jamaica, recommend urgent need to reform the local government and governance structures to ensure sustainable development for Jamaica as a Small Island Developing States (SIDS).

In addition, the Report, among other things, proposed a legal framework that makes local government relevant to current realities and emerging trends regarding local governance and conducive to the achievement of good governance, sustainable development, empowerment of communities, and the active participation of citizens in the local governance process. In essence, the Report encourage decentralized local services provision, revenue collection and decision-making.

These recommendations in the Report are in keeping with several global think tanks approach to decentralization. Chief among them is the World Observatory on Subnational Government Finance and Investment. The World Observatory on Subnational Government Finance and Investment proposes the following definition: 'decentralization consists of the transfer of powers, responsibilities and resources from central government to sub-national governments, defined as separated legal entities elected by universal suffrage and having some degree of autonomy'. This is in line with the view that the level of autonomy of the local government is critical to the decision about the proposed change in the geo-political designation of the Municipality of Portmore.

6. Conclusion

The proposal to change the geo-political designation of the Municipality of Portmore raises several questions about the economic profile of the area and the capacity to enjoy the benefits of urban agglomeration given its location attributes. The efficacy of the proposal was also considered against the background of the competitive advantage of the locality when compared to competing urban spaces, the political economy arrangements, and the legislative and policy environment to support this change. The economic viability of the proposition is at best questionable as the economic base of the municipality is limited and its capacity to generate linkages demand serious considerations. The change would most likely be accompanied by more intensive land use if this can be achieved without compromising the goals of sustainable development. Furthermore, many of these services are funded in part by several agencies but there is no clear picture of the total amount being allocated for these services. A reporting mechanism which emphasizes the services rather than the agency or Ministry will give a clearer picture of how much is being allocated for the particular service.

Another recommendation to ensure clearer understanding of the delivery of the public services is to develop clearer performance measures and a reporting format that is intrinsic to the whole process of the service being delivered. This includes, the management, planning, monitoring and evaluation of the service. Well defined and easily measured goals reported in a concise manner will help the decision makers and the general public has a better understanding of how the public purse is being spent towards the delivery of these services. These clear structures will help decision makers and the average citizens to determine the most appropriate urban agglomeration and the attendant government and governance structure. It is important for local government to concentrate on service delivery and get the necessary support from central government to support these service deliveries rather than an "upgrade" parish status without the requisite autonomy for expanding it economic base for improve service delivery.

Decision-makers are of the view that Portmore's has the potential to attract the kind of economic activities in the quantities that will generate linkages and capitalize on economies, and which will provide job opportunities for its residents without acknowledging that the municipality's current position is based on the relationship that exist within the network provided by the Kingston urban agglomeration. For Portmore to exist above or outside of the existing agglomeration, there is need for the appropriate mix of land uses and production linkages that will trap the benefits of knowledge sharing and the cross-fertilization of ideas, inputs, and outputs, that will expand returns to firms and by extension to the locality. Review of the draft land use plans for the Municipality of Portmore does not suggest that consideration is given to land use management that increase the return on investment to enable the provision of basic urban services.

It is also evident that central government in Jamaica continues to maintain autonomy over revenues and the sources of income and local government at the parish level are not empowered to make the necessary adjustments to manage its provision of public goods and services. As a result, the designation of parish may be accompanied by increased fiscal responsibility but not necessarily by the level of autonomy and financial independence of the local government that is needed for the area to grow at the same level or rate of the larger metropolitan area. More emphasis is needed to make local government relevant to current realities and emerging trends regarding local governance and conducive to the achievement of good governance, sustainable development, empowerment of communities, and the active participation of citizens in the local governance process rather than changing status of an urban areas which is the direct result of the principles of urban agglomeration.

Author details

Carol Archer* and Anetheo Jackson
University of Technology, Kingston, Jamaica

*Address all correspondence to: carcher@utech.edu.jm

IntechOpen

References

[1] Howard E. Garden Cities of To-morrow. London: Swan Sonnenschein & Co., Ltd; 1902

[2] Fang C, Yu D. Urban agglomeration: An evolving concept of an emerging phenomenon. Landscape and Urban Planning, Volume. 2017;**162**:126-136

[3] Geddes P. Cities in Evolution: An Introduction to the Town-Planningmovement and the Study of Cities. London: Williams and Norgate. Ginsburg; 1915

[4] Fang C. Important progress and future direction of studies on China's urban agglomerations. Journal of Geographical Sciences. 2015;**25**:1003-1024

[5] Brueckner. Lectures on Urban Economics. Cambridge, Massachusetts: The MIT Press; 2011

[6] Porter ME. Location, competition, and economic development: Local clusters in a global economy. Economic Development Quarterly. 2000;**14**(15):15-34

[7] Reilly W. The Law of Retail Gravitation. New York, USA: Knickerbocker Press; 1931

[8] Harvey J, Jowsey E. Urban Land Economics. Great Brittain: Palgrave Macmillan; 2004

[9] Gurr T, King D. The State and the City. University of Chicago Press; 1987

[10] Clarke S. The profound and the mundane: Analyzing local economic development activities. Urban Geography. 1993;**14**(1):78-94

[11] Eris S. Local government and local development: Bridging the gap through critical discourse: Evidence from the commonwealth Caribbean. Commonwealth Journal of Local Governance. 2014;**10**:5-31

[12] Peterson P. City Limits. Chicago: University of Chicago Press; 1981

[13] Stone CN. Urban regimes and the capacity to govern: A political economy approach. Journal of Urban Affairs. 1993;**15**(1):1-28

[14] Boyne G, Editor. Managing Local Services: From CC7 to Best Value. London: Frank Cass and Company; 1999

[15] Hero R. The Urban Service delivery literature: Some questions and concerns polity. Polity. 1987;**18**:659-677

[16] Walker B, Davis H. Perspectives on contractual relationship and the move to best value in the local authories. Local Government Studies. 1999;**25**(2):16-37

[17] Ragoonath B. Challenges for local government in the Caribbean. In: Paper Presented at Regional Seminar on Innovative Approaches to Local Development and Management in the Caribbean; October 3-7; Montego Bay, Jamaica. 2005

Chapter 6

Perspective Chapter: Belem and Manaus and the Urban Agglomeration in the Brazilian Amazon

Tiago Veloso dos Santos

Abstract

Exploring the results of research on the relationship between urban agglomeration and the Brazilian Amazon region, this chapter demonstrates, from a comparative study between the two main metropolitan agglomerates - Belem and Manaus -, the different ways of production in the regional space. Based on the organization, systematization, and analysis of data regarding the main elements of the urban structure in both agglomerates, namely: the patterns of housing settlements of the upper, middle, and lower classes; the distribution of industrial zones and the patterns of urban expansion, we sought to highlight the intra-urban differences between the two metropolises, considering their importance in the configuration of the regional urban network. The most recent evidence points to two quite different metropolitan structures, explained both by the distinct nature of the urbanization processes that produced them and the highlighted intra-urban characteristics.

Keywords: urban agglomerate, regional metropolises, Belem, Manaus, Amazon

1. Introduction

The Amazon region, identified throughout history by its biogeographic and morphoclimatic characteristics, has undergone a significant change in its tropical forest natural landscape from the second half of the twentieth century, which is no longer the only visual reference. This change, resulting from the production of the regional space, places cities and the urban as the main territorial reference at the beginning of the twenty-first century.

The image of the urbanized forest as a reference [1] is a symbol of this transformation. Following the urbanization trends in the Brazilian territory, the region had about 70% of its population living in urban areas in 2010, in contrast to the 30% in the 1950s, according to the Brazilian Institute of Geography and Statistics (IBGE). More than the statistical expression of this change in the composition of the regional population, these data show a trend toward the formation of urban agglomerates as a basis for spatial planning.

This arrangement was made possible by the formation of a frontier economy since the formation of cities in the Amazon was part of the implementation of a

IntechOpen

regional urban network, the locus of action of the institutions responsible for the integration [2].

This evolution of urbanization demonstrates the genesis of urban "condensations" since it is related to the increase in the number of cities in certain regions of the settlement system [3]. In this case, it is necessary to consider that, for the study of urbanization processes, the presence of these densities is as important as the increase in the size and number of cities or changes in their hierarchical structure.

However, thinking beyond the urban density in a more recent period, a new quality emerges in the dynamics of the urbanization of the Amazonian space, which can be classified as the emergence of metropolitan agglomerates in line with the metropolization movement of the Brazilian space.

The repercussions of this metropolization process in the Amazonian space arise from the need for expansion toward the frontier and the new patterns of capital accumulation and regional labor market organization, but it also concerns the general movement of urban complexification of Brazilian society [4]. Regional metropolization is associated with a pattern of transition from urbanization of society and territory to a trend toward the metropolization of space. It is from this interpretation that the Amazonian metropolitan agglomerates are presented.

However, if the socio-spatial processes of regional metropolization show up as a revealing trend in Amazonian urbanization, this does not mean that such processes have homogeneous configurations. On the contrary, the region has metropolitan agglomerations that present different characteristics, either to the type of economic-spatial dynamics that potentiates metropolization, or about a particular landscape produced as an expression of deeper processes.

The metropolitan agglomerates of Belem and Manaus, the two main cities in the region, are the references for this analysis since it is about recognizing the spatial manifestations expressed in them from a description of their constitutive characteristics as urban and regional phenomena.

2. Urban agglomerates and regional metropolization

The metropolitan reality is present throughout the Brazilian territory to a greater or lesser extent, and it has been also expressed in the regional Amazonian context in recent decades, in which significant portions of the region currently follow the trends of metropolization. The configuration of this phenomenon in the regional scenario is a consequence of various aspects of the globalization expansion and how this process is presented in the region, considering the insertion of the Amazon in the internationalization of the Brazilian economy since the mid-1970s, through integration and development policies.

The territorial impacts of the world economy unfold in two related manifestations: the ones that act on the intra-urban level and those that express themselves on the regional level around the metropolises. The metropolises of regional projection are parts of this global economic geography, causing transformations that can be synthesized as follows "The current world-system causes a "multi-scale restructuring of capitalist socio-spatial configurations", leading to "qualitatively new geographies of capital accumulation, state regulation, and uneven development" [5].

Considering the official data (**Table 1**), 72% of the population in the Amazon region is in urban centers. Although they can be evaluated according to different degrees of lack and precariousness regarding basic services, the existing urban centers must be considered as constituents of an urbanization model. In addition, urbanization cannot be measured only by the spread of the urban stain or even by the emergence of new cities, but also by the dissemination of its values by society.

Year	Urbanization rate (%)
1950	29.60
1960	35.70
1970	42.70
1980	50.20
1991	57.83
2000	69.83
2010	72.80

Source: IBGE [6].

Table 1.
Brazilian Amazon urbanization.

In this case, it is recalled that the image of the Amazon as an "urbanized forest" had spread as virtuality since the 1980s.

Although it is possible to speak of metropolization, it is important to bear in mind that this process is not hegemonic in the Amazon case. For this reason, a particular type of extensive urbanization [7] is identified in the region, that is, a diffusion pattern of an urban way of life in the territory that does not need an exclusive urban center, but rather that it spreads in the territory of production relations and general living conditions, which have significant urban content to the point of creating demands with metropolitan profiles [8].

This characteristic of urbanization expansion with intensifying metropolization generated the interpretation that the settlement systems configuration in the Amazon is irregular and detached from a general principle of spatial organization. In this case, there is a whole literature stating that regional urbanization would be functionally disjointed from industrial and agricultural regional developments, because while agricultural expansion and industrial growth are limited to specific locations in the Amazon, the growth of the urban population is widespread across the region, leading to the conclusion that the urbanization process is disconnected from local development processes [9].

I assume a different theoretical premise to analyze the process of regional urbanization from the metropolitan agglomerates because in the two metropolises analyzed - Belem and Manaus - the existence of metropolization relates exactly to the networks of relationships that keep the Amazon, in a varied way, connected to global economic forces, which would be a general guiding principle of regional metropolization, and therefore not a reflection of an alleged functional disarticulation.

In Belem, this manifests itself through the expansion of the connection networks of the metropolis with the most dynamic regions of the countryside, through the expansion of the urban network dispersion radius and the logistical infrastructures that follow it.

In Manaus, the location of an industrial hub that connects the city to global networks of production and circulation of goods, with a relevant degree of specialization, becomes a "knot" in the international network of cities.

Therefore, it is relevant to understand that, unlike the urbanization and metropolization process in other regions of Brazil, in which the expansion of the urban area happened along with the process of the conurbation and the creation of territorial mobility networks strongly marked by industrialization, the Amazonian urbanization is characterized by the allocation of a set of the system of objects in the territory and a system of punctual actions - the large objects [10], which provided the regional urbanization for the expansion of the frontier economy.

Based on these aspects, it is assumed that the space metropolization in the Amazon region is inserted into two sets of variables. In the first one, the existence and expansion of metropolization is functional to the new forms of appropriation and capitalist accumulation on a global level, in which the production of value in the urban space leads to the consolidation of new forms of accumulation.

The second variable is linked to the internal structuring axis of the metropolises. The fragmentation of space in these agglomerations demonstrates this unequal reality produced as one of their elements in common, despite their different patterns.

These agglomerates represent an important aspect of regional dynamics in a long historical period, being carriers of a reality that reflects what the process of regional metropolization is nowadays. The importance of Belem and Manaus as the largest agglomerations in the Brazilian Amazon is expressed in **Figure 1**.

Although pre-1960 regional history helps to identify the genesis of these agglomerations, it must be considered that it is the integration and development strategies in the second half of the twentieth century that intensify the current urban occupation pattern, expressed in their population growth (**Table 2**)[1].

According to the data, Belem presents oscillation with a decreasing trend of demographic concentration in its metropolitan space, although the institutional area has been significantly expanded in the analyzed period, with the insertion of new municipalities in the Metropolitan Region of Belem (MRB).

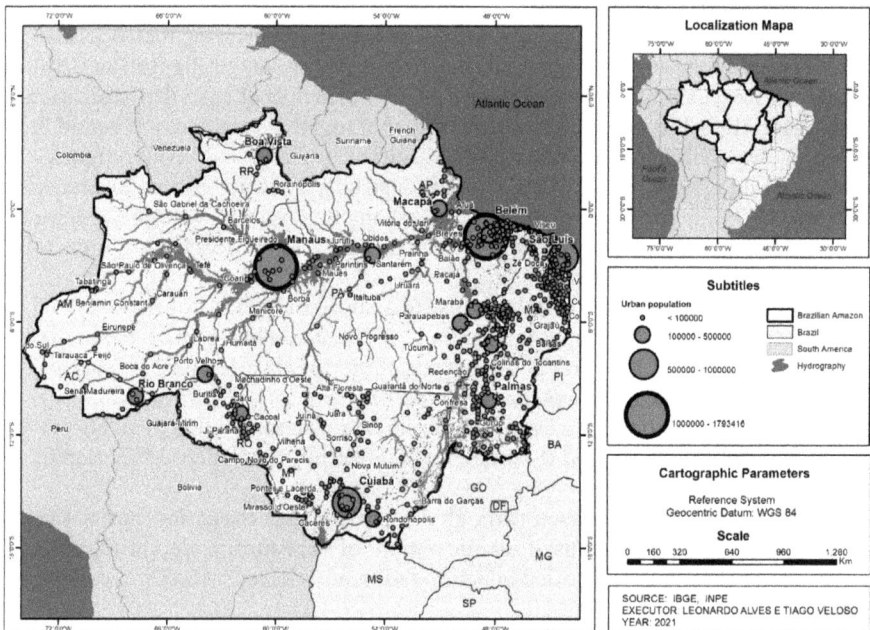

Figure 1.
Brazilian Amazon: cities populations (2010).

---

[1] The percentages of the populations in the agglomerates were calculated considering the contingent of the main municipality and the populations of the metropolitan municipalities. As new municipalities were included, they started to be added to the count, considering the decades in which the metropolitan regions were constituted.

Year	Pop. Belem/PA (%)	Pop. Manaus/AM (%)
1950	22.70	27.16
1960	25.93	24.31
1970	30.00	32.70
1980	28.50	44.34
1991	25.10	48.05
2000	29.00*	49.90
2010	27.90*	61.10*

*It considers the municipalities that became part of the Metropolitan Region.
Source: IBGE.

Table 2.
Belem and Manaus: Metropolitan population in relation to the states.

These municipalities were in part created from divisions and dismemberments of the municipalities that already made up the MRB.

Manaus, on the other hand, has the opposite trend, with increasing demographic concentration. This trend is explained by the significant growth from the implantation of the Manaus Free Trade Zone (MFTZ) and the Manaus Industrial Pole (MIP) in the 1960s. More recently, the inclusion of municipalities at the time the Metropolitan Region of Manaus (MRM) was created explains the population increase, given that the population of the municipalities becomes part of the metropolitan region.

In line with the trend of the participation of the metropolitan population in the state total, a similar aspect is observed when we consider the degree of concentration of the metropolitan Gross Domestic Product (GDP). In the two states (Amazonas and Pará), a significant percentage of wealth is concentrated in the metropolitan region, although this is more intense in Amazonas, with 85% of GDP concentrated in the Metropolitan Region of Manaus. In turn, Pará has important concentration levels, with the Metropolitan Region of Belem participating in 35% of the state's wealth (**Table 3**).

These trends are related to the way in which the integration and urbanization policies of the territory were established, which are part of the differences since there is an overlap between the formation of the metropolises and their regional surroundings.

In this case, there is a set of elements that help to demonstrate such differences, such as the induction of metropolization, the types of connections established with the global plan and the regional scale, the types of circulation systems, and the ways of territorial management that present distinct characteristics in the two agglomerates, making up a mosaic of diversity, much rather than the homogeneous reproduction of the same process. Such elements (**Table 4**) reaffirm the argument of the

State	Metro. Region	State Total	Metrop. Region Total	Core City	Other cities	MR/ State	Core City / MR
			Amount (R$1.000)				Share (%)
Amazonas	Manaus	39.166.314	33.426.618	31.916.257	1.510.361	85.3	95,5
Pará	Belem	44.375.376	15.680.400	12.520.322	3.159.818	35.3	79,8

Source: IBGE.

Table 3.
Amazonas and Pará: Participation in the Gross Domestic Product of the metropolitan region and the state (in current R$ thousand) – 2010.

Elements/Metropolises	Belem	Manaus
Metropolization inducers	Dismantling of old agrarian and riverside structures of the hinterland and modernization of the tertiary sector	Modernizing agglomeration economy of the Manaus Free Trade Zone
Global connections	Large economic project and modern export port system	Assembly industry and globalized tourism.
Main modalities of regional articulation	River – road – airway	River – airway
Growth of the metropolis in relation to the region	The region grows more than the metropolis, with a tendency toward demetropolization.	The metropolis grows more than the region, with a profile of a macrocephalic metropolis.
Intrametropolitan Configuration	Scattered, discontinuous with the presence or absence of conurbated satellite cities	Concentrated, continuous, without satellite cities or conurbation
Segregation Pattern	Concentration of classes with high purchasing power and increasing suburbanization of urban poverty	Concentration of classes with high purchasing power with increasing suburbanization of poverty and formation of selective sectors in pleasant suburbs
Urban fabric Configuration	Urban fabric of double configuration (conurbated and discontinuous) and unified by a system of regular flows	Single, not conurbated urban fabric, with recent and rarefied connections with adjacent municipalities
Metropolitan Region Creation	Older (1970s)	More recent (2000s)
Scope of the metropolitan institutional framework	Smaller than the actual metropolitan agglomerate	Larger than the actual metropolitan agglomerate
Territorial Planning	Limited to local districts and with little intercity permeability	Limited to the municipal district, but with a metropolitan scope

Source: [11].

Table 4.
Metropolitan agglomerates in the Brazilian Amazon: characterizing elements.

regional metropolitan difference and the territorial and content differentiation of the urban forms that the region presents.

The recognition of these elements leads to the conclusion that, in regional terms, these agglomerates can be understood when viewed along with the regional dynamics, which confirms the existence of an articulated complex between metropolis and the region. This complex produces metropolitan spaces that are mirrors of the differentiated sub-regional occupation profile.

These metropolization-inducing elements in different urban-regional realities make up a scenario that must be considered for the understanding of their organizational bases when it comes to the internal structures of the agglomerates. Based on these conclusions, we proceed to the analysis of the internal structure pattern of each of them.

3. The structure of the urban agglomerates: Belem and Manaus in perspective

Although the articulation between metropolitan agglomerates and their regional immediate surroundings is on a scale of understanding the role played by these spaces, it is important to highlight that the metropolis can also be interpreted from its internal structure, that is, the way it is organized according to elements that define its intra-urban, or intrametropolitan space.

In this case, it is necessary to identify the elements that make up the urban structures of the regional metropolitan agglomerates. The goal is to identify the different degrees of differentiation of these structures, considering that they do not mechanically follow the same dynamics. Thus, we will see the elements that mark both Belem and Manaus agglomerates.

3.1 Belem: From urban confinement to dispersed restructuring

From the seventeenth century until the first half of the twentieth century, the trajectory of urban growth in Belem followed the needs of the Amazonian urban network, which still had little need for a complex urban space. Belem rose in the regional and national urban network at some moments in history, such as during the cycle of Pombaline reforms in the eighteenth century and the rubber period at the end of the nineteenth century. But with the end of these cycles, the city returned to its profile with the limited urban fabric.

In a more recent period, it is possible to affirm the configuration of a phase of urban expansion, marked by the dynamics of metropolization, which presupposes the advance of the urban network in relation to the period and the previous phases. This phase "begins in the sixties and is consolidated in the following decades and presupposes the incorporation of cities and villages close to Belem, defining a unique urban network, despite being fragmented" [12].

In this case, it is understood that Belem had its moment of expansion toward the formation of a metropolitan fabric from the 1960s, in the context of alteration of the circulation networks with the construction of the Belem-Brasília highway, the first major axis of road penetration in the Amazon Basin [13]. The highway (**Figure 2**) is one of the fundamental elements to understand the expansion of the urban fabric and the consequent spread of the city because until the 1960s the urban fabric was confined to the perimeter demarcated by its central neighborhoods and immediate peripheries. The stimulus coming from new regional dynamics, such as the introduction of road axes, propels growth toward other districts.

Simultaneously, the limitations on the demands of the new regional configuration stimulated changes of intra-urban spatial nature due to the growth of the city. The existence of a large area destined to state and parastatal agencies, forming an "institutional belt", made the introduction of the road axes to become one of the elements of the intensification of land use, contributing to the formation of a metropolitan core marked by real estate use and making possible to overtake the initial area of the city. This central area or metropolitan core undergoes vigorous densification, followed by a vertical landscape of the central neighborhoods.

On the one hand, if verticalization is the predominant vector in the central areas of the metropolis, on the other hand, the transformations caused changes in the uneven landscape, observed from its slums and baixadas[2], expressions of a metropolization that intensifies a type of urban peripheralization. The formation of peripheries, even within the central area of the metropolis, had a close relationship with this limited urban configuration at the time, since the "existence of institutional areas, bypassing the initial limits, made continuous expansion of the city impossible, making it difficult to access, with few urban services and equipment; a fact that contributed to the population densification in the most central areas, including the baixadas, located below the "institutional belt" [15].

---

[2] The term "Baixadas" derives from the original topographic conditions of certain areas of the city which are floodplain level, constantly flooded or subjected to flooding at certain times of the year and that made up about 40% of the urban site, in the denser area.

Figure 2.
A stretch of BR-316 highway: Boundaries between cities become imperceptible given the conurbation and the intensity of flows. The BR-316 Highway is part of a set of federal highways that connect the capital, Brasília (DF), to Belem (PA), in a connection known as Belem-Brasília. Source: [14].

However, this institutional belt began to be broken in the 1960s, consolidating the spread of the urban fabric toward the two main routes of expansion: the BR-316 federal highway and Augusto Montenegro Avenue, which have guided the directions of expansion since the 1980s (**Figure 3**).

This movement made it possible to expand the metropolis to its immediate periphery in the following decades, with growth toward the peripheral municipalities, configuring the expansion area and shaping the old confinement in a new way. The municipality of Ananindeua reaches a demographic growth of around 18% over the 1990s. In the 2000s, all municipalities in the Metropolitan Region had greater growth than Belem (**Table 5**).

The described movement of metropolitan expansion caused the concentration of the highest-income population in the metropolitan core to undergo some changes in a very recent period. It is representative of a change - which cannot yet be classified as a trend - the fact that Belem receives real estate investments from high-income developments, such as Alphaville[3], but this time located in an area far from downtown, in the district of Outeiro. This district is characterized as an area of low-income housing and leisure, which somewhat contradicts the effort of the upper classes in Belem to remain close to downtown. Otherwise, it reaffirms the trend of dispersion of the metropolitan space, only this time, not from the peripheralization of low-income classes, but rather from the suburbanization of high-income ones.

The industrial areas are also present in the metropolitan area of Belem, even though the urban expansion was not a process derived from the allocation of

---

[3] Alphaville is the brand of a horizontal real estate development exclusively for high-income classes. The brand is a national reference in horizontal projects, planned neighborhoods and urban centers, present in 21 Brazilian states and the Federal District.

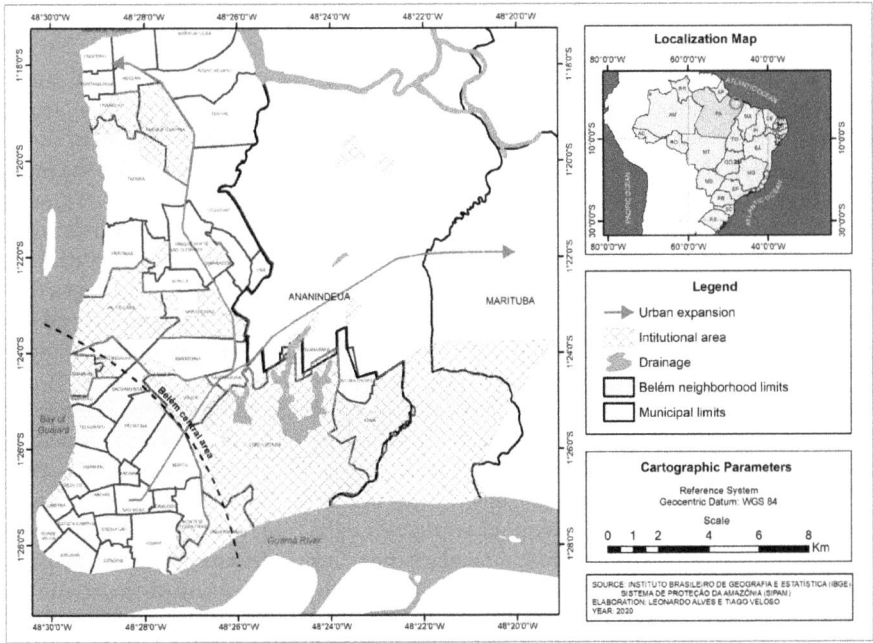

Figure 3.
Belem: Expansion of urban space beyond the metropolitan core.

Municipalities	1980–1991 (%)	1991–2000 (%)	2000–2010 (%)
Belem	2.65	0.32	0.85
Ananindeua	2.68	18.09	1.83
Benevides	10.73	−7.02	3.81
Castanhal	—	—	2.56
Marituba	—	—	3.82
Santa Bárbara do Pará	—	—	4.18
Santa Izabel do Pará	3.01	2.93	3.24

Source: [16].

Table 5.
MRB: Municipalities growth rate (1980, 1991, 2000 e 2010).

industrial capital, because the organization of the metropolitan space in Belem is not structured around industrialization, unlike other Brazilian metropolises.

In this case, we refer to the industrial experience carried out in Barcarena[4], a municipality that shapes the current configuration of the metropolitan space. The installation of a third roadway corridor in the 2000s[5] brings the metropolitan

---

[4] The Barcarena industrial pole was defined as one of the projects of the Grande Carajás Program, whose goal is to implement industrial plants aimed at the processing of aluminum, kaolin and alumina. At approximately 36 kilometers from Belem, it has an energy supply provided by the Tucuruí Hydroelectric Power Plant, a road system and a port with capacity to serve ships of up to 60 thousand tons.
[5] The Alça Viária, PA-483 highway, inaugurated in 2002, is a complex of bridges and roads with 74 km of highways and 4.5 km of bridges, built to integrate the Metropolitan Region of Belem into the countryside.

influence closer to the industrial pole, causing the metropolitan agglomeration to spread, made possible by the rapid flows between Belem and Barcarena.

In addition to the importance of these industrial areas, in economic terms, it is necessary to consider that Belem was historically characterized as the gateway to the Amazon Basin, and still has river navigation in its relationship with the rest of the country and in the diversification of its regional economy. Belem "is one of the most dynamic metropolitan centers in the network, having influence not only over the state of Pará, but also in Amapá, the western portion of Maranhão and the northern Tocantins. This influence of the metropolis has been made possible by the restructuring of the metropolitan area of Belem through the construction of highway axes that connect the capital to the countryside and the region itself, which improved the fluidity and shaped the Pará Integration System, linking the Metropolitan Region of Belem to the other regions of the state" [17].

The existence of ports in the metropolitan space confirms this position. The maintenance of the importance of river navigation combined with a dispersed metropolization has made road-river transportation possible, integrating the industrial and port structure, making the configuration of the metropolitan agglomerate more complex.

If the urban form of the metropolis was confined at first, assuming its dispersed character from the 1980s onwards, nowadays there is an increasing complexity of metropolization at a regional level, made possible by the increase of flows, the implantation of the infrastructure of material circulation and the expansion of the influence of the metropolis on the region. This can be seen on **Figure 4**.

This characterization sets up a metropolitan structure that is not limited to political-administrative limits but rather explained by the fixed points and flows that make up a functioning metropolitan agglomerate. As such, its structure can be thought of from its organization into sectors (**Table 6**).

In this structure, the core of the metropolis is formed by the central neighborhoods of Belem located in the initial perimeter of the city, either by high-income and upper-middle-class neighborhoods, but also by low-income class ones, known for their precariousness in terms of urban services and facilities, despite their

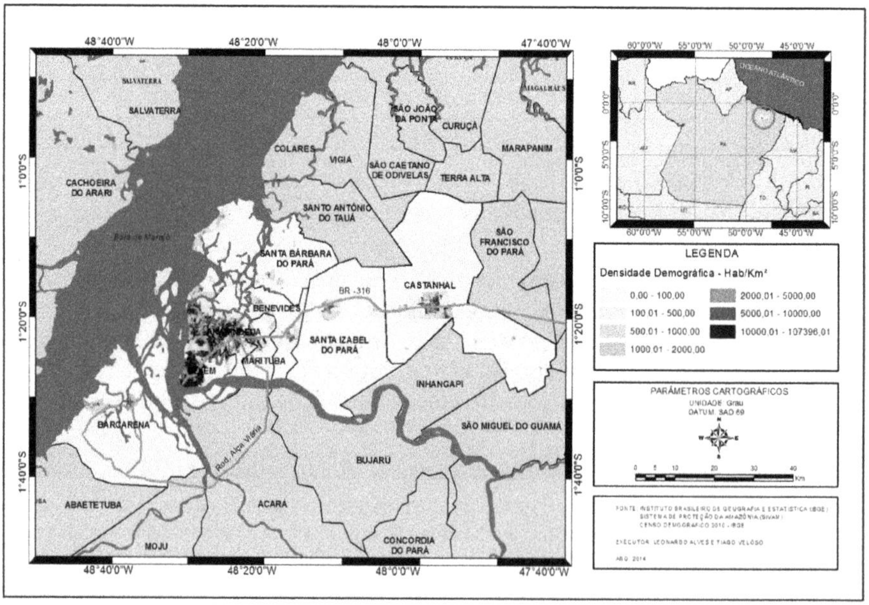

Figure 4.
MRB: Demographic densities.

Sectors	Subdivisions	Characteristics	Municipalities
First Légua Patrimonial (metropolitan central area)	Metropolitan core	Old neighborhoods with commercial, port, service, and residential functions	Belem
	Pericentral Neighborhoods	Old or recent neighborhoods, predominantly middle and upper-class ones	Belem
	Baixadas	Recent residential neighborhoods, low-income classes, and poor infrastructure	Belem
Transition Areas	Institutional Areas	Areas destined for public civil and military institutions.	Belem
	Residential Areas	Recent residential neighborhoods, lower and lower-middle-class ones	Belem
Expansion areas	Vector 1 (Augusto Montenegro Avenue)	Recent and low income industrial or residential sectors; Sectors of lower-middle classes and high-income suburbanization	Belem and municipal districts (Outeiro, Icoaraci, Mosqueiro)
	Vector 2 (BR-316 Road)	Recent and predominantly low income industrial or residential sectors	Ananindeua, Marituba, Santa Bárbara, Benevides, Santa Izabel, Castanhal
	Vector 3 (Alça Viária)	Recent industrial, port, and residential sectors.	Acará, Barcarena, Abaetetuba.

Source: [18].

Table 6.
Belem: Metropolitan structure.

proximity to downtown. This core, the most valorized area of the city thanks to the pattern of concentration of services, jobs, and urban equipment, has been experiencing an increase in density in the form of verticalization [19], with new types of social selectivity, incorporation of sophisticated leisure equipment and high real estate prices, outlining the reinforcement of the trend of segregation for high-income social segments in the central area.

The transition areas are identified by the spaces destined to public and private institutions, which in the past served as a restraint to the expansion of the city and have a reasonable degree of lower-middle-class residential settlements, whose inhabitants still manage to live relatively close to the metropolitan core, counting on the services offered.

Finally, the metropolis expansion areas, which follow the direction of three vectors. The first one, Augusto Montenegro Avenue, which goes toward the peripheral districts, is inhabited mainly by low-income classes. This vector has been the object of recent transformations in urban dynamics because although it remains a vector in which there is the presence of low-income neighborhoods and classes, it has shown qualitative changes caused by the actions of the local real estate sector associated with the national real estate circuit. This expansion has been the scenario of a possible trend of upper-class suburbanization, represented by the arrival of middle- and upper-class developments.

The second vector is the BR-316 highway, which goes toward the peripheral municipalities of the Metropolitan Region of Belem, such as Ananindeua, Marituba, and Benevides, which were the ones that grew the most in the last decades, a growth that was partly due to the metropolitan peripheralization of lower-middle classes and low-income classes, who leave the metropolitan core toward these municipalities. This peripheralization, the main constituent element of the BR-316 vector, took place in a state-stimulated manner via housing policy, through the construction of large housing estates, but it also took the form of "spontaneous" lower class reproduction strategies, with occupations of areas for low-income housing.

More recently, in the 2000s, the expansion of this vector was expanded toward more distant municipalities, such as Santa Izabel do Pará, allowing a territorial discontinuity to happen on a landscape level, but reaffirming the contiguity of the metropolitan network, especially by flows related to the new spaces of low-income settlements that exist in this municipality and the dynamics and demands related to population and urban growth, which ratifies the need for a policy of common metropolitan services [20]. In addition, this BR-316 vector is configured by the existing relationship with the municipality of Castanhal, which, like Santa Izabel do Pará, was recently recognized as a member of the MRB.

The third and most recent vector of metropolitan expansion follows the direction of Alça Viária toward the integration of the metropolis with the closest or more dynamic state sub-regions, as in the cases of Lower Tocantins and Southeast of Pará, respectively, which presupposes the existence of a metropolis more integrated into the region's countryside, hence the affirmation of restructuring the urban-metropolitan network of Belem based on the design of this new structure.

The analysis of these processes of expansion of the metropolitan network allows to conclude the redefinition of the metropolitan dynamics which, enlarged from these different processes, consolidates a more complex metropolitan structure in the regional scenario.

3.2 Manaus: From industrial enclave to concentrated metropolization

The growth of Manaus as a city of regional reference dates to the end of the nineteenth century, when it began to experience the first forms of capitalist interaction under an agro-export basis, because of the exploitation of natural resources (**Figure 5**). This economy enabled the development of an agro-extractive production base, without incentives for the processing of primary products, in the same way, that it triggered the existence of a migratory movement that became workforce for the greater productivity of latex extraction. The end of the period of economic expansion and urban growth caused by the rubber activity until the first decade of the twentieth century was followed by a period of decline in economic, demographic, and urban aspects.

Thus, the movement of little expansion of the urban fabric during the first half of the twentieth century is partly explained by the period of decline, which was only changed from the 1960s with the arrival of regional development programs, when the city's rise to the status of a metropolis began. In the demographic evolution of Manaus (**Table 7**), it is noticeable how there were changes among the highlighted periods: the decline of the population after the 1910s, the slow collapse of the rubber economy, and the growth stimulated by the MFTZ[6] and MIP in the 1960s onwards.

---

[6] The Manaus Free Trade Zone (MFTZ) was established in 1967, with the purpose of creating an industrial, commercial and agricultural center in the countryside of the Amazon, through an import and export free trade area and tax incentives with determined deadlines, in the Export Processing Zones model. Within the MFTZ strategy, a Pole was created to attract industries through reductions in customs fees, tax exemptions, land concessions and infrastructure provided by the State.

Figure 5.
Manaus and the Negro River Bridge: The expansion to the other side of the river. Source: Personal archive to the author.

Year	Total population
1910	85.340
1920	75.704
1940	106.399
1950	139.620
1960	173.343
1970	311.622
1980	633.392
1991	1.011.000
2000	1.405.835
2010	1.802.014
Source: IBGE.	

Table 7.
Manaus: Population evolution (1900–2010).

Following the trend of population expansion, stimulated to grow throughout the Amazon region in the 1970s, an intensification of the urban area and the current configuration of the metropolitan agglomerate can be seen. The evolution of Manaus and the expansion of the occupation of areas further away from downtown can be perceived as causally related to the movements of the region.

While under the influence of Belem, the economic modernization projects contributed to the production of a network of cities in the Amazon's countryside.

In the capital of the state of Amazonas, there was a concentration of urbanization and productive activities in the urban environment. The impact of the industrial enterprise caused the landscape of Manaus to be mediated by the industry.

The initial limits of the city were overcome when Manaus started to receive investments for the improvement of its infrastructure aiming at the implementation of the Free Trade Zone: an international airport was built, the port underwent changes and telecommunications services were implemented [21]. The new urban configuration, brought about through economic activities, marked the transformation of the city, because "with the consolidation of the Free Trade Zone, in the 1970s, the city underwent profound transformations, both in its form and in its social content. Manaus stopped being the "Paris in the Tropics" of the great works of the Rubber Cycle, to become a modern metropolis, with all the economic, social and regional contradictions" [22].

The reformulation of the city's profile since the arrival of the industrial pole is remarkable, with an immediate impact on the production of the space (**Figure 6a** and **b**).

From then on, the metropolis landscape follows the restructuring dynamics of the urban space in the logic of industrial production, which acquires its own economic importance. The urbanization process in Manaus has not stopped since then, and the expansion of the urban fabric was intentionally stimulated by state and market agents, especially the real estate sector.

Due to state actions, the changes in the administrative headquarters of the governments of the state of Amazonas and Manaus city hall, which were in the central area of the capital until the 1990s, were elements that induced the growth of the city toward the west-north vector since the decentralization of some of the administrative structures is consistent with the objectives of metropolitan deconcentration. The areas to which these services were relocated were coincidentally the ones which grew the most during the 1990s and 2000s, a period of the changes described (**Table 8**).

The state action in different administrative spheres has a common goal, to create the conditions for the expansion of the metropolis to one of its sectors, in a structural movement. In this sense, a type of center was created for the middle and upper classes of the city, located in the southern part of the city.

In fact, the production of the manauara space, which is uneven due to the nature of its urbanization, also tends to produce an urban area that has as a characteristic the income inequality, manifested in the city from the forms of land use and housing production. This inequality is constituted in the demographic distribution of

Figure 6.
Manaus: The Industrial Pole in two moments: (a) during its construction, in 1967, with the extension of reserved land in the city; and (b) consolidated in the urban structure in 2012. Source: [23].

Urban Zones	Permanent private housing units total, 1991	Permanent private housing units total, 2000	% of growth (1991–2000)
Central-West Zone	24.880	32.342	29.99
South-Central Z.	20.653	31.739	53.68
East Zone	34.382	76.783	123.32
North Zone	23.463	66.587	183.80
West Zone	38.508	47.952	24.53
South Zone	62.966	68.846	9.34
Source: *IBGE*.			

Table 8.
Manaus: Total households by urban zones (1991–2000).

the population, quite concentrated in Manaus when compared with the extensive territory that constitutes the metropolitan region (**Figure 7**).

It is in this sense that a project of expansion in today's Manaus is conditional on overcoming natural obstacles, such as the river and the forest. This growth model tends to deny the dynamics of nature in its development process and in its representations. As we saw in the metropolitan reality of Belem, the recent arrival of large real estate projects for high-income consumers, such as Alphaville, is representative of this qualitative change in the production of urban space in Manaus.

It happens that, unlike Belem, the location of the enterprise in Manaus is close to the metropolitan core, precisely in the highest income area of the city, the south zone. This location is justified by some specific reasons regarding the urban structure of the city, such as the particularity of the population and economic concentration in the metropolitan core and the existence of green areas, which could not only

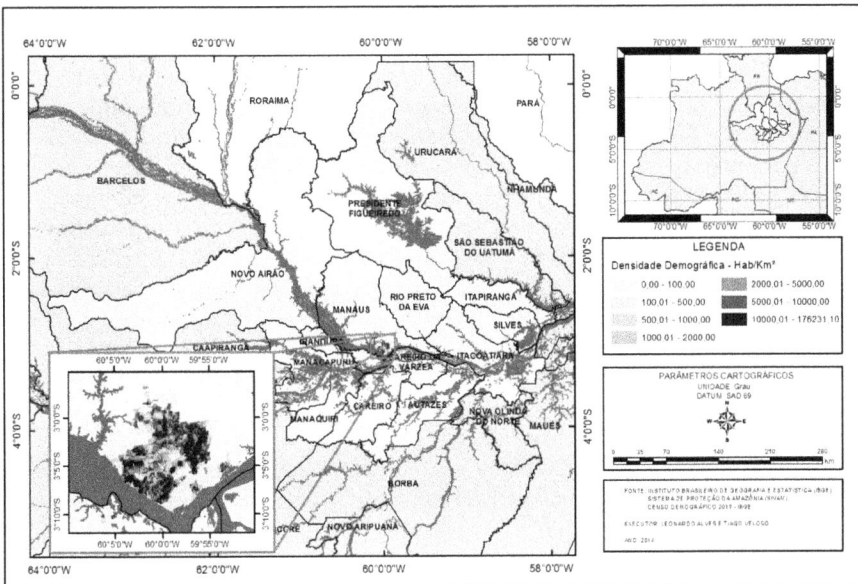

Figure 7.
MRM: Demographic densities.

be used for projects but are also an element of urban marketing when the product is offered to a specific target audience.

Meeting these new landscape realities, the city has also presented urban interventions for areas of lower classes, such as the Igarapés de Manaus Social and Environmental Program (Prosamim), a state government strategy for the housing and health issues in occupations adjacent to Manaus streams, historically occupied by these populations.

In addition, Prosamim is made possible with the possibility of expanding the urban-metropolitan network beyond municipal limits. This expansion has recently been made possible by the construction of the bridge over the Negro River, which provides road access between Manaus and some neighboring municipalities, such as Iranduba and Manacapuru. In the case of Iranduba, there is already a pilot project for Prosamim, reaffirming the influence of the metropolis on the adjacent municipalities.

In the new face of a Manaus-metropolis, the influence of the bridge on the Negro River cannot be minimized, as it composes a new scenario and reinforces the influence of the capital on the immediate region. It is the most important object symbol of the metropolitan landscape, expressing the arrival of the urban on the "other side of the river". This is one of the spatial expressions that large objects tend to intensify, given that in the municipalities close to Manaus the flow of relationships taking place in the metropolis has not yet been established.

Iranduba is a small city in terms of economic dynamics, with little capacity for its activities to add value on a local or regional level [24]. Such conditions occur due to its proximity to Manaus, with integration by highways, which allows it to present a greater quantity and variety of shops and services.

Manacapuru, on the other hand, is classified as a medium city with an intermediary function, since it performs an intermediary function between the other cities and Manaus due to the proximity of the metropolis and the road connection, that is, its importance is not only for the municipality itself but also for those smaller ones around it.

In other words, from the point of view of expanding relations to the municipalities that make up the metropolitan region of Manaus, one can identify a growth trend in these relations, although at present they are not a consolidated fact, due to the concentrating characteristic of the Manaus metropolis.

It is necessary to add that the expansion of these relations is not mediated only by the consolidated urban network between Manaus and the adjacent municipalities. There is a type of movement in the metropolitan space that is made possible by the existence of spaces metropolized[7] by commuting and specific flows, such as those made possible by tourism activities in the municipality of Presidente Figueiredo. This municipality is connected contiguously to Manaus and has flows and commercial activities derived from tourism that cause it to be influenced by a space consumption characteristic of metropolitan areas, including the use of nature elements, such as waterfalls, which are used to enhance tourist activities[8].

Despite the differences in elements of the urban structure, such as economic activities and the resulting socio-spatial impacts, the type of location of the upper

---

[7] The inspiration for understanding the metropolized space comes from Bernard Kayser (1969), for whom "The metropolized space is characterized as a space that is closely and concretely linked to the big city through the flow of people, goods, capital; flows that are quite intense and permanent and that in fact coincide with the major axes of urbanization" [25].

[8] In general, these establishments are privately managed, and there are several of them along the BR-174 highway. In general, they offer a short stay package (one-day long), which includes restaurants, tours and bathing in waterfalls. The use of the English language on billboards is common, characterizing a type of international tourism.

classes, and the metropolitan peripheralization, it is possible to find similarities between at least one aspect of the intra-urban structure of Manaus and what occurred in Belem: the port activities, and in this case, it is necessary to pay attention to the fact that Manaus was favored by its location, since the city was conditioned to perform the port function, concentrating the flow of the hydrographic system of the western Amazon. This urban function became the main force of the city's development, initially driven by the rubber cycle and later by the model of implantation of the Free Trade Zone and the Industrial Pole.

Such characteristics of the metropolitan expansion in Manaus, articulated by urban and regional infrastructures, make it possible to expand the scope of the metropolization toward the state of Roraima, an expanded periphery crossed by the internationalism of the Brazilian northern border, although this increase in regional connectivity provided by the geographical fixed points produces few benefits in terms of quality of life in regions of extensive and peripheral urbanization [26].

We can say, therefore, that the metropolitan structure of Manaus is intrinsically associated with the development policies that produced the Industrial Pole and the Free Trade Zone as pillars of its urban expansion and economic growth.

More recently, over the 2000s and 2010s, this process has been intensified by state-sponsored green entrepreneurship initiatives, which deepen the trend toward metropolization in an uneven geographic development pattern [27].

When we consider these characteristics of the metropolitan space of Manaus and associate it with its main zones and its growth trends and expansion vectors, we have a table that reveals what the metropolitan structure of Manaus is like (**Table 9**).

The main housing areas of the middle- and upper-classes are in the metropolitan core, particularly in the southern part of the city. There is a concentration of urban services, goods, and equipment in these areas. However, the core is not exclusively occupied by these higher-income classes. There are also medium-income sectors bordering its central-south zone, and even low-income populations, especially around the streams. It is in these areas that the main urban intervention programs take places, such as Prosamim, oriented toward interventions aimed at changing the occupation profile, whether in the aspect of the standard type of housing or in the relationship between the city and the river.

The residential areas of the upper-income groups are also located close to downtown, reflecting a scenario of intra-urban segregation, mimicking a particular type of corporate metropolis. So far, there has been no suburbanization of the high-income classes.

The transition and consolidated occupation areas are identified from two occupation profiles. In the central-west zone, residential districts of the middle and upper classes stand out, following the BR-174 expansion vector. In the east zone, industrial areas reserved for the Industrial Pole, as well as recent sectors of housing of low-income nature, predominate. Given the mononucleated characteristic of the metropolis, the low-income residential areas are located within the municipality of Manaus, but in areas relatively distant from downtown, characterized by a low supply of equipment, infrastructure, and urban services, especially in the northern zone.

The mononuclear metropolis characteristic remains fundamental to understand the metropolitan structure of Manaus, although recently there has been an expansion toward neighboring municipalities, such as Iranduba and, to a lesser extent, Manacapuru, particularly oriented by the duplication of AM-070 highway and stimulated by the construction of the bridge over the Negro River, as well as by real estate production and the flows of goods, services, and goods along with it.

Sectors	Subdivisions	Main characterization	Cities
Metropolitan core	South zone (core)	Old neighborhoods with commercial, port, service, and residential functions	Manaus
	South zone Pericentral neighborhoods	Old and recent neighborhoods, predominantly middle- and upper-class ones	Manaus
	Poor areas (near streams)	Recent low-income residential neighborhoods with poor infrastructure	Manaus
	Central-south zone Pericentral neighborhoods	Middle-class residential neighborhoods	Manaus
Transition and consolidated occupation areas	Central-West Zone Pericentral neighborhoods and BR-174 expansion vector	Middle- and upper-class residential neighborhoods	Manaus
	East zone Industrial areas (Manaus Industrial Pole)	Industrial and institutional sectors, such as new housing sectors, predominantly low-income ones	Manaus
Expansion areas	West Zone Vector (AM-070)	Recent middle-, upper- (Alphaville) and low-income class residential neighborhoods	Manaus Iranduba Manacapuru
Expansion areas	North zone	Recent low-income and low-middle class residential neighborhoods (large housing settlements)	Manaus
Metropolized spaces	Vector (AM-010 and BR-174 roads)	Municipalities with intense trade flows and tourist activities	Presidente Figueiredo Rio Preto da Eva

Organization: [28].

Table 9.
Manaus: Metropolitan structure.

4. Conclusions

The general picture of metropolitan structures in Belem and Manaus offers an understanding of these agglomerates from a regional characterization. It is possible to identify, within the scope of the particularities presented, aspects of differentiation in the configuration of the two metropolises. Elements such as the production of industrial, logistics, and port areas; the segregation profile of low-income classes, and the self-segregation of middle and upper classes lead us to conclusions about the pattern of metropolitan agglomerates in the Amazon.

First, metropolization is intensified from regional integration processes via economic ventures. The urban structure of agglomerates is influenced by regional dynamics that interact with capitals internal to cities, which makes it possible to state that in the Amazon case, regional dynamics directly influence the organization of metropolitan spaces. The movement of integration of the region in a frontier dynamic guided by the Brazilian State does not simply cause the structures to present a common pattern. On the contrary, the particularities of the agglomerates take shape when we consider the different ways in which each of the references had the process of induced metropolization.

In Belem, the execution of major development projects in the countryside of the central Amazon indirectly mobilized urban restructuring and the consequent space metropolization. The role of urban reference in the region, combined with the migration movements of the workforce, conditioned its dispersed metropolitan structure.

In Manaus, metropolization was induced due to the implementation of a Free Trade Zone combined with an Industrial Pole, elements that boosted urbanization in the western Amazon, which had been hitherto stabilized in the post-rubber economy period. The industrial core and commercial activities led to the establishment of a concentrated metropolitan structure.

These regional conditions act along with other conditions for structuring the metropolitan space but can be seen through the profile of human settlements, the industrial occupation, the circulation logic promoted by these cities and that articulate not only the intra-urban space of the metropolis, but also connect all the regional environment to which they are related, therefore being product, condition, and means of these regional realities.

Both agglomerates have a socio-spatial segregation profile, although it cannot be affirmed in any way that this is an Amazonian peculiarity. Again, the peripheralization appears as a defining element of the metropolitan structure of the two references analyzed, but in different ways.

In Belem, peripheralization is a more dispersed network due to the trend of occupation of more distant areas by low-income populations. In Manaus, the peripheralization is basically inside the city, but even so, located in the distant periphery (northern sector) of the metropolis.

Finally, they are mononucleated agglomerates. Here, a pattern of similarity is identified, because when considering the profiles of the metropolises, it is assumed that they follow an expansion pattern from downtown, which reveals the type of occupation of their elites around the areas with most urban equipment and services. This tendency to maintain the metropolitan centrality has even caused the urban soil to be increasingly densified in the central and pericentral areas, through verticalization.

Acknowledgements

I would like to thank the Studies and Research Group of "Urbanodiversidade and Territorial Management at Amazon" from Federal University of Pará (GEOURBAM/UFPA) and the Urban and Regional Research Núcleo from State University of Amazonas (NPUR/UEA) for its continued and valuable input and feedback throughout the process of writing and revising this paper. Also, I am very grateful for the research funding provides all those years for my actual institution of work, the Federal Institute of Science, Education and Technology of Pará (IFPA).

Author details

Tiago Veloso dos Santos
Instituto Federal de Educação, Ciência e Tecnologia do Pará, Belem, Brazil

*Address all correspondence to: tiago.veloso@ifpa.edu.br

IntechOpen

References

[1] Becker B. Amazônia. São Paulo: Ática; 1990

[2] Becker BK. Revisão das políticas de ocupação da Amazônia: é possível identificar modelos para projetar cenários? Parcerias Estratégicas. 2001;**5**(12):135-159

[3] Machado LO. Urbanização e mercado de trabalho na Amazônia brasileira. Cadernos IPPUR. 1999;**13**(01):109-138

[4] Lencioni S. Concentração e centralização das atividades urbanas: uma perspectiva multiescalar. reflexões a partir do caso de São Paulo [Concentration and centralization of urban activities: A multiescalar view from the São Paulo city]. Revista de Geografia Norte Grande. 2008;**39**:07-20. DOI: 10.4067/S0718-34022008000100002

[5] Gaspar RC. A economia política da urbanização contemporânea. Cadernos Metrópoles. 2011;**13**:235-256 Available from: https://revistas.pucsp.br/index.php/metropole/article/view/5989

[6] IBGE (Brazilian Institut for Geography and Statístics). Estatísticas do Século XX: Informações dos Censos demográficos brasileiros: 1950, 1960, 1970, 1980, 1991, 2000, 2010. Brasília: Instituto Brasileiro de Geografia e Estatística; 2006

[7] Monte-Mór R. Extended urbanization and settlement patterns in Brazil: An environmental approach. In: Brenner N, editor. Implosions/Explosions: Towards a Study of Planetary Urbanization. 1st ed. Berlin: Jovis Verlog GmbH; 2014. pp. 109-120

[8] Monte-Mór R, Castriota R. Extended urbanization: Implications for urban and regional theory. In: Paasi A, Harrison J, Jones M, editors. Geographies of Regions and Territories. 1st ed. Vol. 1. Massachusetts: Edward Elgar Publishing Limited, 2018. pp. 332-345

[9] Browder J, Godfrey B. Rainforest Cities: Urbanization, Development, and Globalization of the Brazilian Amazon. New York: Columbia University Press; 1997

[10] Santos M. Técnica, espaço, tempo: globalização e meio técnico-científico informacional. São Paulo: Edusp; 2008

[11] Veloso T, Trindade Júnior SCC. Dinâmicas sub-regionais e expressões metropolitanas na Amazônia brasileira: olhares em perspectiva. Revista Novos Cadernos Naea. 2014;**17**:177-202. DOI: 10.5801/ncn.v17i1.1323

[12] Trindade SCC Jr. Formação metropolitana de Belém (1960-1997). Belém: Paka Tatu; 2016

[13] Vicentini Y. Cidade e história na Amazônia [History and city in the Amazon]. Curitiba: Editora da Universidade Federal do Paraná; 2004

[14] dos Santos TV. Metrópole e região na Amazônia: Trajetórias do planejamento e da gestão metropolitana em Belém, Manaus e São Luís (Tese Doutorado). Belém, Brazil: Núcleo de Altos Estudos Amazônicos, Universidade Federal do Pará; 2015

[15] Trindade SCC Jr. A cidade dispersa: os novos espaços de assentamentos em Belém e a reestruturação metropolitana. [Sprawl City: The New Settlements in Belém and it's Metropolitan Changes]. [Thesis]. Brazil: São Paulo University; 1998

[16] Pará. Produto interno bruto dos municípios do Pará. Belém: Governo do Estado do Pará; 2012. pp. 03-53

[17] Huertas DM. Da fachada atlântica à imensidão amazônica: fronteira agrícola

e integração territorial. São Paulo: AnaBlume; Fapesp; Banco da Amazônia; 2009

[18] Trindade Júnior SCC. Grandes projetos, urbanização do território e metropolização na Amazônia. Revista Terra Livre. 2006;**26**:177-194

[19] Oliveira JMGC. A verticalização nos limites da produção do espaço: parâmetros comparativos entre Barcelona e Belém. In: COLÓQUIO INTERNACIONAL DE GEOCRÍTICA. Porto Alegre: Universidade Federal do Rio Grande do Sul; 2007

[20] Cavalcante FC. Metropolização e dispersão urbana na Amazônia: a dinâmica socioespacial do município de Santa Isabel no contexto da urbanização belenense [Thesis]. Brazil: Universidade Federal do Pará; 2011

[21] Oliveira JA, Schor T. Manaus: transformações e permanências, do forte à metrópole regional. In: Castro E, editor. Cidades na floresta. São Paulo: Anablume; 2009. pp. 13-39

[22] Ribeiro Filho V. Novas centralidades em Manaus. In: Oliveira JA, editor. Espaços urbanos na Amazônia: visões geográficas. Manaus: Valer; 2011. pp. 71-89

[23] SUFRAMA. Superintêndencia da Zona Franca de Manaus. Manaus, Amazonas, Brazil: Acervo Fotográfico; 2012

[24] Sousa IS. A ponte Rio Negro e a Região Metropolitana de Manaus: adequações no espaço urbano-regional à reprodução do capital [Thesis]. São Paulo: Universidade de São Paulo; 2013

[25] Lencioni S. Metropolização do espaço: processos e dinâmica. In: Ferreira A, Rua J, Marafon GJ, Silva ACP, editors. *Metropolização do espaço*: gestão territorial e relações

urbano-rurais. São Paulo: Consequência; 2013. pp. 17-35

[26] Kanai JM. On the peripheries of planetary urbanization: Globalizing Manaus and its expanding impact. Environment and Planning D: Society and Space. 2014;**32**:1071-1087

[27] Kanai JM. Capital of the Amazon rainforest: Constructing a global city-region for entrepreneurial Manaus. Urban Studies. 2014;**51**(11):2387-2405

[28] dos Santos TV. Metrópoles amazônicas: Dinâmicas regionais, estruturas urbanas e políticas de planejamento e gestão em Belém, Manaus e São Luís. Belém: Editora Paka Tatu; 2021

Devolution of Decision-Making: Tools and Technologies towards Equitable Place-Based Participation in Planning

Donagh Horgan

Abstract

Neoliberal development has increased spatial inequalities for communities in both urban and peri-urban settlements across in the global north and south alike. The financialisation of property has increased urban development in favour of opaque private and semi-public actors, making it harder for community stakeholders to influence decision-making. Social innovation in which diverse stakeholders collaborate towards sustainability and resilience in the built environment, offers pathways towards place-based policy-making and more inclusive growth, but needs political support and tools to facilitate participation. Using findings from a set of international cases, this chapter considers the effectiveness of participatory approaches to decision-making, and digital tools that facilitate public consultation. Cases consider the effectiveness of mechanisms available to communities in the cities of Moscow, Belgrade and Edinburgh to influence urban development. Literature review and new knowledge is brought together to shine light on whether information and communications technologies are used to provide a veneer of engagement with communities, and whether more bottom-up or insurgent tactics can give citizens a voice to influence more equitable future cities.

Keywords: urban development, neoliberal planning, community participation, platform urbanism, living labs

1. Introduction

In a study looking at a set of cases from both the Global North and South, Horgan and Dimitrijević [1] found that increasing inequality is manifest in the built environment, as trends towards populist governance decreases the ability of communities to influence spatial decision-making. The authors use examples from around the world to illustrate how political ideologies influenced by global capitalism dominate urban planning systems. Even in cases where overarching policy objectives pursue sustainability and resilience, the underlying political system often prioritises economic growth at the expense of more holistic investment. Their research demonstrates that competitive or capitalist values promoted by neoliberal governments are often in conflict with social and ecological priorities, resulting in development strategies that continue to favour the market in practice. This is most evident in the increasing monetisation of property and commodification of home—which has exacerbated acute housing crises in a number of cities and territories [2]. Horgan and Dimitrijević's [1] findings follow warnings from Lefebvre [3] and Harvey [4] on the exclusionary power of a pro-growth politics that produces capitalist forms of spatial development. Lamentably, the research would seem to suggest that decision-makers favour models that align to short term political priorities—that ignore and even inhibit social innovation in communities—in order to reduce risk for those funding development.

In a challenge to financialised development models, Horgan and Dimitrijević [1] identified networked approaches within communities seeking to take control over local development, and how community networks are making steps towards autonomy, self-reliance and resilience. Common across the phases of social innovation observed with those communities was the need for strong supportive governance systems to ensure participation at the grass-roots influences spatial strategy at higher levels of organisation. Tokenistic forms of engagement will only serve to further alienate communities suffering spatial inequality, and encourage citizens to challenge opaque urban governance. The authors conclude that the COVID-19 pandemic has brought new attention to the politics that buttresses spatial inequality, and the precarity of unsuitable and undignified living conditions in cities and peri-urban settlements all over the planet. Similar findings in scholarship which support that capitalist modes of production produce spatial inequalities—through the configuration and allocation of space—inform the context of this research. Semi-structured interviews with key informants provide insights into the nature of participation in decision-making related to the built environment in three urban communities, and provide a lens for further analysis. Case studies from Belgrade, Serbia; Moscow in the Russian Federation and Edinburgh, Scotland are examined—looking particular at opportunities for citizen participation to influence spatial development.

The three cases reveal commonalities in how city governments approach entrepreneurial models of development, albeit within vastly different political regimes. The study confirms a relationship between politics and planning that influences the type of community engagement that accompanies spatial transformation in each case. The chapter provides insights into how communities approach issues related to spatial development within different political contexts, and the mechanisms available to them to influence urban policy in each. Included are relevant themes from literature review that background the political and cultural context of spatial inequality in the chosen cases—with reference for example to previous socio-political systems such as those relying on forms of self-management, as is the case in the former Yugoslavia. The research strengthens findings from previous research on how aspects of ownership and participation in planning are aligned to the nature of the surrounding political ecology. It reveals how different forms of participation—including networked opposition and organisation—can open up decision-making in the medium to long term—and contribute to a sustainable lasting social investment from urban development.

2. Situating hard and soft power in planning and placemaking

The global pandemic has demonstrated the need for resilient social infrastructures, and the highlighted importance of co-creating social value through place-based approaches. Such values are in immediate conflict with pro-growth models of development that prioritise global flows of capital, and the commodification of property and urban assets [1]. At the city scale, pro-growth and competitive modes of urban governance can result in ill-defined forms of collusion between state and market actors—necessitating the need for transparent participation mechanisms for public oversight of planning. Top-down governance—from authoritarian to laissez faire—inhibit nuanced conceptions of public value that promote alternative measures of social capital [5]. In what Swyngedouw [6] calls the post-democratic city, one-way or tokenistic forms of participation in planning spread further disillusionment and disengagement—and eventually post-politicisation of the built environment. In a challenge to increasingly technocratic global governance, agile and insurgent approaches to defend spatial rights are emerging borne out of networking at the grassroots [1, 7].

Within a development context dominated by smart city narratives, and top-down platform approaches to urbanism that rely on big-data, interest has grown around policy-making that engages with the financialisation of the built environment [8]. Funded by the European Union's *Horizon 2020* research and innovation programme, the *Urban Maestro* action involved UN-Habitat and academic partners in order, *"to understand and encourage innovation in the field of urban design governance through a better understanding of alternative non-regulatory ('soft power') approaches and their contribution to the quality of the built environment"* [9]. Collecting best practice from a host of cross-sectoral activities the project was built around a premise that soft coordination mechanisms create a common vision to promote alignment of a large number of stakeholders towards the same objectives. The project looked at the relationship between financial mechanisms and informal tools of urban design governance—looking at how synergies between such tools have the potential to make both approaches more effective in attaining their desired outcomes. A survey conducted for the project among European countries revealed under-exploited potential to use financial tools in urban design governance to reward good behaviour, and discourage poor behaviour. Bento and Carmona [10] found that while the role of the state should be to incentivise high quality development over development for development's sake, many administrations across Europe are *"attempting to do this with one hand tied behind their back"*. The authors concluded their report with a reminder that *"the public sector nevertheless has a special responsibility for creating the conditions within which a high quality built environment can flourish"* pointing at a host of tools already available to governments [10]. These ranged from tools for evidence gathering, knowledge dissemination and proactive promotion, to structured evaluation and direct assistance tools that seek to develop community capacities for placemaking.

It is within this research context that the three cases below are considered—seeking to understand how decision-making is facilitated (or not) within a set of different political environments. Findings reveal persistent concerns around citizen engagement platforms in Russia, while in Serbia an engagement vacuum has activated networked approaches to achieving combined social and spatial rights. Despite a pro-market planning agenda in Edinburgh, an enlightened policy context in Scotland has allowed for sophisticated co-production in the planning process that is more compatible with strategies for an inclusive growth. This chapter offers insights into how cities can look to enhance practices of urban design governance within Europe and beyond, for the ultimate benefit of all citizens delivering better-designed places. It demonstrates how spatial equality can provide a lens for wider democratic deficits, and that while technology allows for participation, it may also give voice to propaganda that mask unsustainable unequal spatial reorganisation.

2.1 Active citizenship in Moscow?

Horgan and Dimitrijević [11] identified that for citizens seeking to participate in spatial decision-making technology can be both a force for good—connecting communities of practice united in the struggle for spatial rights—yet can also be tools to advance neoliberal policies and planning orthodoxies. Their study looked at technology-enabled approaches to engagement, alongside tools to facilitate spatial decision-making in the community setting, such as Scotland's Place Standard [11]. This research follows warnings from a number of scholars that caution how technology is not a panacea to enable participation, and how seemingly 'smart' approaches to urban governance are driven by neoliberal ideals in conflict with the citizens' desired social outcomes [12–14]. That study, which focused on a case study from Moscow in the Russian Federation, revealed a technology-enabled approach that embraces tokenistic participation in planning, allowing city governments to shape dialogue around urban development to suit their own speculative strategic ends. These findings are supported by Wijermars [15], who found that alongside an increase in the number of internet users, the country saw a drastic increase in regulation over Information and Communication Technologies (ICT) which gradually decreased its salience as an object of participatory socio-political construction. Recent years have seen the development of the Internet as a critical form of modern communication infrastructure that supports citizen's lives, as well as being used by governments as a tool of political influence. In their pre-pandemic study, Horgan and Dimitrijević [11] found that through a joint venture, the city of Moscow developed an online engagement platform—*Active Citizen* (Активный гражданин)—to support urban development decisions and others as part of its smart cities strategy. Through interviews with a number of citizens in the city, the authors found widespread suspicion around the platform, and concerns at how it was being used to demonstrate support for large-scale demolition and urban renewal on valuable sites in Moscow. The article noted that alongside questions around how such platforms are used for explicit political means, major concerns exist around aspects of ownership, governance and participation on the platform itself. When decision-making is required on highly impactful issues such as large-scale urban renovation or displacement, there is no substitute for offline face-to-face engagement through incremental and iterative engagement [11].

In an article by the Russian newspaper *Novaya Gazeta*, data analysts found that voters on the Moscow platform had done so in several neighbourhoods at the same time, suggesting some people participated in polls about things without any informed knowledge [16]. The investigation found that rewards for users in casting votes and posting to social media often skewed the voting process, and that in some cases users could claim rewards for simply liking an improvement. The article referred to a number of voting anomalies identified by several sources, including too-regular voting behaviours and seemingly fictitious users. The newspaper cynically suggested that the platform had undergone a tremendous evolution towards emulating the free expression or the will of living people, since earlier well-publicised failures in the system's credibility. Emerging from the success of Active Citizen, a version of the portal is being developed for all cities according to the requirements of the Ministry of Construction of Russia. In fact, cities should procure a platform to comply with the roll out of the Ministry's *Smart City* programme across the federation. A white-label version of the portal sold by LLC "Internet business systems" provides cities with technology to facilitate citizen polls as part of urban renewal projects and policymaking, towards a comfortable urban environment. Already online in cities as diverse as Ulan-Ude and Chelyabinsk, the platform allows citizens the chance to vote on minor aspects of urban development, at a point where any significant influence on urban strategy is impossible.

For example, a scan of posts on a regional platform offered citizens the chance to decide the phasing of a development rather than provide feedback on anything of strategic importance, without revealing anything related to procurement or the tender process. The veneer of transparency around urban development given over by these technologies obscures the relationships between local government and development actors, offsetting public scrutiny. An academic looking at critical area studies, and an expert on Russian *blagostroitsva* (improvement)—initiatives that promote amelioration of the public realm—confirmed that these platforms were an integral part of a strategy to foreground top down decision-making in planning with vague policy rhetoric around the smart city. In a semi-structured interview, the informant explained how the platforms are meant to sit alongside strategies for urban renewal in Russia's regions, and are tied to policies towards promoting a perception of decentralisation of power and constitutional reform to the population. This process, which began under President Medvedev, was piloted by Moscow mayor Sobyanin, and is continued through the work of Russia's Ministry of Construction in other cities.

It has the effect of keeping dissatisfaction with governance at bay within the general populace through the spectacle of improvement, mirroring neoliberal pro-growth approaches to development in other areas of the economy. Undeterred by supply chain disruption during the pandemic, Moscow—and cities across Russia's regions—are still carrying out large scale urban reorganisation at an unprecedented scale, as part of a strictly authoritarian modernisation agenda. The academic informant emphasised how that in Russia, the *Active Citizen* portals follow a model where economic actors conspire to present a speculative development project as a bottom-up intuitive—using participative technology to sanction interventions on public land. Such a model is important following the success of consortia in cities such as New York to convey speculative development in the public as a public good—such as the High Line—while local developers and funders make massive returns on investment on adjacent lots. This approach takes advantage of loosely defined ownership status in Russian law, and is emblematic of how broader privatisation disinherits the community of what was social property under the Soviet system. What is hidden by simulated engagement in the form of what has been described by Asmolov [17] as 'vertical crowdsourcing' platforms, is the incremental transfer of state assets—once part of a socialist commons—to new forms of organisations, and attributed to spatial redevelopment in the public good. While motivations for participation in online polls can be driven by a host of reasons—from incentives to active citizenship—a lack of transparency around the initiators of development proposals, means that users are often willingly championing neoliberal development with little other option. While those can afford to shape how the city is remade, without levers to exercise democratic control, everyday citizens are often passive participants in decision-making that govern place.

Evidently the pandemic has increased tendencies in governments worldwide to substitute online platforms for direct democracy opportunities in real life, the need for greater transparency over online platforms also increases. The likelihood of spatial policies informed by the logic of crisis urbanism means that citizens are likely to become further disenfranchised from decision-making, as city governments use paradigms of austerity and recovery to sanction the absence of public oversight and participation. Writing on urban policy making after the global financial crisis, Theodore [18] found that the financial crisis saw a *"redefinition of the state's role in fostering a globally competitive environment through marketisation, deregulation, and fiscal conservatism"*. In the article, the author shows how the resort to austerity has been highly regressive, socially as well as politically all over the world—aggravating spatial inequalities. For Theodore [18], austerity-led development represents, *"the consolidation of neoliberal urbanism, driven by its underlying logics and deepening its effects on governance arrangements and everyday life"*. This follows authors such as Peck [19], who noted that following a brief period of economic introspection, *"commitments to the antisocial credo of market fundamentalism were soon [...] renewed and reinvigorated"* around the world. There is a growing scholarship documenting insurgent tactics being used by communities to challenge neoliberal orthodoxies in planning, and safeguard the right to the city for citizens [6, 20–22].

2.2 The right to a new Belgrade?

The shock of regime change has often provided an excellent foil for sweeping changes to urban governance, policy and development structures, in particular in relation to the provision of housing [23]. Prior to the collapse of its uniquely itera-tive economic system of market-socialism, housing was confirmed as a social right in Yugoslavia. In building a national state at the confluence of imperial powers, its early federal government saw housing for all as a significant policy objective in bringing about a prosperous and aspirational egalitarian society [24]. As a mecha-nism for equal distribution and management of society, a right to housing was introduced as part of the socialisation of property after the establishment of the socialist state [25]. In fact, the concept of Yugoslavia itself became an inspirational platform for social innovation in architecture—translated into the production of socialised housing informed by an expert evidence base and knowledge culture—supported by high-quality research, technical prototyping and testing on smaller scales [26]. Yugoslav urban planning was a spatial application of a set of ideas for community, providing for genuinely heterogeneous neighbourhoods, the outcome of a distinct economic system of self-management. Within this polycentric gov-ernance system, a unique ideological framework for society was constantly being negotiated [27].

Self-management allowed for successive economic reforms and policy experi-mentation that set Yugoslavia apart [28]. While it did not solve the housing crisis in the federation, it made possible large-scale residential developments with a capacity to house 5000–10,000 inhabitants, managed by locally-elected units [29]. While the virtue of participation emphasised in self-management doctrine did not translate into a genuine participation of citizens in urban or regional planning, the participation of state enterprises encouraged a culture of open innovation [30]. During the 60s and 70s, New Belgrade continued to be one of the largest building sites in Europe, whose housing typologies and designs were as much influenced by socialist thinking in neighbouring countries as the architecture of the welfare states of Northern Europe [31]. Despite its limitations, the system allowed workers to participate in decision-making related to their enterprise that led to improvements in their general quality of life, including housing [28]. Ultimately the Yugoslav system unravelled as unscrupulous actors exploited inconsistencies in its nuanced understanding of Marxism, which was unable to withstand mounting individual-ism at all levels of society. The nationalist parties that emerged to lead the newly independent states introduced a neoliberal capitalist economic system that allowed foreign capital to cheaply purchase production resources in former Yugoslavia, previously owned by the workers [28].

Belgrade is still reeling from the upheavals of the late twentieth century, which saw the stock and production of public housing reach record low levels [32, 33]. An estimated 53% of housing stock in Belgrade was socially-owned apartments in 1991 [34], dropping to less than 1% today [35]. While the politician vacuum and popula-tion movements in the period during and after the war saw a significant demand on housing and space in the city, and overall lack of strategy for is one of the most significant outcomes for the built environment since the introduction of free-mar-ket capitalism in Yugoslavia [36]. Hirt and Petrović [32] (2011) note that globalism and the collapse of the old system are to blame for the spread of gated housing in

Belgrade—evidence of ever decreasing social and spatial solidarity. Projects such as Belgrade Waterfront are emblematic of a speculative approach to planning in the city which prioritises high-wealth individuals in access to public space and spatial decision-makers [37]. In the absence of a coherent strategy put to public consultation, large scale public assets such as the city's main train station are relocated in favour of projects concerned with the exchange value of property rather than social value, and the actual needs of society.

Indeed, movements for spatial rights in Serbia have grown, perhaps in parallel to an ever-growing discontent with urban governance and a litany of unpopular planning decisions and actions across the country. In a study on frameworks for social innovation. Horgan and Dimitrijević [1] identified organisations in the city working to achieve combined social and spatial rights, and address a democratic deficit in decision-making. Spatial activism and organisations such as *Ne Davimo Beograd* (NDB, Do not strangle Belgrade) and *Ministarstvo Prostora* (MP, Ministry of Space) present parallel political movements offer hope for Belgrade as development can be impeded by a lack of strategy and perceived kleptocratic governance [38]. These organisations employ shared methodologies, management and governance to offer ownership to citizens. An interview with the organisers of NDB in a year since the outbreak of the global pandemic, suggests that the top-down spatial development approach in the city is increasingly trending towards monumentalism, following Northern Macedonia's model for statecraft through urbanism in Skopje. While interventions are presented as strategic, *NDB* cannot figure out the pattern in Belgrade's spatial development, which they consider to be harsh and inconsistent, ignorant of the needs of citizens. The group sees no clear divisions between branches of government in Serbia, no independent institutions or transparency over city authority. Since the introduction of a market economy in the 2000s, a privatised state has emerged, producing a stark deterioration of the public sphere, visible in housing issues and everyday life struggles. A perception that political and economic actors are basically one has been behind widespread criticism of monumental projects such as Belgrade Waterfront, which since 2015 has advanced sweeping changes to the city's urban fabric [37].

For spatial activists, the urban development approach combines symbols of raw capital and raw power, that mask a captured state with weak social institutions. Since the coronavirus pandemic the market in Serbia has been stagnant, making it easier to identify opaque relationships in small scale residential developments on key strategic sites in the old city. Often these sites are on formerly state-owned industrial assets which have lain vacant after years of unsuccessful privatisation. It is within this context that *NDB* and their partner organisations have been successful in making an incremental impact towards spatial equality. Their approach has been a mix of hyper-local and regional campaigns that engage citizens on environmental issues such as air pollution and the lack of adequate sewage systems common across peri-urban areas of the former Yugoslavia. Campaigns have focused on topics as wide as forests in the city's development plan to the upgrading of former bomb shelters and other shared public amenities, and included solidarity actions around the pandemic. Tactics have sought to bring about policy change through greater awareness, bringing undisclosed issues into the public domain and mainstreaming dialogue around urban planning—in a country with a polarised and limited media. This is not a linear approach, and its agility comes from an ability to connect public and expert spheres, social and environmental movements within a community of practice.

This networked approach to social innovation is important in driving a rise in new civil society actors in Serbia, and grow participation in spatial decision-making. Without giving these networks the tools to influence development frameworks, and co-produce imaginative social architectures, prevailing political ideology will always stand in the way of change [1]. Belgrade's city government has attempted to develop ownership to citizens through its attempts at participatory budgeting, however the lack of a feedback loop in the process brought attention to the state of the city's badly neglected public services, ultimately derailing the project. For those at the helm of *NDB* the biggest challenges related to a lack of human resources, made worse by an ever-present brain drain from Serbia. This presents challenges for those seeking to grow collective rights, and reinstate some of the innovative forms of ownership of the former socialist federation. Also difficult is the promotion of shared or holistic narratives for spatial development within a system where privatised development is seen to deliver functioning security for those who can afford it.

2.3 Enlightened engagement in Edinburgh?

In a study of the enlightenment in Edinburgh, Scottish author Murray Pittock [39] uses the methodology of urban innovation to describe the civic networks and cultural change at the heart of what made Edinburgh a smart city in 1700. This analysis reveals that civic development produced the innovation and dynamism that made social transformation in the Scottish capital possible and the political power of the gentry and patronage [39]. The current neoliberal turn in the city's mode of governance has been lamented by many, and is visible across a whole host of spatial strategies, from planning for festivals and tourism to economic development. Tracking contested development proposals in the city, Ballard Tooley [40, 41] found that the political and economic logic of urban development in Edinburgh privileges economic growth within a system of interurban competition, by prioritising the needs of business over citizens. The studies explore tensions between the values of community and efficiency in urban development, and reveal how community engagement itself unearthed conflict between political visions of a pro-growth urban renewal agenda with neoliberal realities for local residents. Edinburgh's growth to become the second largest financial centre in the United Kingdom after the City of London has rested on the service sector, is a major contributor to the local economy with culture, consumption and knowledge exchange vital pillars for economic development. Sutherland [42] found that the complexity of devolution in Scotland means that policy priorities, notably inclusive growth and approaches to reducing economic and social spatial disparities differ greatly from that of the Westminster government. For Sutherland [42], evidence of neoliberalism can be found in urban developments, in place-marketing to multinational corporations, and in the competition to attract city investment—efforts which are often at the expense of engagement with citizens, and public service innovation. It is apparent in Edinburgh's ambitions to become the 'Data Capital of Europe', an entrepreneurial framing of the smart city concept, yet difficult to disaggregate in practice. This ambition is anchored by Edinburgh's City Deal—part of a UK government framework to reduce spatial imbalances between London and other parts of the country—a combined investment from the United Kingdom and Scottish governments and match funded through capital investment from universities and other sources [42]. Sutherland [42] notes that Edinburgh's growth faces significant bottlenecks, in the availability of skills and housing, and is almost certainly destined to draw activities and people away from other places, whether in the UK or further afield—necessitating massive spatial reorganisation.

In a study looking at the establishment of a Commission on climate change in the city, Creasy et al. [43] noted a persistently technocratic model of governance in Edinburgh, where a *"fast-tracked conceptualisation of place, instigated from the top-down"*, legitimises a focus on policy-making through 'expert' knowledge. The authors found such an approach to be in direct conflict with Massey's [44] relational interpretation of place, which seeks to engage with citizen's place-based knowledge on the ground. Creasy et al.'s [43] study indicated that in the process of establishing the commission, the city was moving towards more experimental approaches to urban governance, seeking opportunities to unlock new resources and possibilities with new partners, as opposed to *"feeding the zero-sum game of carving out resources from existing local allocations"*. The project has helped to amplify the place-based agency of the city at the grass roots, helping to *"to balance the vital networking and learning opportunities they facilitate with resisting a one-dimensional and static interpretation of the 'places' that they seek to network together"* [43]. In Scotland however, place-based principles are an integral part of a National Planning Framework that mandate levels of community participation in the planning process—in both spatial and community planning—which has encouraged the development of a host of tools and methods that support collective decision making [11]. This means that while economic agendas may drive urban strategy to an extent in Edinburgh, negative impacts and social exclusion may be mitigated against through focused and targeted engagement with specific cohorts. Even as development strategies focused on growth prevail at the policy level, socially innovative place-based approaches to decision-making at the community level—generating ownership over changes in the built environment.

Such an approach is visible in the work of Edinburgh's Living Lab (ELL, part of the University of Edinburgh), whose work takes a holistic perspective that integrates data innovation and place-based practices that put people and sustainable futures at the heart of decision-making. Through community engagement, applied researchers at the ELL help to define these options for the City of Edinburgh Council—providing them with information that allows for a better understanding of multiple perspectives, usage patterns, and aspects of neighbourhood and place—to enable evidence-based decisions. Working with the City of Edinburgh Council, the Living lab produced a report on Data and Design for Property and Planning based around a collaborative placemaking project with a community in the Gracemount area of the city [45]. Through their service design programme the local authority worked with the lab to test an embedded data-and-design methodology to make better decisions about significant changes to the council estate. Project objectives included an audit and analysis of local authority data to address key questions about building use; a process to identify key community values to support the management of assets and service delivery in the area; the definition of future options for the future of buildings; and a set of guidelines for replicating a similar approach in other areas of the city.

The developed placemaking methodology evident in the work of the Edinburgh Living Lab, works to offset impacts of neoliberal economic planning by allowing citizens to participate in decision-making at the neighbourhood scale—building out capacities that may contribute to better urban governance at higher levels of organisation. Their work in Gracemount, Southhouse and Burdiehouse neighbourhoods made better use of data held in the Council to open up the planning process—to explore and validate community perceptions and priorities alongside those of the council, ultimately arriving at a shared vision for social transformation. The planning report found that collaboration between different service, building, and community stakeholders helped build relationships and strengthen project outcomes [45]. Specifically, the process helped the property directorate join up their decision-making with other departments to deliver better services with less resources, making cost savings and optimising the Council estate. Since the project, the report found the narrative had changed from council to community, meaning that communities are more engaged in decision-making, participate in making difficult decisions, understand and are therefore more likely to accept outcomes. An active engagement methodology facilitates ongoing engagement with communities, while spatial data from multiple sources provides more of a holistic insight into access to and exclusion at an appropriate scale.

The methods developed by the ELL in Gracemount have informed another project in Edinburgh which combines citizen engagement and co-design with urban data and research to help support the high street to become a more successful and liveable place as communities emerge from the global pandemic. Scaling tactics developed with other communities the 'Future of the High Street' project led by the Edinburgh Futures Institute (in collaboration with Edinburgh Living Lab) will develop a toolkit of 6 possible ideas to tackle common high street challenges through digital co-design workshops with local businesses and other stakeholders in Gorgie-Dalry high street and Dalkeith town centre—with two ideas selected to be rapidly prototyped. Limited by social distancing requirements, wider digital engagement with residents, young people, local organisations and other stakeholders is designed to focus more holistically on the high street as a 'place'—helping to understand how the high street may be adapted or accommodate new and innovative uses. This helps to address common challenges experienced by local independent businesses and residents that would make the high street more liveable and successful—leaving an actionable legacy.

The Future of the High Street project incorporates use of urban data through baseline assessments of land use, character and business data alongside existing reports and reviews of previous consultation—to respect both this prior work and citizen time and input that had already gone before, in former community engagement. Important here is to emphasise a feedback loop—and two-way relationship between community, the project team and council—and to put forward a nuanced understanding of challenges and opportunities that can inform a robust evaluation of pilots, across a set of common indicators. As ELL continue to refine the methodology use in the Gorgie-Dalry pilot, legacy impacts are already visible in changes to the built environment and public realm at the neighbourhood scale—which involve the prototyping public realm improvements, and artistic installations—based on learning form the pilots. It is hoped that these tactics will influence historically top-down planning with respect to the public realm—and wider infrastructure—in Edinburgh. For many communities on the periphery of Edinburgh's growth, their participation offers a safeguard against social exclusion and spatial inequalities that accompany top-down planning approaches. For communities on the edge of more speculative developments closer to the centre of Edinburgh's financial core, it remains to be seen if the council will offer them the same opportunities to influence development decisions. This raises important questions regarding the political ideologies that govern offers of participation, and whether ownership over such processes is ever truly devolved to citizens.

3. Conclusions: towards a collective urban governance?

As recovery from the great slowdown necessitates capacities for more sustainable social, economic and environmental resilience, urban governance must seek to achieve a balance in the organisation of settlements to support equitable work, housing and economy. Concluding in early 2021, the *Urban Maestro* project found that while "*European cities have developed sophisticated laws and regulations ('hard power') to secure diverse public interest objectives through the governance of urban design, the quality of the resulting urban places can be disappointing [...] often outcomes are not aligned with commonly shared objectives such as creating environmental sustainability, human scale, land use mix, conviviality, inclusivity, or supporting cultural meaning"* [9]. Thus, decision-making related to the built environment must become much more developed to citizens, allowing communities to generate their own set of indicators, and base decision on informed empirical evidence—presented in a way they can use effectively. Towards a useful typology of tools for urban design governance, means recognising that the urban environment is shaped by various interventions and policy decisions over time and reflects the collective work of multiple stakeholders. *Urban Maestro's* recommendations promote extensive discussions, whatever the local circumstances, and identify 6 fundamental factors for improving the quality of the built environment to be based on: culture; capacity; coordination; collaboration; commitment and continuity. These factors are important considerations for cities seeking to open up decision-making to the public, private and community actors who co-produce the city. Importantly, the project assembled a wealth of best practice that supports how soft power—and social innovation—can influence levers of hard power present in cities by focusing on the process of carrying out an urban project rather than the end product itself.

The findings in this chapter suggest that for the methodologies promoted by Urban Maestro to have impact, city governments must divorce the planning process from politics, and allow for feedback loops in the system through honest engagement processes. Short term political or economic advantage is often at the expense of longer-term sustainability, blocking processes that encourage the development of capacities for resilience in communities. We should not underestimate the role of visions, narratives and cultivated propaganda in the governance of urban design—and the need to challenge established political ideologies that inhibit social innovation in the built environment [1]. Launching the European Union's *New European Bauhaus* initiative, Commission president Ursula Von der Leyen announced it was not just an environmental or economic project, but a new cultural project for Europe. The policy calls for, *"a collective effort to imagine and build a future that is sustainable, inclusive and beautiful for our minds and for our souls"* and *"sustainable solutions that create a dialogue between our built environment and the planet's ecosystems"* [46]. In working towards these objectives, cities in Europe and around the world need to provide meaningful opportunities for participation in decision-making that allow stakeholders at all levels to influence urban development, while keeping the barrier for entry low. This means a more concerted effort to design and deliver opportunities for co-production that devolve new capacities to the communities who participate—without further perpetuating processes of social and spatial exclusion [47, 48].

Notes/thanks/other declarations

The author would like to thank all informants who gave their time or interviews in each case study.

Author details

Donagh Horgan
Department of Work, Employment and Organisation, Strathclyde Business School,
University of Strathclyde, Glasgow, Scotland

*Address all correspondence to: donagh.horgan@strath.ac.uk

IntechOpen

References

[1] Horgan D, Dimitrijević B. Social innovation in the built environment: The challenges presented by the politics of space. Urban Science. 2021;**5**(1):1. DOI: 10.3390/urbansci2010013

[2] Rolnik R. Urban Warfare. London: Verso Trade; 2019

[3] Lefebvre H. State, Space, World: Selected Essays. Minneapolis: University of Minnesota Press; 2009

[4] Harvey D. From managerialism to entrepreneurialism: The transformation in urban governance in late capitalism. Geografiska Annaler: Series B, Human Geography. 1989;**71**(1):3-17. DOI: 10.2307/490503

[5] Mazzucato M. The Value of Everything: Making and Taking in the Global Economy. London: Hachette UK; 2018

[6] Swyngedouw E. Promises of the Political: Insurgent Cities in a Post-Political Environment. Cambridge: The MIT Press; 2018. DOI: 10.7551/mitpress/10668.001.0001

[7] Horgan D. Placemaking. London, United Kingdom: Elsevier Health Sciences; 2019. DOI: 10.1016/b978-0-08-102295-5.10680-8

[8] Aalbers MB. Financial geography III: The financialization of the city. Progress in Human Geography. 2020;**44**(3):595-607. DOI: 10.1177/0309132519853922

[9] Urban Maestro. New Governance Strategies for Urban Design—Urban Maestro. 2021. Available from: https://urbanmaestro.org/wp-content/uploads/2021/04/urban-maestro_new-governance-strategies-for-urban-design.pdf

[10] Bento J, Carmona M. Informal Tools of Urban Design Governance,

the European Picture. Urban Maestro: New Governance Strategies for Urban Design. 2021. Available from: https://urbanmaestro.org/wp-content/uploads/2021/04/um_survey-report.pdf

[11] Horgan D, Dimitrijević B. Frameworks for citizens participation in planning: From conversational to smart tools. Sustainable Cities and Society. 2019;**48**:101550. DOI: 10.1016/j.scs.2019.101550

[12] Odendaal N. Smart city: Neoliberal discourse or urban development tool? In: The Palgrave Handbook of International Development. London: Palgrave Macmillan; 2016. pp. 615-633. DOI: 10.1057/978-1-137-42724-3_34

[13] Grossi G, Pianezzi D. Smart cities: Utopia or neoliberal ideology? Cities. 2017;**69**:79-85. DOI: 10.1016/j.cities.2017.07.012

[14] Cardullo P, Kitchin R. Smart urbanism and smart citizenship: The neoliberal logic of 'citizen-focused' smart cities in Europe. Environment and Planning C: Politics and Space. 2019;**37**(5):813-830. DOI: 10.1177/0263774x18806508

[15] Wijermars M. The digitalization of Russian politics and political participation. In: The Palgrave Handbook of Digital Russia Studies. Cham: Palgrave Macmillan; 2021. pp. 15-32. DOI: 10.1007/978-3-030-42855-6_2

[16] Zayakin A, Smagin A. Продам свой голос за свитшот (in Russian). Novaya Gazeta; 2018. Available from: https://novayagazeta.ru/articles/2018/09/21/77914-prodam-svoy-golos-za-svitshot

[17] Asmolov G. Vertical crowdsourcing (Russia). In: The Global Encyclopaedia of Informality: Towards an

Understanding of Social & Cultural Complexity. London: UCL Press (Taylor & Francis Ltd.); 2018. pp. 463-467

[18] Theodore N. Governing through austerity: (Il)logics of neoliberal urbanism after the global financial crisis. Journal of Urban Affairs. 2020;**42**(1):1-7. DOI: 10.1080/07352166. 2019.1623683

[19] Peck J. Situating austerity urbanism. In: Cities under Austerity: Restructuring the US Metropolis. Albany: State University of New York Press; 2018

[20] Miraftab F. Insurgent planning: Situating radical planning in the global south. Planning Theory. 2009;**8**(1):32-50. DOI: 10.1177/1473095208099297

[21] Roy A. Why India cannot plan its cities: Informality, insurgence and the idiom of urbanization. Planning Theory. 2009;**8**(1):76-87. DOI: 10.1177/1473095208099299

[22] Huq E. Seeing the insurgent in transformative planning practices. Planning Theory. 2020;**19**(4):371-391. DOI: 10.1177/1473095219901290

[23] Madden D, Marcuse P. In Defense of Housing: The Politics of Crisis. London: Verso Books; 2016

[24] Krstić I. The housing policies in Yugoslavia. In: Knežević V, Miletić M, editors. We Have Built Cities for You: On the Contradictions of Yugoslav Socialism. Belgrade: Centre for Cultural Decontamination; 2018

[25] Sekulić D. The Ambiguities of Informality. The Extra-Legal Production of Space in Belgrade During and After Socialism. Eurozine; 2018. Available from: https://www.eurozine.com/the-ambiguities-of-informality/

[26] Alfirević Đ, Simonović AS. Urban housing experiments in Yugoslavia 1948-1970. Spatium. 2015;**1**(34):1-9. DOI: 10.2298/spat1534001a

[27] Le Normand B. The contested place of the detached home in Yugoslavia's socialist cities. In: The Cultural Life of Capitalism in Yugoslavia. Cham: Palgrave Macmillan; 2017. pp. 173-190. DOI: 10.1007/978-3-319-47482-3_10

[28] Uvalić M. The Rise and Fall of Market Socialism in Yugoslavia. Berlin: DOC Research Institute; 2018. Available from: https://doc-research.org/2018/03/rise-fall-market-socialism-yugoslavia/

[29] Topalović M. New Belgrade: The modern city's unstable paradigms. In: Belgrade: Formal/Informal: eine Studie über Städtebau und urbane Transformation. Switzerland: ETH Studio Basel, Scheidegger & Spiess; 2012. pp. 128-228

[30] Bojić N. Social and physical planning: Two approaches to territorial production in socialist Yugoslavia between 1955 and 1963. Architectural Histories. 2018;**6**(1):25, 1-14. DOI: 10.5334/ah.309

[31] Stierli M, Stephanie E, Vladimir K, Tamara BK. Toward a Concrete Utopia: Architecture in Yugoslavia, 1948-1980. MoMA, New York: Museum of Modern Art; 2018

[32] Hirt S, Petrović M. The Belgrade wall: The proliferation of gated housing in the Serbian capital after socialism. International Journal of Urban and Regional Research. 2011;**35**(4):753-777. DOI: 10.1111/j.1468-2427.2011.01056.x

[33] Vilenica A. Contemporary housing activism in Serbia: Provisional mapping. Interface: Journal for and about Social Movements. 2017;**9**:424-447

[34] Petrović M. Post-socialist housing policy transformation in Yugoslavia and Belgrade. European Journal of Housing Policy. 2001;**1**(2):211-231. DOI: 10.1080/14616710110083434

[35] Vilenica A. Contradictions and antagonisms in (anti-) social (ist)

housing in Serbia. ACME: An International Journal for Critical Geographies. 2019;**18**(6):1261-1282

[36] Mojović D. Serbia. In: Gruis V, Tsenkova S, Nieboer N, editors. Management of Privatised Housing: International Policies and Practice. New Jersey, USA: John Wiley & Sons; 2006. pp. 211-228. DOI: 10.1002/9781444322613.ch11

[37] Perić A. Public engagement under authoritarian entrepreneurialism: The Belgrade waterfront project. Urban Research & Practice. 2020;**13**(2):213-227. DOI: 10.1080/17535069.2019.1670469

[38] Jakovljević A. Fighting corruption with pyramids: A law and economics approach to combating corruption in post-socialist countries: The case study of the Republic of Serbia [doctoral dissertation]. Staats-und Universitätsbibliothek Hamburg Carl von Ossietzky

[39] Pittock M. Enlightenment in a Smart City: Edinburgh's Civic Development, 1660-1750. Edinburgh, Scotland, UK: Edinburgh University Press; 2018. DOI: 10.3366/edinburgh/9781474416597.001.0001

[40] Tooley CB. Competition and community in Edinburgh: Contradictions in neoliberal urban development. Social Anthropology. 2017;**25**(3):380-395. DOI: 10.1111/1469-8676.12419

[41] Ballard Tooley C. "Beauty won't boil the pot:" Aesthetic discourse, memory, and urban development in Edinburgh. City & Society. 2020;**32**(2):294-315. DOI: 10.1111/ciso.12279

[42] Sutherland E. Edinburgh: Data Capital of Europe. 2019. Available from: SSRN 3504740. DOI:10.2139/ssrn.3504740

[43] Creasy A, Lane M, Owen A, Howarth C, Van Der Horst D.

Representing 'place': City climate commissions and the institutionalisation of experimental governance in Edinburgh. Politics and Governance. 2021;**9**(2):64-75. DOI: 10.17645/pag.v9i2.3794

[44] Massey D. The responsibilities of place. Local Economy. 2004;**19**(2):97-101. DOI: 10.1080/0269094042000205070

[45] City of Edinburgh Council. Data and Design for Property Planning [Internet]. 2019. Available from: https://issuu.com/edinburghlivinglab/docs/edinburghlivinglab_dataanddesignforpropertyplannin

[46] European Union. New European Bauhaus: Our Conversations Will Shape Our Tomorrow [Internet]. 2021. Available from: https://europa.eu/new-european-bauhaus/index_en

[47] Arnstein SR. A ladder of citizen participation. Journal of the American Institute of Planners. 1969;**35**(4):216-224. DOI: 10.1080/01944366908977225

[48] Lund DH. Co-creation in urban governance: From inclusion to innovation. Scandinavian Journal of Public Administration. 2018;**22**(2):3-17

Section 2

Studies and Researches on the Conformation of Spaces and Their Agglomerations

Chapter 8

Impact of Urban Open Spaces on City Spatial Structure (In Case of Isfahan)

Ghazal Farjami and Maryam Taefnia

Abstract

Public spaces can be considered as important elements to improve the quality of the environment and increase the sense of citizenship. On the other hand, the cohesive network of urban spaces shows the integrated structure of a city's spatial organization, in which not only the connection of form and function is considered, but also meaning finds its place in a complex urban system. Since the spatial structure of the traditional Iranian cities is ingrained in geographical factors and culture of the settlements, the evolution of this structure in Isfahan as one of the most famous historical cities in Iran is examined. The aim of this study is to answer this question: How do urban open spaces impact city spatial structure? The research method is descriptive-analytical, which has been concluded in a process of content analysis. The development of Isfahan's structure over time and role of urban spaces in its formation has been studied. Entrances, key points, roads, and water edges as main urban spaces impact on city structure direction. The structure has changed from linear-nuclei to central-radial and finally, an integrated network to the Safavid era, but cohesive nature of the structure has changed from the Pahlavi period with multiple sections of streets.

Keywords: city structure, spatial structure, urban open spaces, urban context, Isfahan

1. Introduction

Looking at city structures, both open and closed parts are considered to form the shape of a city. Therefore, cities are not just about masses and not mere open lands but the combination of these two make the city structure, which differs in different regions based on so many factors.

The structure of traditional Iranian cities has a special physical-spatial cohesion and order that is guaranteed by their richness and physical quality. One of the most important features is the continuity of the city and neighborhoods through the centers, main passages, and the bazaar, which has led to the formation of a clear and legible structure in the city and the continuity of components and elements of the city. The composition and construction of the city in the past of Iran have been such that the main passages and the bazaar have been responsible for the connection between the important elements of the city.

However, since the first years of the present century, when the street has emerged as the dominant and decisive element in the city, urban cohesion has undergone serious changes. The street runs through the city, presenting itself as the powerful lips within the city, and from the integrated structure of traditional cities, only residential contexts remain, such as islands cut off from the arteries of urban life. At the same time, the spaces and elements of the communication network must establish an organized and regular relationship with the components and structures of the city and the current activities in it. Because the formation of urban spaces and elements along the roads is influenced by the behavioral patterns, culture of the people, and the economy of the society [1].

In contrast to the modern Iranian series, which are simply copies of the contemporary diffused European and American cities, the traditional Iranian city is concentrated and how much genius in its buildings combining diverse land uses in a tight relationship with each other. In this way, three main factors affecting the early compact Iranian cities may have been the physical environment of the Iranian plateau, trade and historical events, and the socio-political structure of the country [2].

One of the important but forgotten elements affecting traditional cities is urban open spaces formed among the compact masses of buildings. Therefore, this research is an attempt to investigate influential factors on city structures and identify the role of open spaces on spatial city organization. In traditional Iranian cities, the urban structure was based on the geographic characteristic of the surrounded environment. Since a vast area of this country is covered with desert and hot and arid climate, cities were shaped in a very compact and dense form. Besides the central courtyard of individual buildings, urban open spaces emerged as a joint element among the masses. On the basis of the carried out research, the paper analyses the historical development of Isfahan as one of the historical cities of Iran with a very compact context affected by geographical conditions, while urban open spaces still emerged as key elements in a very unique form.

2. City structure

The Latin root of the word structure "struere" means to build, grow, and evolve. Hence, the structure means "working together continuously to evolve." For example, living features grow and evolve in a continuous, purposeful, and highly organized movement. In this way, each structure has its own function and shape, which plays an important role in facilitating the function of the structure [3].

The structure is a complete set of relationships in which the elements may change but remain dependent on the whole and retain their meaning. The whole is independent of its relations with the elements. The relationships between the elements are more important than the elements themselves. Elements are interchangeable, but relationships are not [4].

Against the views of those who believe that structures are formed based on functions and goals, there is also the opinion that structure is determined by its elements and their combined features, regardless of the specific function and purpose [5]. The structure is a set of interdependent processes and interrelationships of elements or a network of relationships between elements' positions that are plotted within the external appearance of the object, the shape [6].

Some believe that the structure is made by the human mind and then projected onto the shape of the city. Researchers seek to discover the subconscious mind structures that are common to all societies. The latter group aims to find the relations and rules that have been effective in the formation of these structures to use them in the emergence of subsequent structures [7].

From all these definitions, it can be concluded that structure is a set of inter-dependent elements, in which the necessary and simultaneous mutual relations or partnerships between components take place abstractly and objectively and depending on the purpose, within a certain range.

The spatial structure of the city shows the order and the relationship between the physical elements and the uses in the city [8]. In other words, the spatial structure refers to a set of communications resulting from the urban form and the gathering of people, the transportation and flow of goods and information [9]. Alain Bertaud combines the spatial structure of a city into two components, namely the spatial distribution of population and the pattern of people's travel from where they live to the various destinations and places, where an important social activity or interaction takes place, such as the workplace, and knows the location of social gatherings [10].

Therefore, city structure includes various elements coming together creating a city with its own characteristic. These elements not only work individually but also generate unity resulting in a homogenous city structure. It does not matter if it is open or close space, but it is important to form in a way that integrates the whole structure. However, urban open spaces are dominant features reflecting the story of the city and residents' culture and lifestyle.

3. Urban open spaces

Open space, on the one hand, refers to a space that is relatively open, less closed, and has more limited space, and on the other hand, refers to a space opened by the masses to the majority of people. This refers not only to landscapes, such as parks and green spaces but also to streets, squares, alleys, and courtyards [11].

"If we want to clarify the concept of urban space without imposing aesthetic criteria, we have to consider the spaces between buildings in cities and other places as urban space," says Rob Carrier. This space is geometrically surrounded by various symbols. Only the clarity of its geometric features and esthetic qualities allows us to consciously consider the outdoor space as an urban space. Outdoor open space is defined for outdoor mobility and is divided into public, semi-public, and private [12].

Zucker considers urban space to be an organized, neat, and orderly structure physically for human activities and based on specific and clear rules; These rules are: the relationship between the shape and the body of the buildings enclosing the same shape and uniformity, with their diversity, the absolute dimensions of the bodies to the width and length of the space between them, and the angle of passages or streets to the square, and finally the location of historical monuments, fountains, and slabs or other three-dimensional elements that can be emphasized [13].

Bruno Zevi considers space to be the essence of architecture, and follows the same definition of urban space, stating that streets, squares, parks, playgrounds, and gardens are all empty spaces that are limited or defined as space [14].

Urban space in a general sense is a kind of interrelation between relationships and behaviors, while the place is adjacent to individual identities, in urban life, it is the most important factor of authentication and affects human behaviors. In addition, urban space, as a public area, is the place of emergence and revitalization of individual and social thoughts and desires, which is why it has a very important role in the development of societies.

Since the emergence of cities and the beginning of urban planning and urbanization is closely related to the need for interaction and the sociality of human beings, undoubtedly, these relationships need their own spaces. Cities are known as places of the emergence of human social relations throughout history, and even the type and quality of urban spaces have been quite effective in the manifestation of these relations. Therefore, one of the most important elements of the urban context is the city structure formed and evolved based on human lifestyle in different periods [15].

The changes experienced in modern cities are reflected in the urban space, and this leads to the gradual extinction of public spaces in the urban structure. Increased urbanization and migration are leading to a loss of integration of public open space in city centers.

4. Urban open spaces in iranian city structure

The physical morphology of the traditional city of Iran is to a large extent a cultural-historical response to the natural environment, especially, the climatic conditions of the Iranian plateau. Its extreme climatic conditions are characterized by a shortage of water, high evaporation than precipitation, intense solar radiation, high seasonal temperature ranges, and damaging dust and sand storms [2].

The structure of traditional Iranian cities has a special physical-spatial cohesion and order that is guaranteed by their richness and physical quality. One of the most important features is the continuity of the city and neighborhoods through the centers, main passages, and the bazaar, which has led to the formation of a clear and legible structure in the city and the continuity of the components and elements of the city. The composition and construction of the city in the past of Iran have been such that the main passages and the bazaar have been responsible for the connection between the important elements of the city [16].

One of the important features of the old context of Iranian cities is its division into several neighborhoods because the historical city as a whole is composed of components in the form of a neighborhood [17]. In general, in Iran, the city was complex. Consisting of homogeneous and homogeneous neighborhoods that are integrated into a specific place based on relations, forms, and affiliations of ethnic, religious, professional, or territorial, and have kept their identity and originality in this way for years and until the new development. The city was considered as the main constituent units or as the cells of the city, the residence of a particular ethnicity, religion or group, and more than any other urban unit, within itself solidarity, unity, ethnic, family, and sometimes administrative, professional and class [18].

What has always been prominent in the construction and establishment of neighborhoods are the aspects of social, cultural, religious, or economic commonalities [17] and in the meantime, the separation of neighborhoods on the basis for differences in religious or ethnic beliefs and characteristics is more visible among large cities with larger and more diverse populations and in nomadic cities. For the emergence of each neighborhood, a limited and coherent geographical area, social interdependence between a specific group, and a specific city design were required for the spaces and houses of the neighborhood, the existence and permanence of the neighborhood depend on their existence [18].

The structure of ancient cities is known as the most obvious and complex part of the physical system that shows the social structures of the city along with its internal contradictions. Dynamics in the design of this structure causes logical relationships between urban components and systems and their function and process together.

One of the ways to organize the space in the historical cities of Iran was to connect the building mass continuously. This method can be seen at the micro level, such as neighborhoods, and at the macro level, as the whole city. For example, the bazaar, as the main and central street of the country, has been an important tool for the continuous growth of the city [19]. Next to, or along, some of the major bazaars in major cities were an urban or regional square. The bazaar was the most important road in the city and in most cases, it was connected to an urban square.

The main bazaar of cities are often linear and formed along the most important urban road. For this reason, in many historical cities of Iran, the most important part and the main element of the context is the main direction of its bazaar. A bazaar order was formed in its simplest form with shops located on either side of it. Many bazaars were gradually built and developed, and for this reason, the extension of the direction of these bazaars, following the natural shape of the passages, has been indirect and organic. Various guilds were stationed along the main bazaar line, thus placing various activity groups in different parts of the mainline. In some large cities, two or more main directions appeared in parallel or intersecting.

One of the main features of past spaces is their centrality and confinement. Each spatial area is central to its surroundings. Gradually, the construction method of the central building replaced the central space. The part of the building that could not be designed due to the connection with the adjacent building was exposed from all sides by being located in the middle of the space, and the necessity of designing all aspects of the building was raised. Each building peaked independently of adjacent buildings in height so that the horizontal connection gave way to the vertical connection [20].

One of the historical cities of Iran is Isfahan, located in the hot and arid area close to the desert while a river is passing through the city. Isfahan has a very special city structure based on various environmental issues. However, urban open spaces play an essential role in city structure.

5. Spatial structure of Isfahan based on urban open spaces

Isfahan is located in 32°38′30″ N latitude and 51°38′40″ E longitude, about 340 km south of Tehran and the capital of Isfahan Province (**Figure 1**) [22]. The main factors of the prosperity of Isfahan during the time have been the Zayandehrud River and the location of the city in the center of the Iranian plateau. So, throughout its history, it has been either the capital or one of the most important parts of Iran [23].

Figure 1.
Location of Isfahan in Iran [21].

The spatial-physical structure of each city is closely related to its history. Therefore, a review of historical periods can enlighten how the city is organized during the time. Most of the old cities of Iran had a specific structure of the main urban elements and functions such as palaces, bazaars, squares, mosques (After Islam), schools, etc. The physical characteristics of the evolution and development of the main structure of Iranian cities up to the contemporary era were mainly in harmony with the growth of the city [24].

The city of Isfahan has been continuously evolving for more than 2000 years. Until the early Islamic centuries (750–1258), Isfahan consisted of two districts, Jay and Judea (**Figure 2**). During the Sassanid Empire, Jay was the administrative and governmental center and included urban elements, such as squares and bazaars. In contrast, Judea and the rural agricultural areas in the north and south of the Zayandehrud River were inhabited [26].

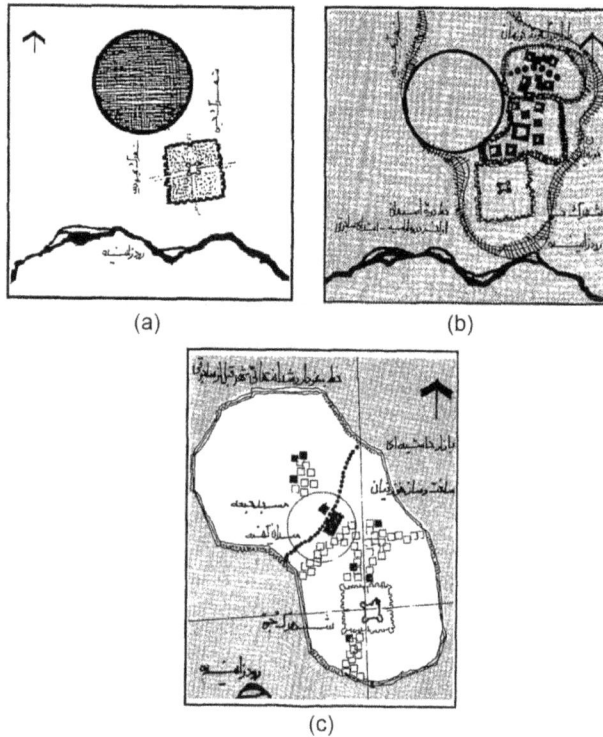

(a)　　　　　　　　　　　(b)

(c)

Figure 2.
Isfahan in the late Sasanian and early Islamic periods (Abbasid era) [25].

After the Arab invasion of Isfahan, in the Abbasid era, Jay gradually became a ruin, while Judea survived. The physical form of the city in the pre-Islamic era included three distinct parts: the governmental area, the central city, and the outer city, but in the Islamic time, the past structures underwent changes, the most important of which was the Grand Mosque (Jame Mosque), as a characteristic of the urban element [27]. Rural groups connected with lines of communication and formed an urban body (**Figure 2**) [28]. The structure of Isfahan in the Seljuk era (1037–1194) was a combination of linear and centralized patterns. Due to the comprehensive development of the city, the central position of the structure was located around the Old Square as the main center of access. The linear part of the city structure has continued in the form of a bazaar to the gates, which has provided the possibility of development in the future [29]. Therefore, the most important urban spaces in this period have been squares, bazaars, and transportation routes which are created the structure of the city (**Figure 3**) [31].

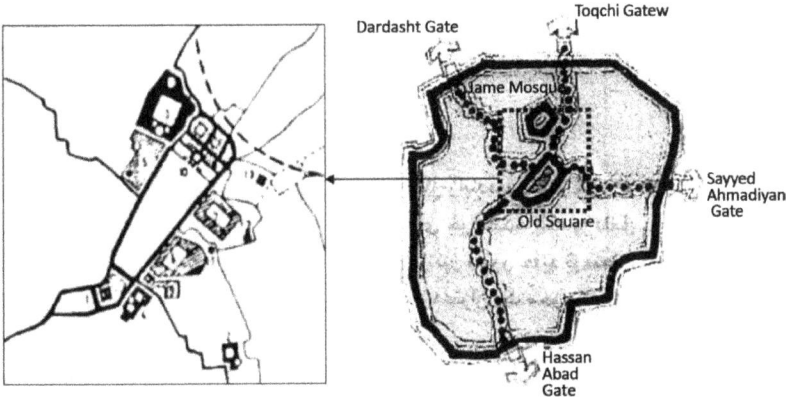

Figure 3.
Structure of Seljuqid Isfahan. Modified by authors [30].

After the selection of Isfahan as the capital of Iran in the Safavid Empire (1501–1722), the main structure of the city was formed. During this period, Chaharbagh Street, Naghsh-e Jahan Square, and its connection to the Old Square by the bazaar was one of the most important measures in urban spaces. Naghsh-e Jahan Government-Ceremonial Square caused the future development of the city to be drawn to this direction and then to Hezarjarib gardens on the other side of the river (**Figure 4**) [33].

During the Qajar period (1789–1925), the empty space of the Old Square began to fill and lost its importance as a reference point in the structure of the city [34].

Over the Pahlavi period, modernism and its developments by ignoring the context, history, and structure of the city, introduced a kind of intervention in historical areas that led to spatial isolation and destruction of traditional structures in the city.

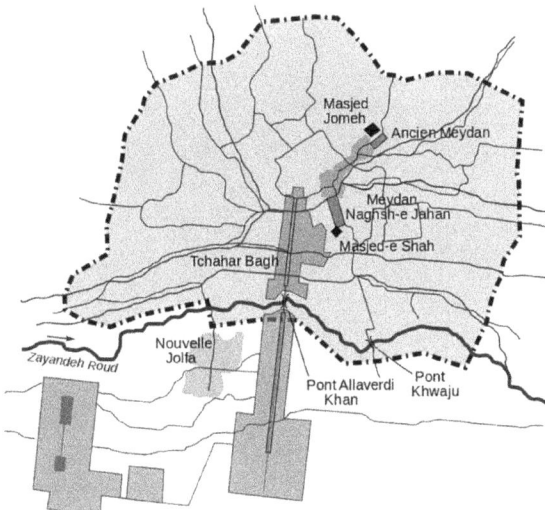

Figure 4.
Structure of Isfahan in Safavi era [32].

During this time, the structure of the city was physically changed from a linear-nuclei model to a network structure, so that the old structure gradually faded in the minds of the people and lost its physical-structural value and reputation (**Figure 5**) [25].

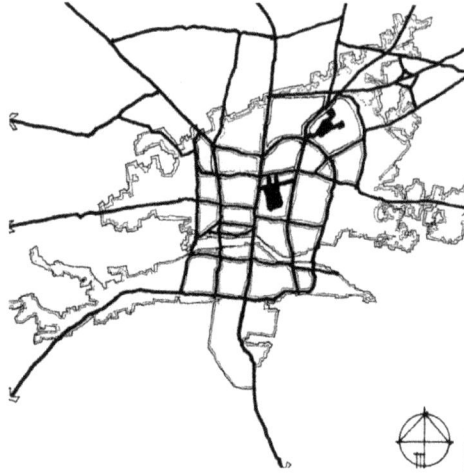

Figure 5.
Structure of Isfahan in Pahlavi period [25].

The first planning measures in this period were street plans in the old contexts and their continuation to the outside based on the grid-system pattern and separation of urban functions, which led to the fragmentation of the old context of Isfahan [35]. This kind of intervention has led to the apparent separation of the main old parts of the city and the destruction of its traditional structure, which led to the complete decline of historical centers in the 20th century. Therefore, it was necessary to prepare master plans. Modern major urban planning began in Iran in the 1950s and 1960s when the first master plans were prepared for some important cities like Isfahan [36]. Isfahan has three main master plans in 1960, 1971, and 1988 (**Figures 6–8**). Then, detailed plans were prepared based on the regions of the city,

Figure 6.
First master plan of Isfahan [35].

Figure 7.
Second master plan of Isfahan [35].

Figure 8.
Third master plan of Isfahan [35].

but with the non-implementation of more than 70% of the comprehensive plans, the strategic development plans were replaced. City Development Strategy (CDS) is a comprehensive flexible planning framework designed to empower urban communities to control and manage the consequences of rapid economic change and increase the growth of economic and social inequalities [37].

6. How urban spaces of Isfahan impact city structure?

The first question that should be answered is: what is the city structure? The structure of the city is a set consisting of the main axis and an interconnected network of land uses and urban elements that integrates the whole city and extends hierarchically in all parts of the city on a proportionate scale (**Figure 9**). This complex is the foundation of the spatial-physical organization of the city and indicates the general and common characteristics of the city [25].

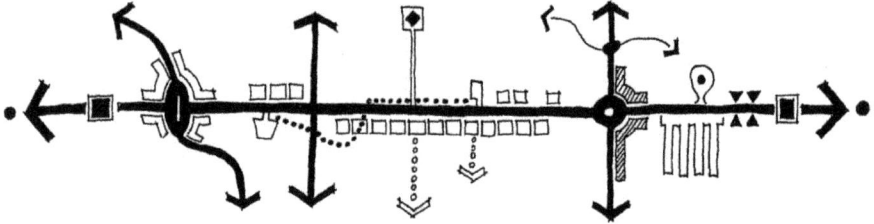

Figure 9.
City structure: Main axis and an interconnected network of land uses and urban elements [25].

In other words, this complex as a linking structure includes parts of the city that are in public use, including movement structure (main roads, public transport cores, and main walking routes), interaction places, gathering places, and public buildings. City context with its specific physical and social characteristics is formed and organized by the city structure. This structure breaks the experience of the city into pieces with spatial locations and at different scales that make the city legible and conceivable. It changes over time and the elements that remain unchanged create the cultural landscape of the city. This structure can also be linked to the natural landscape (**Figures 10** and **11**) [38].

The second question is what is the urban open space? Urban space is the scene where the story of social life begins. It is a space that allows all people to access and work in it. Based on researches, there are different points of view about urban open space typologies (refer to [39]) but the focus of this study is based on five main categories: entrances urban nodes especially squares, paths, water edges, and urban stairs. The entrance is a joint for connecting two places. The entrances of the cities and the

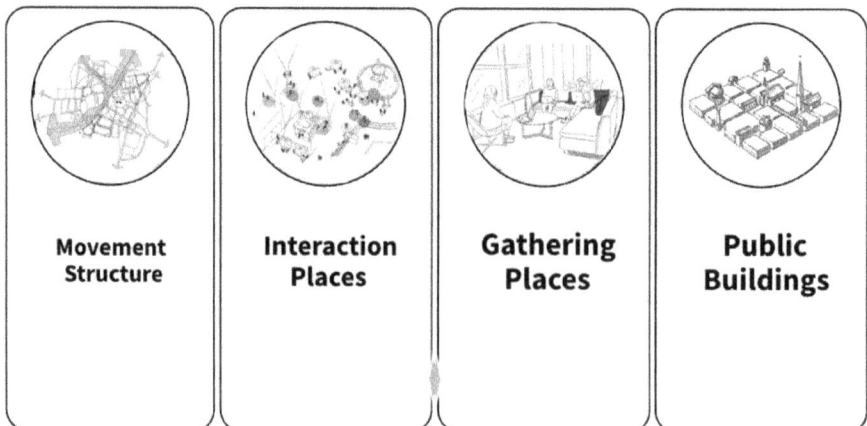

| Movement Structure | Interaction Places | Gathering Places | Public Buildings |

Figure 10.
City structure elements (authors).

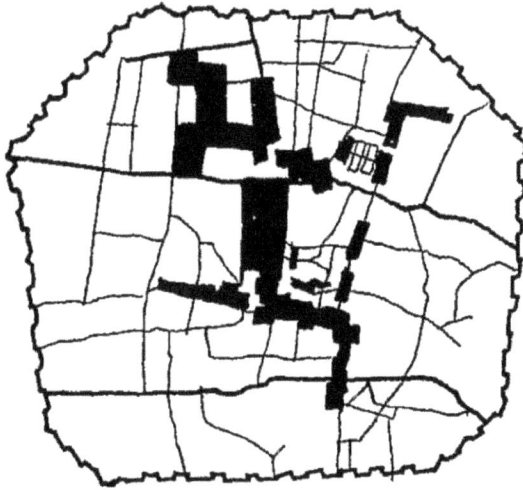

Figure 11.
City structure: Movement structure, interaction and gathering places, and public buildings [25].

neighborhoods entrances are public spaces that play the role of urban space. Squares are the most influential urban spaces in the mental image of citizens. They can be on an urban, local scale, or play as a ceremonial place. In people's minds, paths are not only the lines that enable the connection of different parts of the city, but also the spaces that accommodate the most social life. They have the largest share of other urban spaces and are manifested in the form of urban streets, passing streets, local streets, boulevards, alleys, dead ends, and pedestrian ways. Water in the city can play a key role. The water's edges can be the basis of different social happenings. The last one is urban stairs which can be a place of social events in addition to the physical role of access (**Figure 12**) [40].

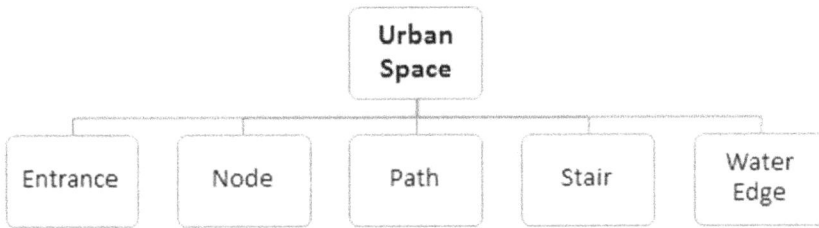

Figure 12.
Urban space typology.

According to the above issues, the last question is what is the role of urban open spaces on the structure of Isfahan? The following diagrams show the evolution of changes in the city structure over time and the interaction of these two main factors.

As mentioned, Isfahan initially consisted of two main cores and the dominant activity model of the people of the city was gardening. These gardens were mainly located on the banks of the Zayanderud River, and people had learned to use the River to irrigate their gardens, thus "Madi's were formed. This pattern of residential activity may be the answer to the question of why the early settlements of Isfahan were formed at a distance from the Zayanderud River. Supplying water through wells was much easier than supplying water to gardens, in addition to the fact that the river was not permanent. People created branches (Madi) from the Zayanderud to deliver water to the gardens in a controlled manner [41].

In the Sassanid period, Zayanderud, Madies, two main cores (Jay and Yahudiyyah), and scattered points of residences created the basis of Isfahan's structure in the multiple nuclei model. The river and Madies, as the first urban open spaces, played a significant role in locating the centers. In addition, the settlements around Yahudiyyah were organized by the Madies in a linear-nuclei connection (**Figure 13**).

In early Islam, the Isfahan spatial organization remained in linear-nuclei type, but the residential areas around Yahudiyyah joined together and organically formed in central organizing. This area is the foundation of the development of settlement as a city in the next years. The oldest neighborhood of Isfahan is in this part of the city and at the same time, functions such as bazaar and mosque were formed next to the palace. Zayanderud and Madies played their role as previous years in the structure of the city (**Figure 14**).

Figure 13.
Isfahan structure in late Sassanid. The structure is linear-nuclei. Edited by authors [42].

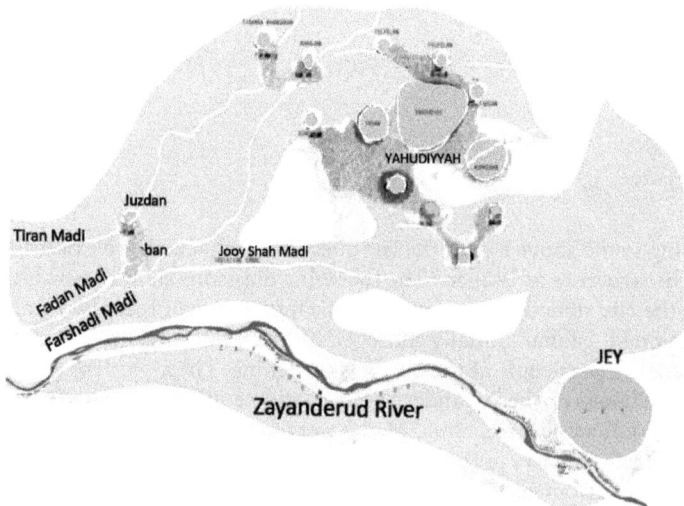

Figure 14.
Isfahan's structure in early Islam. The structure is linear-nuclei. Edited by authors [26].

During the Seljuk period, the foundations of the Iranian-Islamic city emerged and the first square of the city was formed at the linkage of Joybareh, Dardasht, and Karan neighborhoods and next to the bazaar. The square and the bazaar, as the main urban spaces, formed the core structure of the Seljuk city along with the paths leading to the city gates. The city gates, as key points of crossing the city wall, strengthen the structure. In this period, the structure of the city core is central-radial with a predominant orientation northeast-southwest and on a larger scale with the Madies and the river is as a linear-nuclei (**Figure 15**).

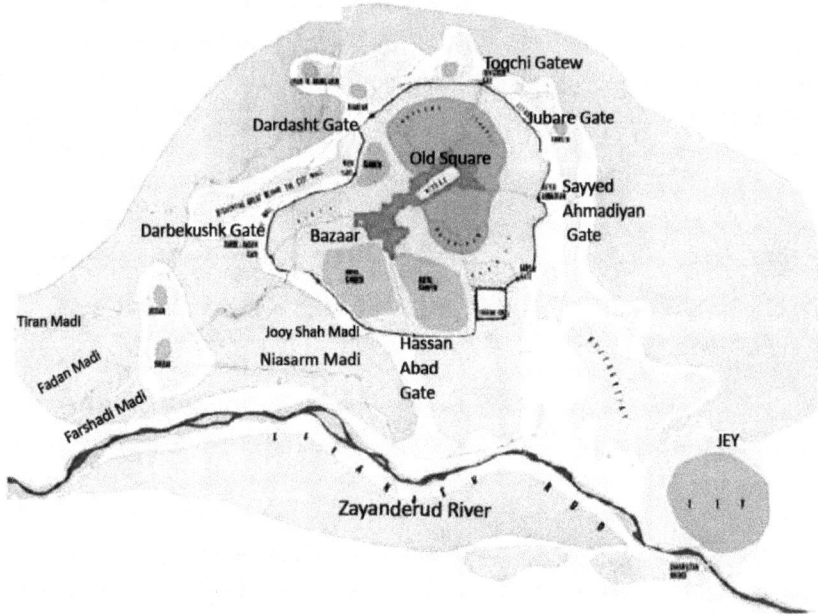

Figure 15.
Isfahan structure in Seljuk era. The structure is central-radial in central of city and linear-nuclei on a larger scale. Modified by authors [26].

During the Safavid period, with a rapid increase in population, four gardens in the middle of the city became residential areas, and the government decided to create new gardens instead of ones that had changed their use, and so the gardens appeared around Chaharbagh Street [41]. Thus, the structure of the city was drawn to the south under the influence of the street route. New Square (Naghsh-e Jahan) was built in linkage to the bazaar between Faden and FarshadiMadies, and following the connection of Khajoo Bridge to Naghsh-e Jahan Square, another part of the city structure was directed to the southeast (**Figure 16**). The crossing of Chaharbagh over the river towards HezarJerib gardens and the axis of Khajoo towards the Takht-e-Foolad Cemetery brought the river to the heart of the city structure. These intersections designed the structure of the city as an interconnected network (**Figure 17**).

It is worth noting that before the Safavid era the growth of the city was organically based on Madies and the river but at this time, the city was developed according to the designed plan (**Figure 18**).

During the Pahlavi era with the aim of renovating the worn-out contexts left from the Qajar period, street construction continued based on the previous structure (**Figure 19**).

Figure 16.
Naghsh-e Jahan Square [43].

Figure 17.
Unban open space along the Zayandehrudriver [44].

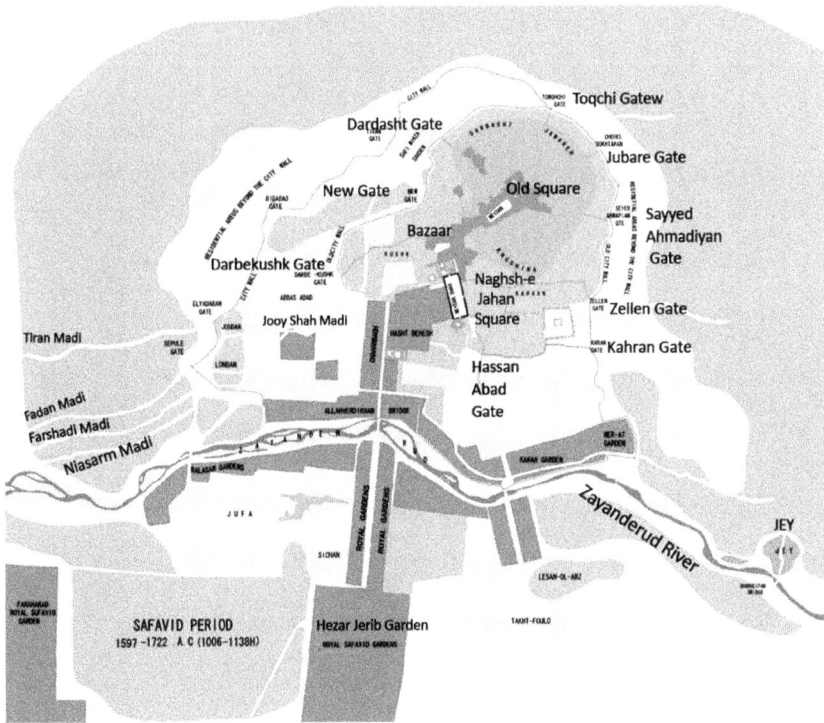

Figure 18.
Isfahan's structure in Safavid era. The structure is an integrated network. Modified by authors [26].

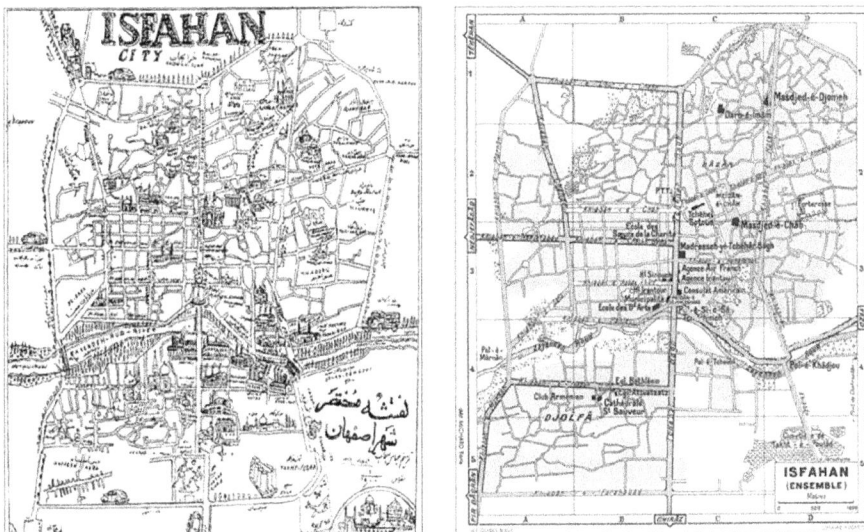

Figure 19.
Isfahan's structure in Pahlavi era (1956) [45]. The structure is networked.

During this time, the new structure expanded its network by passing through the old texture, regardless of the size and orientation of the context pattern. From this period onwards, the streets are the main public open spaces that shape the structure of the city (**Figure 20**).

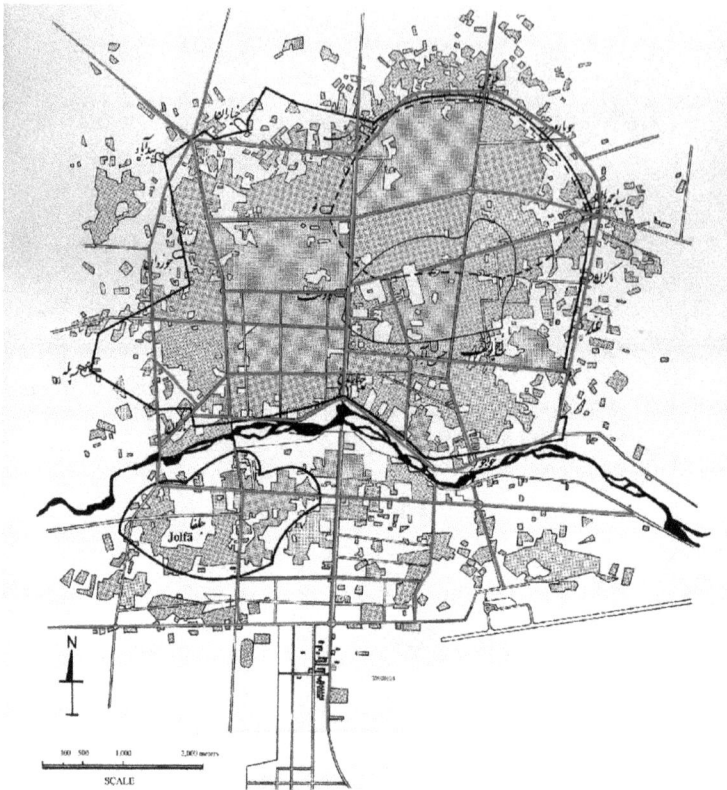

Figure 20.
Isfahan structure in 1970's and 1986 [46]. The structure is networked.

Figure 21.
Old Square [47].

With the regeneration of the Old Square in the contemporary era, it returned to the structure of Isfahan and along with the bazaar and Naghsh-e Jahan Square, physically organized the historical core of Isfahan (**Figure 21**).

So as a result, the evolution of the structure of Isfahan over time is as follows (**Table 1**):

Time	Structure type	Open spaces affecting the structure
Late Sassanid	Linear-nuclei	Nodes (Residential areas)
		Water edges (Zayanderud and Madies)
Early Islam	Linear-nuclei	Nodes
		Water edges (Zayanderud and Madies)
Seljuk Era	Central-radial in central of city linear-nuclei in larger scale	Entrances (Gates)
		Paths
		Nodes (Old Square)
		Water Edges (Zayanderud and Madies)
Safavid	Integrated network	Entrances (Gates)
		Paths (Chaharbagh/Bridges)
		Nodes (Old Square and Naghsh-e Jahan square/HejarJerib Garden/Takht-e-Foolad)
		Water edges (Zayanderud and Madies)
Pahlavi	Network	Paths (Chaharbagh/Bridges)
		Nodes (Naghsh-e Jahan Sauer/JolfaSquare/Takht-e-Foolad)
		Water edges (Zayanderud and Madies)
NOW	Network	Paths (Chaharbagh/Bridges/Main streets)
		Nodes (Naghsh-e Jahan square/Sofe Mountain/Takht-e-Foolad)
		Water edges (Zayanderud and Madies)

Table 1.
Development of Isfahan structure and the main urban spaces affected over time.

7. Conclusion

Today, the viewpoint of natural and indigenous conservation refers to the fact that by maintaining and strengthening the indigenous structure, the social capacities of the place can be formed [48]. Urban open spaces are the main components and the most basic elements in the physical structure of a city. By identifying them, as well as determining their role in space and connecting their functions, we can take action to revitalize the ossification of traditional cities. This strategy is reinforced by defining a multifaceted role for them and a new skeleton is expected to be formed in the city. With such an approach to changing the structure of the city and strengthening the urban joints that connect the past and history to the present and the future and diverse activities to each other and citizens to civic life, the quality of urban places and spaces is improved and sense of richness and belonging strengthen.

Urban open spaces as vital factors play an important role in connecting the constituent elements of the city. The old context of cities, due to the preservation of their original structure, has appropriate models for recognizing and analyzing life-giving open spaces. These open spaces generate hierarchical space organization; breathing spaces among solid parts and city livability. Regarding the modernization process of cities, these valuable spaces were faded while mass spaces are mostly considered. It has resulted in very massive urban contexts affecting social interaction, legibility, city image, etc. Isfahan as one of the historical cities of Iran is well-known because of its urban open spaces which create specific city structure.

As mentioned, entrances, key points (nodes), roads, and water edges are the main urban spaces that in each period in the form of city gates, squares, and Madies routes and the river have strengthened the structure of the city. During the Safavid period, these elements in an integrated connection cause the expansion of the city to the south. With the passage of Chaharbagh through the Zayanderud River, the river finds a central role in the structure of the city, and these two artificial and natural axes form the foundation of the city's later expansions. During the Qajar period and after that, the Old Square and the Madies lost their role in the structure of the city. With the construction of several streets during the Qajar, Pahlavi, and contemporary eras, the structure of the city expands in the form of a network and the roads are the main elements of the city.

Today, with the revitalization of the valuable historical structure of the city, such as regeneration of the Old Square and also rehabilitation of Madies green network, their role in the structure of the city has regained its importance.

Author details

Ghazal Farjami[1*] and Maryam Taefnia[2]

1 Department of Architecture, Daneshpajoohan Pishro Higher Education Institute, Isfahan, Iran

2 Department of Urban Development, Daneshpajoohan Pishro Higher Education Institute, Isfahan, Iran

*Address all correspondence to: qazalfarjami@gmail.com

IntechOpen

References

[1] Alalhesabi M, Jabari M. Investigation and functional-physical strengthening of ossification of Qazvin city with emphasis. Armanshahr. 2011;**6**:27-34

[2] Kheirabadi M. Iranian Cities: Formation and Development. Syracuse: Syracuse University Press; 2000

[3] Bohm D. Fragmentation and Wholeness. Jerusalem: Van Leer Jerusalem Foundation; 1976

[4] Luchinger A. Structuralism in Architecture and Urban Planning. Stuttgart: Karl Kramer Verlag; 1981

[5] Mosso L, Mosso L. Self-generation of form and the new ecology. Ekistics. 1972;**34**:316-322

[6] Grichting WL. The meaning of social policy and social structure. International Journal of Sociology and Social Policy. 1984;**4**:16-37

[7] Kurzweil E. The fate of structuralism. Theory, Culture and Society. 1986;**3**:113-121

[8] Cheng J, Turkstra J, Peng M, Du N, Ho P. Urban land administration and planning in China: Opportunities and constraints of spatial data models. Land Use Policy. 2006;**23**(4):606-616

[9] Rodrigue J-P. The Geography of Transport Systems. London: Routledge; 2020

[10] Bertaud A, Malpezzi S. The Spatial Distribution of Population in 48 World Cities: Implications for Economies in Transition. Madison, WI, USA: The Center for Urban Land Economics Research, The University of Wisconsin; 2003

[11] Yang J, Zhang F, Shi B. Analysis of open space types in urban centers based on functional features. In: International Symposium on Architecture Research Frontiers and Ecological Environment (ARFEE 2018). 2019

[12] Krier R. Urban Space. Michigan: Rizzoli; 1993

[13] Zucker CP. Town and Square. Vol. 4. New York: Columbia University; 1959. pp. 12-19

[14] Zevi B. Architecture as Space. Michigan: Horizon Press; 1993

[15] Nezahd Ebrahimi A, Farshchian A, Khoshrokh P. Investigating the theoretical framework of urban space and the effects of existing forces in the formation of aesthetic urban attitudes. Green Architecture. 2015;**9**:61-69

[16] Soltanzadeh H. Urban Spaces in Traditional Iranian Contexts. Tehran: Cultural Researches Office; 1993

[17] Tavasoli M. Construction of the City and Architecture in the Hot and Arid Climate of Iran. Tehran: University of Tehran; 2002

[18] Khaksari A, Shakibamanesh A, Ghorbanian M. Urban Neighborhoods in Iran. Tehran: Institute of Humanities and Cultural Studies; 2001

[19] Frieden RA, Mann BD. New Influence on Persian Cities: Case Study of Lerman, Iran. London: Paul Elek; 1986

[20] Tavallaei N. Integrated Urban Form. Tehran: Amirkabir Publishing Corp; 2008

[21] Mirbag A, Shokati Poursani A. Indoor radon measurement in residential/commercial buildings in Isfahan city. Journal of Air Pollution and Health. 2019;**3**(4):209-218

[22] Assari A, Maghreby S, Mousavi NM. Investigation of smart growth in traditional Islamic culture: Case study of Isfahan city in Iran. Journal of Geography and Regional Planning. 2017;**10**:47-56

[23] Taefnia M. Revitalization and Restoration of the Historical Interconnection Route between Ali Mosque and Holy Ismail. Isfahan: Art University of Isfahan; 2010

[24] Habibi SM, Cite a La Ville DL. Analyse historique de la conception urbaine et aspect physique. Tehran: Tehran University; 2018

[25] Hamidi M, Sabri RS, Habibi MH, Salimi J. Structure of Tehran City. Vol. 1. Tehran: Technical and Engineering Deputy of Tehran; 1998. p. 273

[26] Omrani M. In Search of Isfahan Urban Identity. Vol. 1. Tehran: Ministry of Roads and Urban Development; 2005. p. 406

[27] Yousefifar S. Towns and Villages in the Middle Ages of Iranian History. Tehran: Institute of Humanities and Cultural Studies; 2010

[28] Taghavi A, Golabi M, Asghari B. Study of the role of religious tendencies in formation and expansion of the city of Isfahan from the era of the Abbasid Caliphate to the end of the Seljuk rule (750 to 1198 AD). Historical Research. 2014;**6**(1):71-84

[29] Sadeghi S, Ghalehnoee M, Mokhtarzade S. The analysis of the effects of contemporary urban development plans on the spatial structure of the north of Isfahan's historical core. Quarterly Journal of Urban Studies. 2012;**2**(5):3-12

[30] Asadi L, Soltanzadeh H. The role of government in shaping the spatial structure of Isfahan. Space Ontology International Journal. 2019;**8**(2):19-32

[31] Azarm Z, Ghalani Z, Ranjbar E. Transformation of public spaces and changing pattern of mobility in a historic city, case study: Isfahan, Iran. In: 22nd International Conference on Urban Planning and Regional Development in the Information Society GeoMultimedia, Vienna, Austria. 2017

[32] Ayvazian S, Diba D, Revault P, Santelli S. Maisons d'Ispahan. Paris: Maisonneuve & Larose; 2002. p. 250

[33] Shafaqi S. Geography of Isfahan. Vol. 1. Isfahan: University of Isfahan; 2000. p. 678

[34] Ghasemi Sichani M. Primary core and the formation process of the city of Isfahan throughout history. Danesh Nama Monthly. 2005;**14**(124-125):6-10

[35] Qureshi SAR. Reasons and factors for non-implementation of physical plans in Isfahan. Journal of Danesh Nama. 2006;**69**(133-134):69-81

[36] Karimi K. Urban planning of Isfahan in the past and present. Journal of the Urban Development and Organization Haft Shahr. 2009;**2**(27-28):10-19

[37] Pour Ahmad A, Darban Astaneh A, Pourghorban S. The role of city development strategy in tourism development management (case study: Hormoz island). Quarterly Journal of Urban Studies (Motaleat Shahri). 2016;**5**(19):37-56

[38] Roberts M, Greed C. Approaching Urban Design (The Design Process). 3rd ed. Tehran: Iran University of Science and Technology; 2015. p. 263

[39] Woolly H. Urban Open Spaces. New York: Spon Press; 2003. p. 259

[40] Pakzad J. Theoretical Principles and Urban Design Process. Tehran: Shahidi Publications; 2006. p. 234

[41] Namdarian A-A, Behzadfar M, Khani S. The network of Madis and the urban development of Isfahan along the Safavid era. Journal of Iranian Architecture Studies. 2017;**5**(10):207-228

[42] Falahat S, Shirazi MR. Spatial fragmentation and bottom-up appropriations: The case of Safavid Isfahan. Urban History. 2015;**42**(1):3-21

[43] Mashregh. Mashregh [online]. 20 July 2020. Available from: https://www.mashreghnews.ir/photo/1094422/ [Accessed: 4 June 2021]

[44] Kheshte Aval. Kheshte Aval [online]. 5 January 2021. Available from: https://avalkhesht.ir/ [Accessed: 5 June 2021]

[45] Entekhabi H. Hossein Abad and Its Dead Ends. Isfahan; 2020

[46] Engineers BC. Aerial Photo of Isfahan 1986. Isfahan; 2005

[47] Seiri dar Iran. Seiri dar Iran [online]. 4 May 2013. Available from: https://seeiran.ir/ [Accessed: 4 June 2021]

[48] Peters K, Elands B, Buijs A. Social interactions in urban parks: Stimulating social cohesion? Urban Forestry & Urban Greening. 2010;**9**:93-100

Dialectics of Mainstreaming Agriculture in Urban Planning and Management of Cities of the Global South

Nkeiru Hope Ezeadichie, Vincent Aghaegbunam Onodugo and Chioma Agatha John-Nsa

Abstract

Most cities in the global south have evolved overtime with significant organic changes in their wake. One of the noticeable changes is the emergence of pockets of city-based agricultural activities, a previously rural-based activity. There are varying interpretations behind this new trend. With increased agglomeration arising from rural-urban migration, residents resort to farming as a panacea to urban challenges. Even employed urban residents resort to agriculture for supplementary income. This emerging scenario has generated debates, dialectics, and polemics among stakeholders as to the propriety or otherwise of this development. This chapter, therefore, takes a panoramic view to all the sides of the issue through review scoping of desktop research method. Specifically, it examines the scope of increase in urban agriculture (UA), the types and nature of UA; urban planners' attitude towards UA, and then propose the management strategies such as promoting agriculture-friendly urban plans for access to agricultural land and practices. The findings revealed that UA takes place on residential land, undeveloped private/public lands, and riverbanks. The prominent UA activities are animal husbandry, aquaculture, cultivation of food and cash crops, etc. The urban-planning measures for integrating UA into the urban environment include inculcating UA-responsive policies in broad plans.

Keywords: urban agriculture, agglomeration, urban planning, urban management, global south

1. Introduction

The increased urbanization of the world cities has thrown up varying opportunities and challenges in its wake. A combination of factors such as population increase, a rapid climate change situation and higher incidences of extreme weather conditions, energy limitations, water scarcities and food security concerns, have bequeathed some air of uncertainty on the evolution of global cities ([1] *c.f* [2]). Due to the current unparalleled city growth, urban areas now offer living space for more than half of the world population that needs to be fed [3] and it is projected that by 2050, 70% of the world population will reside in cities ([4] *c.f* [2]).

Adedeji and Ademiluyi [5] suggest that population increase adversely impacts food security from both ends. It increases food demand on the front end and also indirectly decreases its supply through environmental deterioration, construction of buildings and marginalization of food production at the back end. The implication is that of the marked increase of agricultural activities in areas hitherto that was not known for such. Available statistics show that the need to feed a rapidly growing urban population has driven above 800 million people on a global scale to practice urban agriculture (UA) [6]. UA is a pervasive phenomenon that is found in both advanced and developing countries ([7] and Mlozi [8] *c.f* [5]). Further challenges associated with food security in an urban setting are the continuous increase in rural–urban migration, environmental implications of large-acre commercial agriculture and overall access to safe and nutritious food [2].

Urban and peri-UA is the 'growing, processing and supply of food' and related products through the cultivation of plants and occasionally raising of livestock in and about cities to feed the local populace ([9] *c.f* [6]). UA comprises small to large areas in and about cities, like community gardens, vacant plots, balconies, farms or gardens on the rooftop, indoor farms and greenhouses [6]. In developing countries, UA is a progressively vital livelihood activity, which adds considerably to both family livelihood arrangements and the urban informal economy [10]. Studies of urban and sub-urban agricultural systems in West Africa are few [5]. However, Game and Primus [6] noted that more than 20 million people in West Africa engage in UA. A broad range of production ventures is found in the agricultural sphere spanning family subsistence to extensive commercial farming. Korir, Rotich and Mining, [10] suggest that the supply and distribution cost from rural foods or the cost to import food for the urban area has been on the increase and it was estimated that urban food insecurity would rise if nothing is done to reverse the trend. Thus, in some parts of Lagos city, unauthorized farms are mostly found along wetland areas used for the cultivation of perishable goods particularly vegetables (carrot, lettuce, spinach, cabbage, etc. while in parts of Ikoyi and Lekki, home-plots are used by the people to cultivate vegetables, do poultry farming for chickens, ducks and turkeys) [5]. Urban farming in most developing nations is embarked on by two groups, recent migrants and traditional farmers, who have been immersed in urban development [10]. The practice of UA in different countries revolves about four broadly distinct farming systems: animal husbandry, aquaculture, agro-forestry and horticulture [10]. Meanwhile, the various parts of the city where agriculture is practiced have been researched by scholars. UA commonly occurs around or in-home spaces, a large expanse of undeveloped private or public land [11], private residential land, roadside borders, river banks and other public lands [10].

UA is harnessing the gap left by rural agriculture (RA) which is the primary producer of urban food as it failed to attain urban food security [6]. UA complements RA in relation to self-provision, marketing and supply flow of food [6]. Furthermore, there is a rising fear that rural agriculture will constrain access to land for accommodation purposes in the rural areas (through land grabbing) and provoke migration to cities thus decreasing rural populations ([12] *c.f* [6]). Nevertheless, UA is doubtful to make any urban area or most of its households entirely self-sufficient in their food requirement ([13] *c.f* [6]).

Presently, the absolute and relative increase in food insecurity and urban poverty is becoming a challenge due to the rising urbanization facing many parts of the world. However, UA is gaining prominence as a mitigating approach to challenges of many cities of the world [10]. Urban farms are mostly on former vacant or underused spaces in the city and are then converted into attractive, safe and useable areas [14]. According to United Nations Development Programme (UNDP) ([15] *c.f* [16]), the international policy area of UA, which addresses poverty, was emphasized at the HABITAT agenda

1996 in Istanbul. In the same vein, the UNDP published an influential volume on UA, which highlights the activity's importance for the creation of a job, eeding urban dwellers and creation of an environmentally sustainable urbanization. United Nations Habitat affirmed that UA in many cities plays a significant role in sustaining environmental integrity and adding meaningfully to the achievement of self-reliance in relation to food. This is achieved through the enhancement of livelihoods of the urban poor, and by growing of varied range of crops and breeding of livestock with considerable yields [10]. The physical environment where people play, work and live, has significant impacts on their health. Areas with clean and safe outdoor spaces for recreation, meetings and exercises, impact positively on the health of the dwellers [14]. Hagey, Rice and Flournoy [14] show that the gains accruable to communities due to the presence of urban farms include: offering beautiful, safe, and welcoming areas for neighbors to come together and play; increasing a sense of communal living and making safer environs.

Korir, Rotich and Mining, [10] revealed that urban households can enhance both food consumption (better access to a low-cost source of protein) and food quality (as poor households in cities who participate in farming consume more fresh vegetables than other households in similar class). Households who take part in community gardening are capable of offsetting usually 30%-40% of their yield needs by consuming food cultivated in their own gardens (Seattle Department of Neighborhoods in Hagey, Rice and Flournoy, [17]). The import of UA to nutrition and food security is likely its most vital advantage [10]. UA is promoted as it is widely seen as contributing to food security, the dispensation of providing creative income opportunities and as an approach mainly geared towards supporting the poorest of the poor in the urban population [16]. Korir, Rotich and Mining, [10] revealed that many claims have shown that the principal motive people engage in UA in cities is in reaction to unreliable, insufficient and irregular access to food supplies and the absence of purchasing capacity. Urban farming offers many gains to struggling societies: neighborhood revitalization, better access to healthy food, workers' training and occupational development [14].

The foregoing has riveted the attention of scholars towards assessing the broad implications of UA. This chapter joins the growing number of works to consider the many sides of the debates on this subject. The authors adopted a more explorative than prescriptive approach to the discourse. The desktop research method through scoping review proposed by Arksey and O'Malle [18] as cited in De Beer et al. [19], was employed in the exploration; and online resources were extensively utilized in the discourse.

2. Methodology

The study employed the desktop research method. For a comprehensive study, a scoping review of the literature was applied to explore the nature and dynamics of UA and the planning panaceas for mainstreaming the same into the urban environment. The methodological structure employed was modified by Levac, Colquhoun and O'Brien [20] as cited in De Beer, Gaskin, Robbertse and Bardien, [19], from its initial exposure by Arksey and O'Malle [18] as cited in De Beer et al. [19]. The scoping appraisal for this chapter trailed the first five phases of this methodological structure.

2.1 First step: defining the objectives of the review

The review objectives include to:

1. Examine through available literature the scope of increase in UA in the world

2. Examine the nature and types of UA as published in journals

3. Examine urban planners' attitude towards UA

4. Identify the means of mainstreaming UA in the city

2.2 Second step: identifying appropriate search strategy for the study

Studies relevant to the study were identified by examining electronic catalogs through the use of various search engines, especially google.com. This opened up electronic databases such as MEDLINE, Schimago, EBSCOhost Web of Science, Scopus, Google Scholar, etc. related to the study objectives. The search word or terms include: UA or farming in cities, urban planning and UA, types of UA, etc.

2.2.1 Eligibility criteria

Articles published from 1995 to 2020, were sought. This was to cover the duration of time perceived to be when UA began to gain scholarly attention. Another reason for choosing from 1995 to 2020, was to avoid selecting materials that are obsolete and do not reflect the present technological and societal changes already in place and further evolving. All pertinent peer-reviewed educational materials and those written in the English language were selected.

2.3 Third step: selection of material for the study

The screening of the title was done by the three researchers using the eligibility criteria. Where it seemed that articles were suitable for the theme, one of the researchers screened the abstracts in relation to the inclusion criteria. The articles were retained if they met the inclusion or exclusion criteria, that is, they answered the research question(s) or tendered towards meeting the objective of the study.

2.4 Fourth step: extracting and charting the data

This study followed Arksey and O'Malley's [18] suggestions concerning a standard data extraction procedure. The procedure employed for this review comprised details of author, study location, methodology, publication year, language and communication requirements and other relevant discoveries.

2.5 Fifth step: collating, summarizing and reporting results

The relevant articles were collated and relevant information on the study objectives was retrieved and reported. A total of 12 articles were considered relevant and retrieved, from which 16 other scholarly information was retrieved. This makes a total of 28 referenced materials for this review study. Those utilized in this study, were articles related to African countries like Nigeria, Kenya, Zimbabwe and a little of other articles centred in developed countries like America.

3. Results and further discussions

With the growing interest in sustainability as an essential concept in urban planning, localized systems of food production ought to be encouraged and assisted [21]. Regardless of the divergent opinions about the import of UA, it continues to rise worldwide ([22] *c.f* [2]) and is positively impacting the course of survival and poverty alleviation ([23] *c.f* [2]) of over 800 million persons with 200 million out of them being urban farmers [2]. The World Commission on Environment and Development (WCED [24]: 254 *c.f* [16]) report advised all governments in the developing world to "*consider supporting urban agriculture*".

In contrast to the valued importance of UA, its negative impacts are also projected such as its situation in close proximity to populated regions and livestock farming contributing 30% of greenhouse gas emissions [6]. Moreover, Korir, Rotich and Mining, [10] opine that UA must, thus, be viewed as a permanent and dynamic aspect of the urban ecological and socioeconomic system, utilizing normal city resources, contending for water and land with other urban activities, affected by urban plans and policies, and supporting urban economic and social advancement. Furthermore, Rogerson, [16] shows that UA is not a safety net for 'the poorest of the urban poor' since the proportion of families in the ultra-poor partaking in agriculture was significantly lesser than higher-income groups.

There is presently a universal recognition of the significance of UA in most nations of the world and particular in the African continent. Existing literature reveals that over the past ten (10) years, fast growth in concern and activity in UA has risen greatly ([25] *c.f* [10]). The renewed interest in UA in the late 1980s and early 1990s demonstrates its importance for creating more sustainable rapid urbanization globally [2]. A study in Kenya indicates that about 64% of households in the city engage in some form of UA ([26] *c.f* [10]). An appraisal done in Cape Town of an urban food gardens initiative showed that UA provides gardeners prospect to get involved in a development scheme that holds great potential and can grow into a commercial venture if enough attention is paid to agricultural development, policy issues, land restructuring and livelihood creation [16]. It is expected that 200 million city dwellers provide 15%-20% of the world's food [10]. Broadly speaking, there is a propensity towards more system of intensive production that better meets the rising urban food demand [5].

UA still gets the least importance in several countries, especially in the aspect of development planning. A critical step in the development of UA is the integration of the same into the urban land use structure and the formation of an enabling policy environment [10]. Decision-makers and urban planners are confronted with the problems of acknowledging the significance of UA production to city sustainability and adjoining areas and several planners believe that urban planning and agriculture are relatively unrelated [5]. For this reason, UA is frequently informal and tends to be segregated to the cities' peripheral, far away from infrastructure and markets without evaluating the environmental, economic and links with other facets of the city [5]. In Kenya, policies on UA have not been considered in the past by the government as a worthwhile livelihood alternative. In Nairobi, regardless of the high level of practice, UA is not an accepted urban land use and it is not classified in land use zoning [27].

In the present situation, introducing UA would be a feasible means of attaining sustainability that tackles structural variations engendered by globalization to groups, their food systems and the value of life for urban dwellers. That is the reason for the attempt to reimage 'the city as a farm' by some urban designers. The American Planning Association also recognizes the significance of integrating UA into the planning of various land-uses in cities ([28] *c.f* [2]). The neglect of UA, in city planning, has created several problems in most cities of developing countries. Some of these problems include physical chaos and its associated challenges of unsustainable city growth and environmental insufficiencies which are a clear demonstration of inadequate and inappropriate land use planning that acknowledges and integrates certain facets of the city into the process of land-use allocation, and mark-out specific tools of administering them. It consequently necessitates that contemporary urban planning approaches should recognize the current urban realities and needs and accept urban livelihood approaches such as urban farming as part of the basis for urban planning and management [5].

Interest in urban and sub-urban agricultural production is largely low amid policy, hence, a consistent approach to UA is hardly available ([29] *c.f* [5]). Thus, the management strategies to UA demand the following points as observed from literature:

- Increase financial allocation for programs that educate urban farmers and provide technical support [14].

- Authorize resolutions, schemes, and laws promoting UA [14].

- Offer safe and accessible land that has access to solar energy and irrigation source [21].

- Improve urban networks and transportation systems to efficiently convey produce to consumers [21].

- Make provisions for the collection and transportation of organic wastes from landfills to urban farms [21].

- Change some open spaces and vacant lands within residential areas to community farms, gardens and other features that permit socializing [21].

- Build networks to connect farmers, with labourers, and markets to help maintain and advance new ventures [21].

- Inculcate UA-responsive policies in broad plans and implement UA-friendly zoning programs [14] to combat heat islands and other unfavorable climatic conditions [21].

There is a pressing need to appraise zoning decisions and land-use planning and embracing more flexible guidelines to assist the urban poor advance UA instead of eliminating it [5]. Furthermore, urban planners should make effort to uphold multifunctional land use, and bigger public participation in the administration of urban open spaces as a means, incorporating UA as a vital model in programs of urban development. And likewise enable negotiation between diverse stakeholders for consensus building on UA [5]. Thus, UA must be incorporated into the urban master plan and a comprehensive modification of the urban regulation needs to be undertaken to add UA as a formal part of urban land use. The "New Urbanism" paradigm aims at correcting the trend of "Urban Sprawl" by utilizing insight gained from traditional urban development methods and thus preserving green areas for active recreation, natural habitat and useful agriculture [5].

4. Conclusion

This chapter discusses the growing phenomenon of UA, which emergence is more out of necessity than careful systematic planning. The growing urbanizing of global cities generally and especially global south cities in particular, largely due to migration and population increase has exacerbated food insufficiency and security. Consequently, UA evolved to fill the gap created by the food demand and supply conundrum from the rural setting. The findings revealed that UA takes place around or in-home spaces, a large expanse of undeveloped private or public land, residential land, roadside borders, river banks and other public lands. The prominent UA activities are animal husbandry, aquaculture, agro-forestry, horticulture, cultivation of perishable crops carrot, lettuce, spinach, cabbage, etc. Traditional urban planning as usual is foot-dragging to acknowledge the organic emergence of UA and integrate it into its planning. Instead, it prefers to view it as an abuse of urban space and dwells more on the nuisance it poses to the environment and constraints it gives to land availability for residential and other formal purposes [30]. This chapter posits that there should be a holistic examination of the pro and cons of UA. It submits that, since, UA has come to stay, it offers employment and food lifeline to a teeming population of urban residents, and it may be mainstreamed in urban planning to maximize the benefit while minimizing the negative side effects. Some of the suggested measures that can enhance the achievement include promoting agriculture-friendly urban plans for access to agricultural land, irrigations, manure and transportation network for distribution of produce, offering safe and accessible land, changing some open spaces and vacant lands to agricultural use, inculcate UA-responsive policies in broad plans.

Author details

Nkeiru Hope Ezeadichie*, Vincent Aghaegbunam Onodugo
and Chioma Agatha John-Nsa
University of Nigeria, Enugu, Nigeria

*Address all correspondence to: nkeiru.ezeadichie@unn.edu.ng

IntechOpen

References

[1] Erickson DL, Lovell ST, Méndez VE. Identifying, quantifying and classifying agricultural opportunities for land use planning. Landscape and Urban Planning. 2013;**118**:29-39

[2] Sarker A, Bornman JF, Marinova D. A framework for integrating agriculture in urban sustainability in Australia. Urban Science. 2019;**3**(50):1-15. DOI: 10.3390/urbansci3020050

[3] Mendes W, Balmer K, Kaethler T, Rhoads A. Using land inventories to plan for urban agriculture: Experiences from Portland and Vancouver. Journal of the American Planning Association. 2008;**74**:435-449

[4] Population Reference Bureau. Urbanization. 2019. Available online: http://www.prb.org/Publications/ Lesson-Plans/HumanPopulation/ Urbanization.aspx [Accessed on 1 March 2017]

[5] Adedeji OH, Ademiluyi IA. Urban agriculture and urban land use planning: Need for a synthesis in metropolitan Lagos. Nigeria, Journal of Geography and Regional Planning. 2009;**2**(3):043-050

[6] Game I, Primus R. Urban Agriculture. New York: GSDR 2015 Brief; 2015. pp. 1-13

[7] Gbadegesin A. Farming in the urban environment of a developing nation - a case study from Ibadan metropolis in Nigeria. The Environmentalist. 1991;**11**(2):105-111

[8] Mlozi MR. Impacts of urban agriculture in Dar es Salaam, Tanzania. The Environmentalist. 1997;**17**:115-124. DOI: 10.1023/A:1018599916581

[9] Kulak M, Graves A, Chatterton J. Reducing greenhouse gas emissions with urban agriculture: A life cycle assessment perspective. Landscape and Urban Planning. 2013;**111**:68-78

[10] Korir SCR, Rotich JK, Mining P. Urban agriculture and food security in developing countries: A case study of Eldoret municipality. Kenya, European Journal of Basic and Applied Sciences. 2015;**2**(2):27-35

[11] Madden JP, Chaplowe SG, editors. For all Generations: Making World Agriculture more Sustainable. Glendale: OM Publishing; 1997

[12] Wallimann I. Personal Communication about 'Urban Agriculture' by Ibrahim Game. Syracuse: and Richaela Primus; 2014

[13] Mougeot L. Growing Cities, Growing Food: Urban Agriculture on the Policy Agenda. Vol. 2000. Feldafing, Germany: Deutsche Stiftung fürInternationaleEntwicklung (DSE); 2000. pp. 1-42

[14] Hagey A, Rice S and Flournoy R. Growing Urban Agriculture: Equitable Strategies and Policies for Improving Access to Healthy Food and Revitalizing Communities, Policy Link Online Report, 2012. 1-49

[15] United Nations Development Programme UNDP. Urban Agriculture: Food, Jobs and Sustainable Cities. New York: United Nations Development Programme; 1996

[16] Rogerson CM. Urban agriculture and urban poverty alleviation: South African debates. Agrekon. 1998;**37**(2): 171-188

[17] Seattle Department of Neighborhoods. "P-Patch". 2012. Retrieved from: http://www.seattle.gov/ neighborhoods/ppatch/aboutPpatch.htm

[18] Arksey H, O'Malley L. Scoping studies: Towards a methodological framework. International Journal of Social Research Methodology.

2005;**8**(1):19-32. DOI: 10.1080/
1364557032000119616

[19] De Beer A, Gaskin A, Robbertse A, Bardien F. A review of the communication needs of persons with stroke within the African context for application within the clinical setting. In: Louw Q, editor. Collaborative Capacity Development to Complement Stroke Rehabilitation in Africa (Human Functioning, Technology and Health Series Volume 1). Cape Town: AOSIS; 2020. pp. 57-96. DOI: 10.4102/aosis.2020.BK85.02

[20] Levac D, Colquhoun H, O'Brien KK. Scoping studies: Advancing the methodology. Implementation Science. 2010;**5**(69):1-9. DOI: 10.1186/1748-5908-5-69

[21] Lovell ST. Multifunctional urban agriculture for sustainable land use planning in the United States. Sustainability. 2010;**2**:2499-2522

[22] Moragues-Faus A, Morgan K. Reframing the foodscape: The emergent world of urban food policy. Environment and Planning A: Economy and Space. 2015;**47**:1558-1573

[23] Chipungu L, Magidimisha HH, Hardman M, Beesley L. The importance of soil quality in the safe practice of urban agriculture in Zimbabwe, Kenya and South Africa. In: Land-Use Change Impacts on Soil Processes: Tropical and Savannah Ecosystems. New York, NY, USA: CAB International Publishing; 2015. ISBN: 9781780642109

[24] WCED SWS. World commission on environment and development. Our Common Future. 1987;**17**(1):1-91

[25] Urban Harvest. Development Focused Research Partnerships in Urban and Peri-Urban Agriculture (UPA). 2008. http://www.urban harvest.info

[26] Foeken D, Owuor SO. To Subsidize my Income: Urban Farming in an

East-African Town. Leiden and Boston: Brill; 2006

[27] Musonga H. Incorporating UPA in urban land use planning. In: Ayaga G, Kibata G, Lee-Smith D, Njenga M, Rege R, editors. Policy Prospects for Urban and Periurban Agriculture in Kenya. Workshop Proceedings. Nairobi. Kenya: Kari Headquarters; 2004. http://www.cipotato.org/urbanharvest/documents/pdf/policy-brief-kenya.pdf

[28] Lovell ST, Johnston DM. Designing landscapes for performance based on emerging principles in landscape ecology. Ecology and Society. 2009;**14**:44

[29] Food and Agriculture Organization of the United Nations. Urban and Peri-urban Agriculture and Urban Planning. Discussion paper for FAO-ETC/RUAF Electronic Conference "Urban and Peri-Urban Agriculture". August 21-September 30, 2000. Prepared by Drescher A. 2000

[30] Onodugo V, Ezeadichie N, Onwuneme C, Anosike E. The dilemma of managing the challenges of street vending in public spaces: The case of Enugu City, Nigeria. Cities. 2016;**59**:95-101

Chapter 10

Reusable Cities: A Circular Design Approach to Urban Regeneration through Materials Reuse

Serena Baiani and Paola Altamura

Abstract

The "circular city" reduces its environmental impacts from many points of view: from construction-related CO_2 emissions to energy production. The key areas for the implementation of an "urban policy for transition" are mainly oriented toward the reuse and recycling of materials from the building processes and urban value chains (urban mining), also through reuse practices of the existing building stock. The results of the research activities, reported in the present contribution, demonstrate the possibility of integrating studies on the environmental benefits of recycling in the building sector, with investigations on the potential of reuse to increase the overall eco-effectiveness of building interventions, interpreting the urban built environment in the perspective of a "reusable city." The hypothesis for a real reduction of raw material consumption in the construction sector is to combine the use of secondary materials with the reuse of building components, resulting from partial or total deconstruction and of materials from other waste streams, not belonging to the construction sector. Therefore, the research sought to understand to what extent the reuse of architectural components and waste materials from other industries can contribute to increasing resource efficiency at the local scale, reducing the consumption of materials, land, and energy.

Keywords: circular cities, urban mining, reuse, building materials, Harvest Map

1. Introduction

1.1 Circularity and climate neutrality: the role of cities

"Rapid urbanization and increased consumption have led to economic growth in many parts of the world, but have also created unprecedented amounts of waste" [1]. The linear economy paradigm, the so called "buy-use-dispose" model, as adopted globally in particular in cities, is no longer sustainable, especially because of the growing production of waste, which is becoming unmanageable in many contexts. The UN Agenda 2030 [2], in fact, within Goal 12 of the Sustainable Development Goals, "Ensuring sustainable consumption and growth patterns," calls for significant waste prevention, reduction, recycling, and reuse by 2030. As the demand for natural resources is constantly increasing along with waste production, circular economy and material resource efficiency represent the only approaches that can help to face the challenge of decoupling growth from resource

IntechOpen

consumption, tackling the "dual issue of increasing waste and decreasing resources by incentivizing actors throughout the value chain to extract maximum use from both existing products and the elements within them" [1].

Implementing the circular approach means reconfiguring all material flows within the city (building materials, water, solid waste, electronic waste, and even heat and energy), in order to avoid waste. Urban areas, in this perspective, represent an ideal environment to implement circular economy, starting from the resources embodied in the built environment and in particular in the existing building stock. Cities around the world are already moving in this direction, experimenting actions and interventions to promote interactions among different value chains and stakeholders, which can effectively foster circularity, urban mining, and sharing economy. This the case for Montevideo, for instance, where less than 2% of the solid urban waste goes to landfill via the waste collection system and, thanks to the support of ARUP, circular economy is being implemented as a strategic approach to enhance the city's resilience [3].

Indeed, cities play a central role in the transition toward sustainability and circular economy: urban agglomerations contribute significantly to climate change and the overexploitation of resources, with impacts including land use, soil consumption, pollutants due to mobility, water and energy consumption, air quality, waste. Nevertheless, with their high concentration of resources, people, capital, data, cities offer excellent chances for cross-collaboration between all key actors (individuals, companies, government, civil society, research, etc.) that to take action and "lead to a more sustainable and livable future for the next generation of urban dwellers" [1].

In fact, as stated in 2020 by the European Commission in the updated Circular Economy Action Plan [4], circularity is a prerequisite for climate neutrality, having an important impact on climate change mitigation and adaptation and on greenhouse gas emission reduction, through carbon removal. Actions can be nature-based, including through restoration of ecosystems, forest protection, afforestation, sustainable forest management, and carbon farming sequestration, or based on increased circularity, for instance, through long-term storage in wood construction, reuse, and storage of carbon in products such as mineralization in building materials.

1.2 A vision of a circular city

Transitioning to circular cities entails defining a vision. According to the Ellen MacArthur Foundation [5], a circular city embeds the principles of a circular economy across all of its functions, establishing an urban system that is regenerative and restorative by design. In such a city, the idea of waste is eliminated, with assets kept at their highest levels of utility at all times and the use of digital technologies a vital process enabler. A circular city aims to generate prosperity and economic resilience for itself and its citizens, while decoupling value creation from the consumption of finite resources [6]. Amsterdam, one of the leader cities in the application of circular economy concepts to city governance, follows seven principles in its transition toward a circular economy, as elaborated in a report commissioned by the city government [7]:

- Closed loop: all materials enter into an infinite cycle (technical or biological).

- Reduced emissions: all energy comes from renewable sources.

- Value generation: resources are used to generate (financial or other) value.

- Modular and flexible design of products and production chains increases adaptability of systems.

- Innovative business models: new business models for production, distribution, and consumption enable the shift from possession of goods to (use of) services.

- Region-oriented reverse logistics: logistics systems shift to a more region-oriented service with reverse logistics capabilities.

- Natural systems upgradation: human activities positively contribute to ecosystems, ecosystem services, and the reconstruction of "natural capital."

The abovementioned principles can be extended to define a vision and an action roadmap for circularity in cities.

The "circular city" reduces its environmental impacts from many points of view: from construction-related CO_2 emissions to energy production. In the construction sector, a circular city would allow a 10-fold reduction in CO_2 emissions and a 75% reduction in soil consumption, with a 30–50% saving in construction costs. Circular construction allows the saving of natural resources, considering that the building and infrastructure sector consumes 1/3 of the world's raw materials, releases 11% of global emissions, and produces 40% of municipal solid waste within demolition processes. It is to be considered that the use of recycled building materials would reduce CO_2 emissions by 40–70% [8].

At the global level, the building sector is in fact a crucial one for the implementation of circular strategies, as demonstrated by the survey on the status of the circular economy in 34 cities and regions documented in the Report "The Circular Economy in Cities and Regions" by the OECD [9], where 61% of involved cities and regions declared to have a circular economy initiative including buildings (**Figure 1**).

Planning and design of urban areas and buildings should draw inspiration from the circular processes that occur in nature, by promoting a closed-loop use of resources, and therefore defining flexible, multipurpose spaces, using reused/recycled and recyclable materials, designing for deconstruction, so as to prevent the production of waste. In fact, in order to successfully deal with the problematic disposal of residues, the very concept of waste must be erased from our design and technological point of view [10], a new circular approach to the design, production, and procurement of materials has to be defined, with the involvement of all the stakeholders, including industry and waste operators.

By resorting to these processes, reproduced in an industrial key, and exploiting the synergy between urban and periurban areas (preferably industrial, to be redeveloped), it is possible to reduce the energy and environmental impact of these areas, rebalancing the impacts of cities.

From the point of view of territorial and urban policies, cities and regions are putting into practice a multiplicity of experimentations of systems and technologies for saving, reusing, and recycling. However, these are largely sectoral practices, still far from the adoption of integrated management and programming models for functions. This integration will increasingly have to bring the various phases of production and management of material and energy flows into coherence and coordinate the activities of the various territorial actors: public administrators, territorial management bodies, producers of goods and services, distributors of goods and distributors of services, users, and workers. The process of adopting integrated models of development and circular management therefore can and must

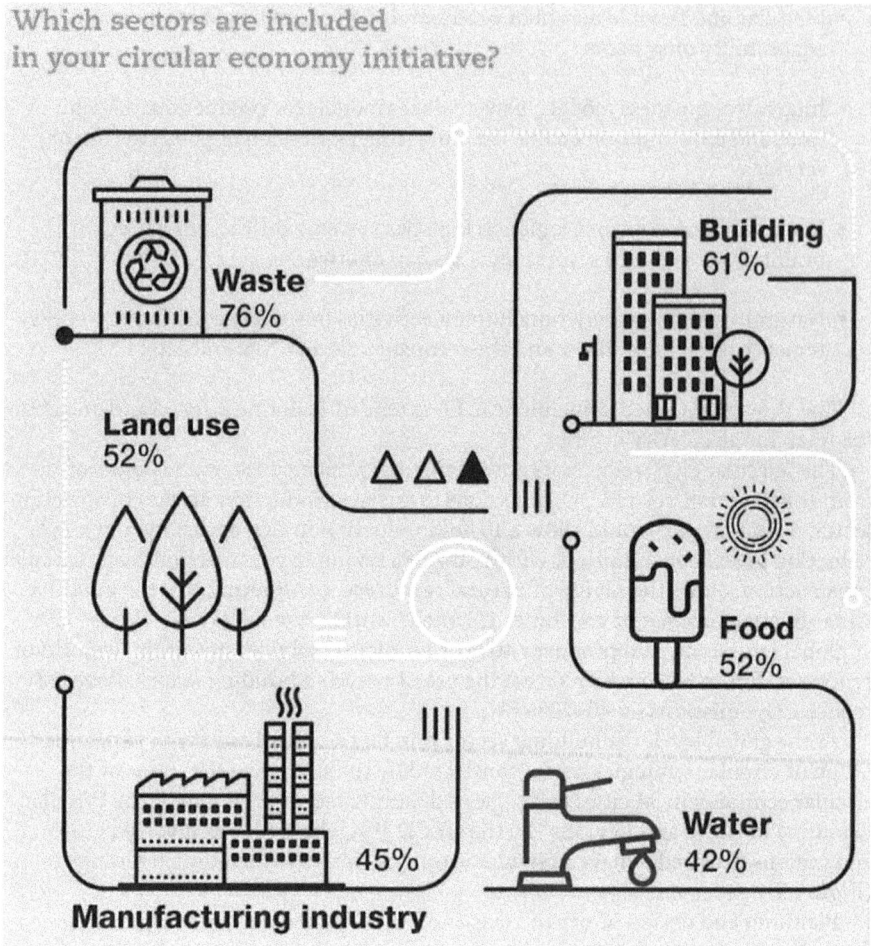

Figure 1.
Fifteen out of 34 cities and regions have a circular economy initiative, where buildings are included in 61% of cases. Source: OECD [9].

be increased and made more effective through coordinated and decisive support by public governance at the level of all sectors of the national production chains, but above all through the organization and the efficient management of the territory as a generator of economy and consumption in a circular sense [8].

Cities and regions are implementing territorial and urban policies oriented toward a multiplicity of experiments with technologies for reduction, reuse, and recycling, but these practices today tend to be sectoral, far from the adoption of integrated management and programming models for functions. Such integration could, in fact, make the various phases of production and management of material and energy flows coherent, by coordinating the activities of the various actors (public administrators, land management bodies, producers of goods and services, distributors of goods and services, users, and workers). The process of adopting integrated models of circular development and management can be increased and made more effective through a coordinated and decisive support of public governance at the level of all sectors of the national production chains, but above all through the efficient organization and management of the territory, as a generator of economy and consumption in a circular sense [8].

What is needed, therefore, is a decisive acceleration in the change of perspective toward circularity: it is necessary to overcome the sectoral, vertical, and fragmented nature that characterizes the circular interventions at the urban scale in the current panorama. Instead, circularity should be considered as central to the eco-systemic and economic functioning of cities and also in their interaction with peripheries, which should be systematically reorganized by putting circularity at the basis of all processes and exchanges of resources that take place at the urban level (food production and consumption, buildings and infrastructure construction, energy production and use, water use and recovery, etc.).

2. The importance of materials reuse in the construction value chain

The key areas for the implementation of an "urban policy for transition" are mainly oriented toward the reuse and recycling of materials from the building processes and urban value chains (urban mining), through the creation of materials management and recovery hubs and the adoption of reuse practices for the existing building stock [8].

At the EU level, the impacts related to construction activities are even higher than those cited at the global level: the building and infrastructure sector uses nearly 50% of the materials in EU by weight; buildings consume 40% of the EU energy and are responsible for 35% of EU GHG emissions [11]. Indeed, the level of material resource efficiency in the European building sector needs to be improved, in order to increase the contribution of the built environment to decarbonization and circularity, tackling climate change and resource scarcity. Through the Roadmap to a Resource Efficient Europe, already in 2016, the EU emphasized the severe impact of the consumption of raw materials in the construction industry, which represents 50% of excavated materials each year. In addition, the total amount of Construction and Demolition (C&D) waste produced annually in the EU represents almost half of total waste, with a recovery rate that is quite high for many member states, but much uncertain for many others. The necessity to significantly boost the closing of production cycles in the building sector was stated by the EU Dir. 98/2008 on Waste, which called for the increase of reuse, recycling, and material recovery of C&D waste to a minimum of 70% by weight by 2020. This target, which has been achieved by many Member States, among which are Germany, Netherlands, and the United Kingdom, for some countries it is particularly ambitious. The Italian situation, apparently in line with the EU threshold for C&D waste recovery (78.1% in 2019, not considering small quantities of C&D waste that do are not counted and the fly tipped waste) [12], is hindered by the lack of complete and reliable data—due to a partial traceability system—on which to develop an efficient C&D waste policy, by planning appropriate strategies and infrastructures.

However, even the virtuous countries must face a new challenge, highlighted by the abovementioned EU Directive, which places reuse above recycling in the waste hierarchy. In fact, in order to close building materials cycles reducing both energy and material consumption, it is necessary to integrate the two strategies, promoting reuse over recycling whenever possible [13]. At present, while high-quality recycling of C&D waste begins to spread, prevention and reuse, notwithstanding their great environmental and energy potential, are still rare. Both reuse and recycling are valid strategies, but their environmental benefits must be considered on a case-by-case basis. While in the future we ought to use only recyclable or biodegradable materials in buildings, so that they can be infinitely regenerated in a closed-loop model [10], as far as existing buildings are concerned, reuse is often the best option in environmental terms [14]. This is particularly true for clay bricks, stone slabs and blocks, steel elements, and other components with high embodied energy and a low performance decay. The Olympic Park in London represents a best practice in this sense [15]. Reuse, despite being well spread in the past, was almost completely abandoned by the construction industry. It only endures in restoration interventions, particularly in countries such as Italy, which often resort to reuse in the preservation of historical buildings. Nevertheless, in the contemporary circular cities' visions, the closed-loop construction value chain—as envisaged, for instance, in the Amsterdam case (**Figure 2**)—a crucial role is played by all the processes that are needed to allow reuse: deconstruction, selective demolition, separation and stocking of reusable components, eventual remanufacturing, repurposing within other construction sites. This model interprets the urban built environment in the perspective of a "reusable city," with buildings meant as "material banks," a concept deeply investigated in the recent H2020 Research Project BAMB (Buildings As Material Banks) [16].

Moreover, in order to favor a sustainable management of building materials and a higher resource efficiency, there are three crucial factors. Firstly, an accurate quantification of the potential supply and demand of secondary materials on an appropriate area (regional/local scale) is needed. This can help in promoting secondary sources of building materials within the urban planning and in forecasting the withdrawal of resources (such as sand, rocks, and aggregates) from the environment. Secondly, a wide range of tools supporting the operators of the building sector can factually help to implement the eco-effective management of waste materials, such as pre-demolition audits or software for the monitoring of waste production on large construction sites. A third factor, which will pay off in a longer term, is the mapping of secondary sources of materials not coming from building sites but rather from the industry, not necessarily from value chains directly linked to the building sector.

The quantification of supply and demand of inert waste and recycled aggregates at the regional level, experimented in a few studies in literature in the last decade

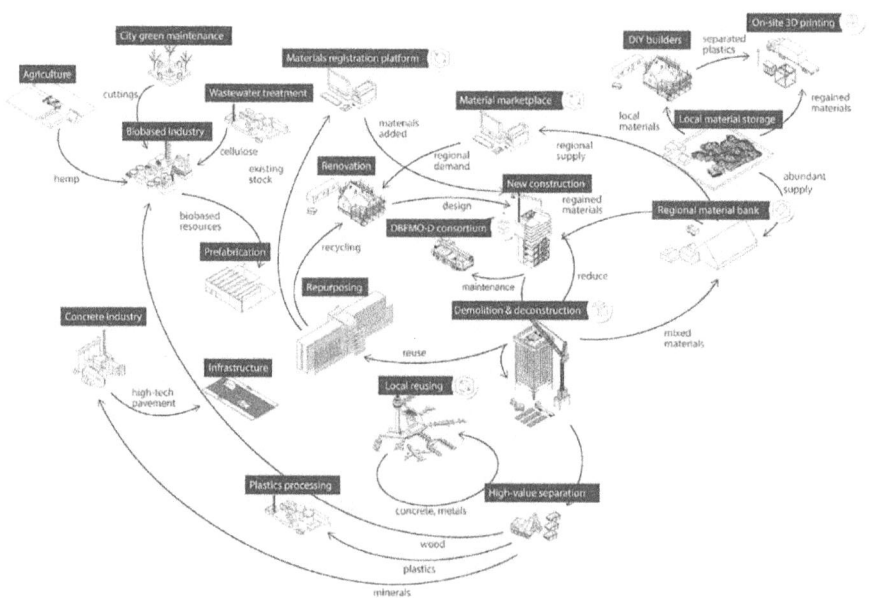

Figure 2.
The vision of a circular construction value chain at the urban/regional scale. Source: Circle Economy, TNO and Fabric [7], redesigned in World Economic Forum [6].

[17, 18], suggested the possibility to investigate, with a similar approach, the *resource conservation potential* deriving from the reuse of building components and waste materials coming from other industry sectors. This research approach will be described in Paragraphs 3 and 4, while the next sections describe in detail the three above mentioned factors through state-of-the-art experiences.

2.1 The estimation of the stock of materials in the existing buildings on an urban scale: Top-down and bottom-up approaches

Quantification, at a neighborhood/urban district scale, of the sources for the potential procurement of secondary building materials is a challenging task, to which European countries are starting to approach, in different ways. A good example is Germany, where different valid methodologies are applied, in order to correctly plan the economic and infrastructural development of the recycling industry. Data can derive from statistical analysis on building stock (top-down approach), such as Material Flow Analysis applied to regional level [17]. Data collection can also be carried out by surveying the materials constituting individual buildings (bottom-up approach) and aggregating the data for homogeneous portions of the building stock [18].

Further research experiences on the topic were illustrated within the SBE19 Brussels BAMB-CIRCPATH International Conference, conclusive of the cited H2020 Project BAMB (Buildings As Material Banks) (2019). The most interesting three are illustrated below.

The REBUILD (REgenerative BUILDings and products for a circular economy) Project, coordinated by Exeter University (UK) [19], addressed the possibility of creating value from the remanufacturing of components from buildings that have reached the end of service life (EoSL), creating new construction products destined to buildings to be realized according to the design for deconstruction approach, allowing in the future the potential new reuse of the same components. A key step in this research, in this sense, is the quantification of the stocks of bricks, steel, and concrete in the existing building stock at the district level, as well as the analysis of the related barriers for recovery and reuse. The project focused in particular on the analysis of bricks, with the development of a new technique for the deconstruction of the masonry with the reclamation of the single integral brick in a mechanized way, complemented by a study of the possible transformations of the element itself and of its use in new building components.

Another research, conducted at the Technical University of Munich [20, 21], has instead developed a dynamic GIS/BIM model to evaluate the stocks of materials in urban areas and the relative flows of materials activated by the construction of residential buildings. The research addressed both the classification of materials stocks (land registry of raw materials incorporated in residential buildings) and the identification of future flows of demolition waste, in order to predict potential sources of secondary raw materials, establish recovery strategies and more suitable control mechanisms. The potential supply of reusable/recyclable materials was therefore compared with the demand, in order to identify the degree of self-sufficiency achievable in a given territorial area, reducing the use of primary raw materials and transport. The developed assessment model was validated by applying it to the Munich-Freiham district, one of the main urban developments in Germany, demonstrating that a self-sufficient supply of steel (from 2036) and recycled aggregates for the production of concrete (from 2031) can be achieved for the construction of residential buildings.

Finally, another interesting research, developed in Belgium by the Hasselt University with the real estate company Essencia [22], experimented the use of existing databases as a tool to explore the potential of the building stock as a bank of materials. The research reports an estimate of the quantities of materials present in the residential building stock in the Flanders region, based on the combination of two existing databases: one relating to the energy performance certificates of buildings, belonging to the Flemish Energy Agency (VEA), with the general characteristics of the buildings, such as volume, type, surface, and information on the envelope of over 1 million assets in the Flemish region; the other one, developed by Essencia Marketing, containing general characteristics, geometric data, and materials on nearly 6000 new residential buildings distributed throughout Belgium. The research examined both databases and defined methods for combining data and for assessing the (future) potential of the existing building stock as a bank of materials.

2.2 Tools for the quantification of waste materials at the building site level

The tools supporting designers and operators (construction and demolition companies) in the estimation, monitoring, and exchange of waste materials in the design and construction phases play a crucial role in optimizing the level of material resource efficiency and circularity in the construction sector.

In the United Kingdom, for example, the share of C&D waste diverted from landfill has significantly grown thanks to, among many regulatory instruments, the mandatory introduction of Site Waste Management Plans in 2008. These Plans, containing an estimate of the waste that will be produced, as well as the accounting of waste actually produced, provide an accurate data collection in real construction/ demolition projects, which can effectively integrate statistical surveys. In England, the collection of data is facilitated by the SMARTWaste program by the BRE [23], an online tool supporting operators in the preparation of waste management plans and in the monitoring of waste on site. The SMARTWaste database enables operators to verify and increase their resource efficiency over time, while simultaneously providing valuable information to public authorities for the optimization of this sector. Such instruments could help those countries, such as Italy, still uncertain on the real quantity of C&D produced/recovered.

Another interesting initiative, with a view to the digitalization of the management of building materials' recovery processes—aimed at optimizing their environmental and economic sustainability—was the publication in Italy of the Reference Practice UNI/PdR 75: 2020 "Selective deconstruction—Methodology for selective deconstruction and waste recovery from a circular economy perspective" (February 2020) [24]. This technical pre-standard aims to define a macroprocess for deconstruction that favors the recovery of C&D waste and is oriented toward the compatibility with the digital management of the process itself and of the material-related information. The envisaged process takes into consideration both existing buildings to be refurbished or demolished and new constructions: for the former, through a pre-demolition audit, a database of materials intended for recycling and reuse is built and used during the intervention; for the latter, it is necessary to compile the database of the materials foreseen by the design project. The deconstruction process is divided into three phases: planning, operational, updating the database/final list of the materials used in the building. The aim of this procedure set by the UNI/PdR is to overcome the difficulties of the current construction waste tracking and management system, which in Italy appears to be a barrier for a concrete practicability of circular strategies. Within the UNI working group, the GEOWEB company offered its contribution on the subject of digital support tools for operators, creating a mock-up that collects instruments and functions covering the following phases: survey; modeling of the three-dimensional geometry of the building; design, planning, and execution of the deconstruction intervention.

The SaaS (Software as a Service) platform supports an end-to-end workflow in which all waste management technical and administrative phases are acquired, processed, planned, validated, and certified. Furthermore, the platform integrates an operational network of services (transport, waste treatment, storage, sale of products from secondary materials) offered by local companies, thus providing information and interoperability tools enabling stakeholders to monitor the process on the territory, to enforce policies, to define capacity planning processes, and finally, to promote incentives for the implementation of circular practices.

Another example of a tool to support the actors of the construction process for enhancing the use of secondary materials is the DECORUM platform, developed in Italy by ENEA [25]. The Platform, in support of all the actors of the supply chain in the decision-making phases, aims at ensuring the compliance of construction/renovation works with regulatory and environmental requirements, in particular with the mandatory national Green Public Procurement Minimum Environmental Criteria (GPP MEC) for Buildings (Ministerial Decree 11/10/2017) defining minimum thresholds of recycled content for the different building materials/products. Moreover, the marketplace section of the Platform gives space to the availability and reliability of recycled materials, promoting their wider diffusion in public works contracts.

Finally, the recent French initiative, which saw the development of the Démoclès platform, should also be mentioned: Démoclès is a traceability model for construction waste, whose methodology is being tested in France to be then disseminated abroad. In fact, the feasibility study [26] of the platform established a European benchmark to identify best practices in terms of traceability for—but not only— construction waste, demonstrating that there are two types of possible tracking systems: those similar to certifications and those that physically track streams accompanied by documents and a "third party guarantor." Furthermore, the study identified the building industry's needs in traceability and allowed the definition of specifications for a specific system, revealing that only a physical traceability of streams would be able meet stakeholders' requirements and enabling the construction of the system and of the Démoclès platform.

2.3 Harvest Maps at the urban district and city level

As mentioned, the most innovative research experiences concerning the design of buildings with a low consumption of raw materials today focus on the reuse of materials and components, not necessarily from the building sector, following an urban mining approach. In particular, some research studies have proposed and experimented the mapping of local sources of reclaimed materials, suitable for architecture but coming from other waste streams, before the designing of the building itself [27], those promoting the valorization of residues *through* the design solution. Indeed, reuse provides not just cultural and esthetic benefits, but concrete environmental and economic advantages, whose potential deserves to be thoroughly investigated.

The process set out by Superuse Studios (NL, formerly 2012 Architecten) is a fundamental reference for the circular project, because it involves sourcing all types of waste materials locally and enhancing their potential through a new design and construction process: the demolition of an existing building can represent the first source of materials; then, sources of other types of waste are sought in the proximity of the project area, opening up to the flows of discarded materials at the urban district and the city level. The project experience shows how it is possible to identify different mines of materials, each one characterized by its own dynamics, referring to different types of residues: End of life cycle products/materials (waste), Construction and demolition (waste), Dead stock (new), Production waste (new), Fast-life (short use). All of these potential sources are geo-referenced creating a graphic map with an overview of the residual materials reused/reusable in the project and their original locations. The map is called "Harvest Map" and its use, within the design process, allows many benefits [28].

The scouting process of waste materials [27] suitable for use in architecture (by-products, defective products, dead stock, leftovers processing waste, C&D waste, etc.) and available in the area adjacent to the intervention site, within a limited distance—on average a radius of 25 km—allows the enhancement of local waste by design, with actions of "superuse" rather than simple reuse, where materials acquire a more relevant technical and esthetic value through the design process. Moreover, this process allows to reduce the energy and carbon embodied in the materials used for the intervention, to avoid consumption and emission for the production and transportation of "new" materials, as well as at to activate small-scale circular economy processes. The research experiences described in the next paragraphs investigate the implications of this early mapping of the materials available on the site of the project, both in terms of optimization of resources use and of material characterization of architecture and the potential for transposing this strategy into a highly repeatable technical option [29].

3. Hypothesis and research objectives

The present contribution reports the results of research activities whose aim is to supplement ongoing studies on environmental benefits of recycling in the building sector, by investigating the potential of reuse to increase the overall eco-effectiveness of construction interventions. The hypothesis to be tested is that for a real reduction of primary materials consumption in the building sector, we need to place side by side the use of secondary materials with two other modalities of supply: the reuse of building components resulting from the partial or total deconstruction of buildings, and the reuse of materials from other waste streams not belonging to the construction sector. Therefore, the reported research activities have tried to understand to what extent the reuse of architectural components and waste materials from other industries can help to increase the resource efficiency at the local scale, reducing the consumption of materials, land, and energy. Final aim of the research is to decline the "circular city" in a specific perspective that of the "reusable city," where the built environment represents a resource to be reused in closed loops of material flows. In order to understand to what extent reuse might integrate C&D waste recycling (in particular that of inert waste, representing the main fraction) and to define the potential for resource conservation related to reuse itself, it is necessary to analyze real contexts to understand the actual availability of discarded materials and components suitable for architecture. In this sense, the points at issue are: the frequency of partial and complete demolitions in town; the instruments that a designer or a contractor can adopt to search for reusable materials; the tools that can be used to signal the availability of materials; the possibilities of activating new and different flows. Given the variety of types of materials from other sectors, which might be adopted in construction, it is necessary to focus on stable flows, constant in time and space, which in some cases can be even more regular than those coming from demolition activities, typically not constant in time and not completely predictable. Specific industrial activities might instead represent a constant source of by-products for the building industry.

The specific research goal is to investigate how reuse can contribute, at the scale of an urban district—and then scaled up at the city level—to reduce raw materials consumption, waste production, energy, and emission in the production and transportation of building materials and components. The scale of investigation has been chosen in order to minimize the impacts of the transport of building materials. Therefore, the research studies have focused on case studies of urban regeneration at the district level, testing the potential impacts of reuse on the life cycle first of a single building, then of a small group of buildings, and identifying, in the end, the factors that make it possible to scale up the results to the urban level. The chosen building has undergone a complete technological breakdown in order to understand, within the deep retrofit scenario, which technical elements are more suitable to be renovated with reused materials. Using a life cycle thinking approach, the average length of the service life of various technical elements, the average frequency of replacement of components, and the duration of the service life of the building as a whole have been taken into consideration. Then the materials requirements haven been considered in order to identify potential secondary/reusable materials available at the local level, taking into account the embodied carbon indicator in order to identify the best option in environmental terms, by quantifying the reduction of CO_2 emission due to the avoided extraction, production, and transportation. This is useful also in order to compare multiple scenarios: the use of primary raw materials, of reclaimed components and recycled materials, of only recycled materials. After evaluating the benefits on the single building, it is necessary to assess the possibility of extending reuse to the building stock at the urban scale, in order to maximize its environmental potential.

Final aim of the research is to develop a procedure (and related verification indicators) supporting the design phase. Thus, the scope in the long term is to facilitate the adoption of reuse as a strategic technical option. These objectives require an interdisciplinary and multiscalar approach, combining different scales of investigation (from the city to the building to the component level) and multiple methods, in order to respond to a new and broader approach to resource efficiency in the building sector.

4. Research methodology

The research methodology is divided into three main phases carried out with specific methods. In order to test the hypotheses defined above and to reach the mentioned objectives, it is necessary to start the analysis at the urban level. The adopted model works on the multiscalar dimension of urban districts, with the aim of redefining the environmental, energy, and social performance of existing quartiers to be turned—through urban regeneration—into circular districts, characterized by high resource efficiency and closed-loop flows of material and immaterial resources, in line with the objectives of decarbonization and climate neutrality. In this approach, the renovation interventions aim at a high level of material resource efficiency in the optimization of the environmental performance of the existing settlements, in order to limit the consumption of raw materials, favoring the supply of "zero km" and/or locally sourced building materials and products and at the same time minimizing the volume of C&D waste through circular strategies, thus reducing the level of embodied carbon in the materials used (**Figure 3**). The renovation of existing buildings themselves is a strategic action in order to reduce the need for materials and limit environmental impacts, both in the short and long terms adopting the Design for Deconstruction strategy.

The first phase of assessment of the adopted methodology involves the identification of the building components and materials that make up the building, the estimation of their volume and weight, and the calculation of the carbon

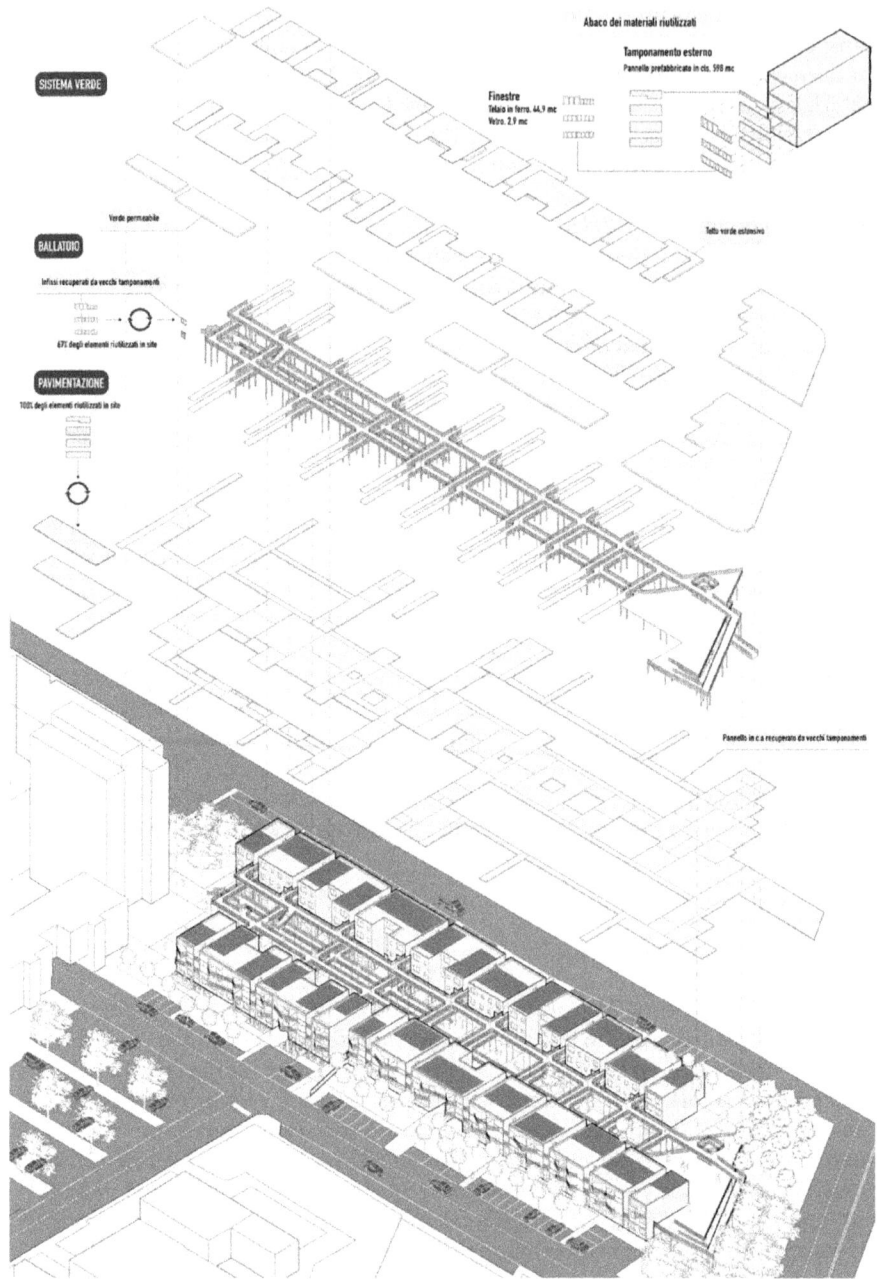

Figure 3.
Circular design process implemented in the design of the linear buildings and the outdoor spaces in the Torrevecchia District, Rome (IT). Source: Research Studies, S. Baiani, P. Altamura with M. Battiata, G. Schiavon, A. Sofi (2021).

embodied in the single materials, starting from a relevant reference database. At the same time as defining the design solutions for the rehabilitation of existing buildings, the volume and weight of the materials destined to be removed from the various technical elements and the relative estimate of the embodied carbon are also quantified.

From a methodological point of view, the evaluation of the level of material resource efficiency achieved is identified through consistent quantitative indicators that allow the measurement of the effectiveness of the choices. In particular, it is possible to measure the recycled content of the materials chosen for the intervention (one of the criteria of the GPP Minimum Environmental Criteria for Construction, compulsory at national level since 2016 for the entire public built heritage); the rate of landfill diversion of materials removed from the existing building; the amount of material recovered on-site; the amount of embodied CO_2 preserved by avoiding the demolition of existing buildings and that preserved through the recovery of materials intended for disposal, in particular on-site reuse or recycling that avoids energy consumption and emission due to transport.

For individual components and materials, potential circular technical options to avoid landfilling are assessed, with reuse as the preferred scenario over recycling and on-site reuse as the optimal solution. The different technical options are compared considering environmental, technological, and economic costs and benefits (**Figure 4**).

The process outlined leads to the integration of three different ways of supplying the materials needed for the deep retrofit intervention: the identification of components that can be recovered from the renovated buildings, during the selective demolition phase (e.g., external and internal fixtures), and that can be subject to remanufacturing and reuse or to recycling and reuse in situ; the identification of sources of waste materials/components/products from buildings or industries in the surrounding area; the selection, to cover the remaining needs, of new renewable and certified materials, which support the objective of reducing environmental impacts and intervention costs, while also ensuring the future reusability and recyclability of materials: "Changing the way products and materials are selected, manufactured and used in the built environment can lower environmental impacts as well as costs. Biological nutrients and sustainable, renewable materials can replace materials that are heavily processed, and hard to reuse and recycle" [30].

It is possible to define the mapping of the sources of waste materials coming from other supply chains (Harvest Map, **Figure 5**), built through an online survey and direct contact with companies, through the provision of questionnaires and inspections aimed at viewing the stocks of materials (surplus, waste, defective products, processing residues, etc.) potentially recoverable in the redevelopment intervention [31].

As part of the experimentation in the urban district of Torrevecchia, in Rome (Italy), online surveys were, as a priority, conducted to identify potential local mines, which were subsequently investigated directly, in collaboration with the respective operators, in order to identify potentially reusable materials. The research led to the definition of a GIS-based map, which identifies potential sources with their inventory of materials, their performance characteristics, and potential uses in relation to the technical elements identified by the project.

The experimentation has identified different typological systems characterizing public housing (ERP) assets (towers and linear buildings) on which the mass flow balance has been developed, considering all the inputs and outputs of materials expected to occur in rehabilitation and maintenance interventions during the whole life cycle.

Through a technological breakdown, with direct surveys and archive research, the technical elements that, on the basis of the project, can be replaced/integrated with recovered components being identified. The evaluation was supported by comparison with projects and experiments that have adopted, with a similar approach, mixed systems containing recovered materials. Through the comparison with case studies, the elements for which the application of reusable components

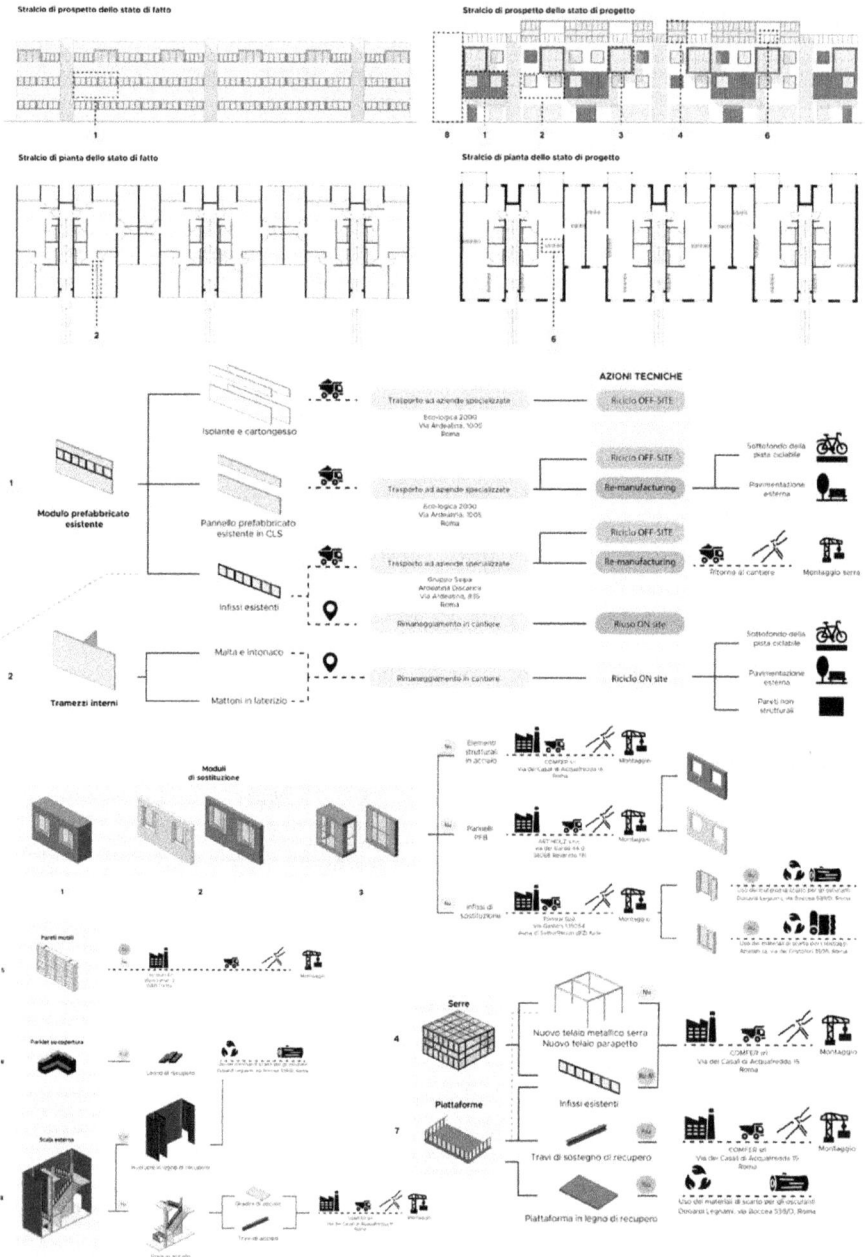

Figure 4.
Torrevecchia District in Rome: Circular building process and its quantitative verification. Research Studies, S. Baiani, P. Altamura, with N.D. Belforte and C. Fabrizio (2021).

is technically, economically, and environmentally more feasible were selected and analyzed in terms of technical requirements and potential performance (**Figure 6**).

An important step in the experimentation is the possible identification of resources on an urban and local scale, starting from demolition materials, reasoning on other types of waste, working on the production of energy at a local level and the reduction of transport emissions, due to the limited size of the district.

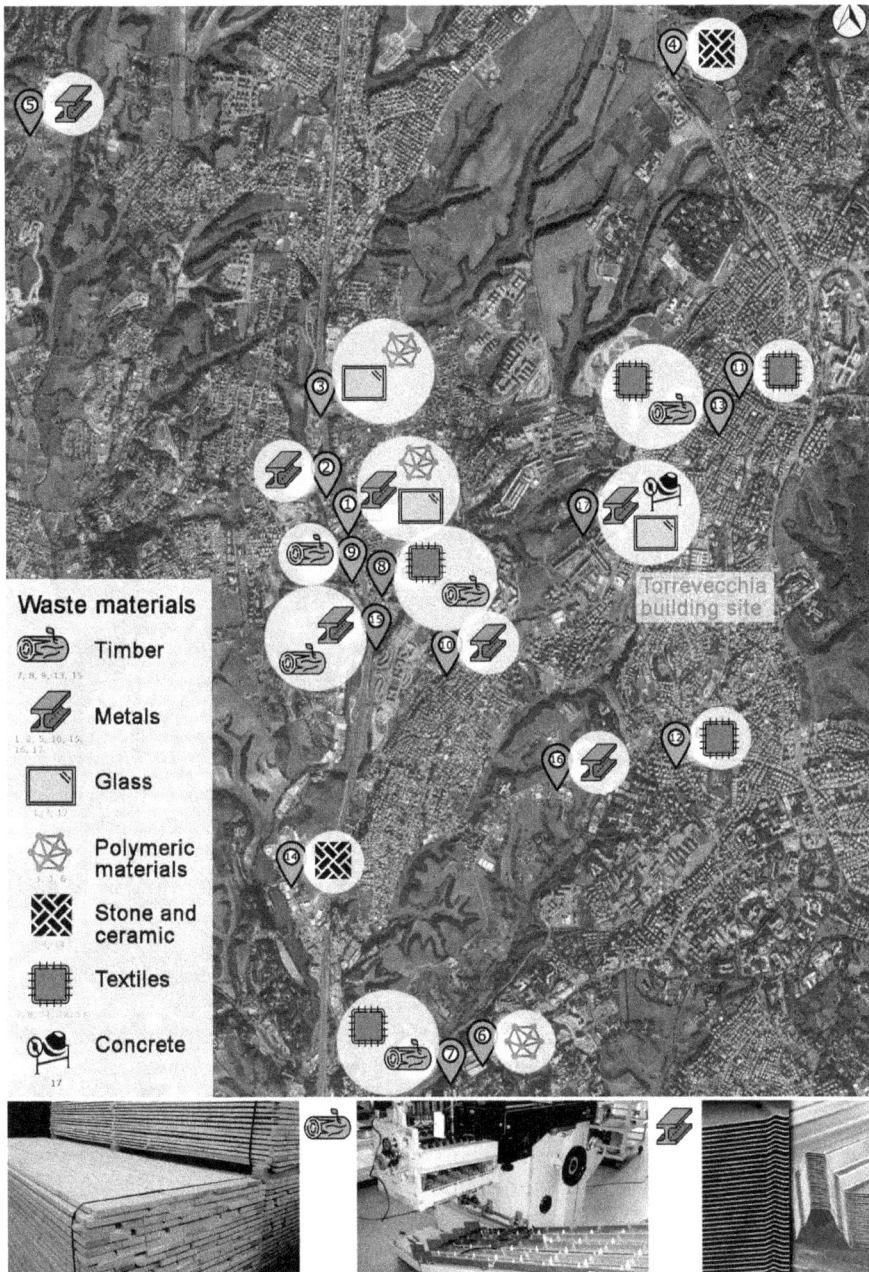

Figure 5.
Harvest Map around the Torrevecchia District, Rome: Map of the companies identified as sources of waste materials around the regeneration site of the Torrevecchia District, within a radius of 6 km. In red, the companies whose waste materials have been chosen for the project. Source: Research Studies, S. Baiani, P. Altamura with M. Rossi and S. Urbinati (2019).

The innovative character of the project lies in the way it verifies the feasibility of reuse in an urban area—and not on an experimental architectural project—to build a dataset that can be used by designers and can be increased by individual users, through shared tools such as the open-source Harvest Map platform. An initial mapping, available to operators in the building industry for the sector, could

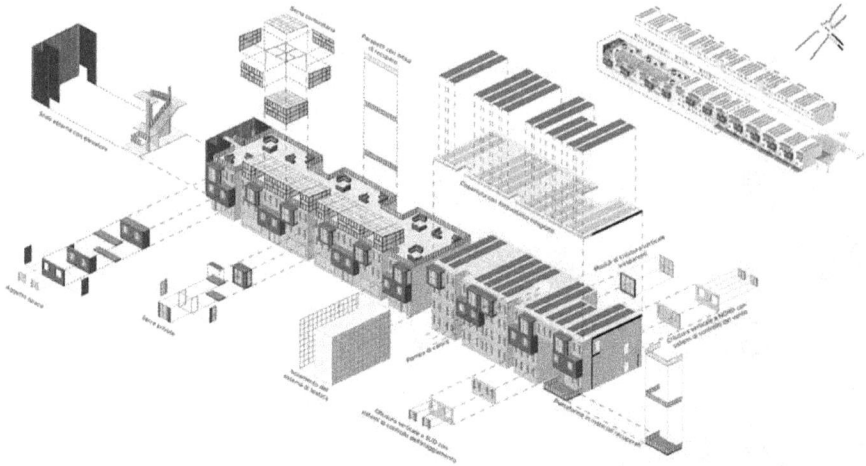

Figure 6.
Technical systems and subsystems of one of the renovated linear buildings in Torrevecchia, Rome: Disassembly of the elements built with reused and recycled materials. Source: Research Studies, S. Baiani, P. Altamura, with N.D. Belforte and C. Fabrizio (2021).

in the future make it possible to direct the methods of intervention, representing a picture—continuously updated—of the material resources available, with significant spin-offs in terms of innovation, involving all the operators in the process.

The experimentation phase, carried out on real cases, examined the potential sources of reusable materials in the city, starting from large construction, demolition, or redevelopment sites. The screening was carried out in an area with a radius of 10–20 km around the project sites, extending the research to more distant areas only in case of specific project characteristics. In the area of the former industrial site Papareschi in Rome (MI.REUSE Project, 2018) [31], for example, the project—aimed at the recovery of the former Miralanza factory with the use of waste materials sourced on site—applied a process that from the scouting phase led to the creation of a Harvest Map, to the redefinition of functions and spaces and the technological design of reversible building components with reused materials [29].

The results achieved in the different contexts, in terms of circular management of building materials in the intervention phase, denote a potentially very high level of circularity achievable through the management of building materials, deriving from partial demolitions and supplied for rehabilitation interventions. Interesting data, derived from experimentation on an public housing urban district in which for each building about 50% of the existing materials are conserved and the remaining half are destined for selective demolition, demonstrate the possibility, through the integrated action of several technical options for the end of life of materials, to reach a recovery quota of materials destined for demolition of about 90% by weight (higher than the 70% threshold of the EU Dir. 98/2008, which GPP MEC have adopted as a criteria) that guarantees to preserve about 80% of the embodied carbon of the materials intended for demolition, which replace new materials for the sub-bases of the external paving, whose environmental impacts are avoided [32].

The investigation involved gathering knowledge about the site in terms of the changes and transformations that led to its current state. Evaluating the building's evolving use has highlighted a series of transformations, which have affected the existing structure at different points in its life cycle. These changes are mainly related to past needs to expand overall living space. A building's life cycle can be analyzed by reading and understanding its construction system. This also makes it possible to understand its peculiarities and limits. In the Torrevecchia District in Rome, in terms of the architectural and construction aspects of the building system, it was made using a heavy and prefabricated system in reinforced concrete (**Figure** 7). This was completed with panels made off-site, limited interior insulating materials and plaster finishes.

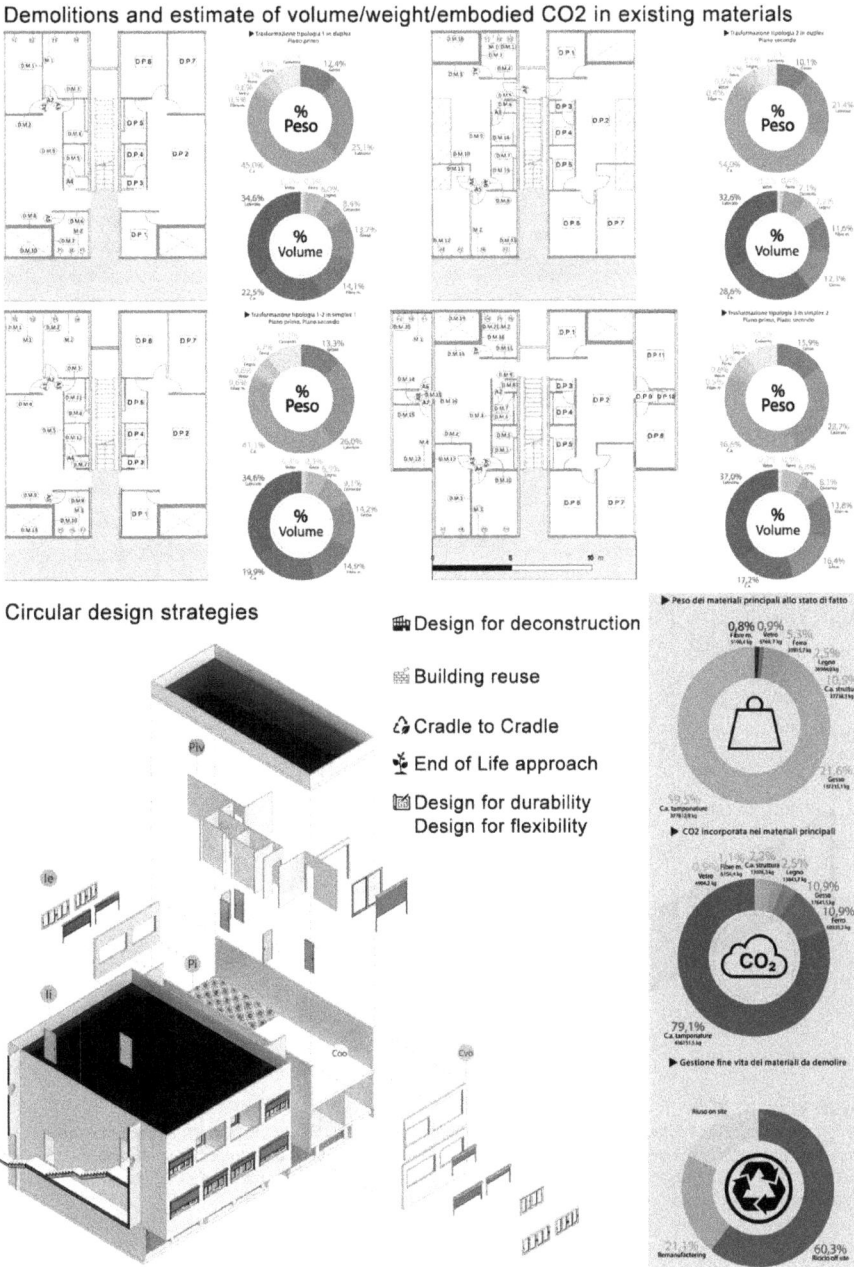

Figure 7.
"Ante operam" state of a renovated building in Torrevecchia, Rome, with the quantification of demolished materials in weight, volume and embodied carbon and the identification of circular design strategies and estimate of the recovery rate. Source: Research Studies, S. Baiani, P. Altamura with F. Ianiri, G. Massaroni, N. Taddei (2020).

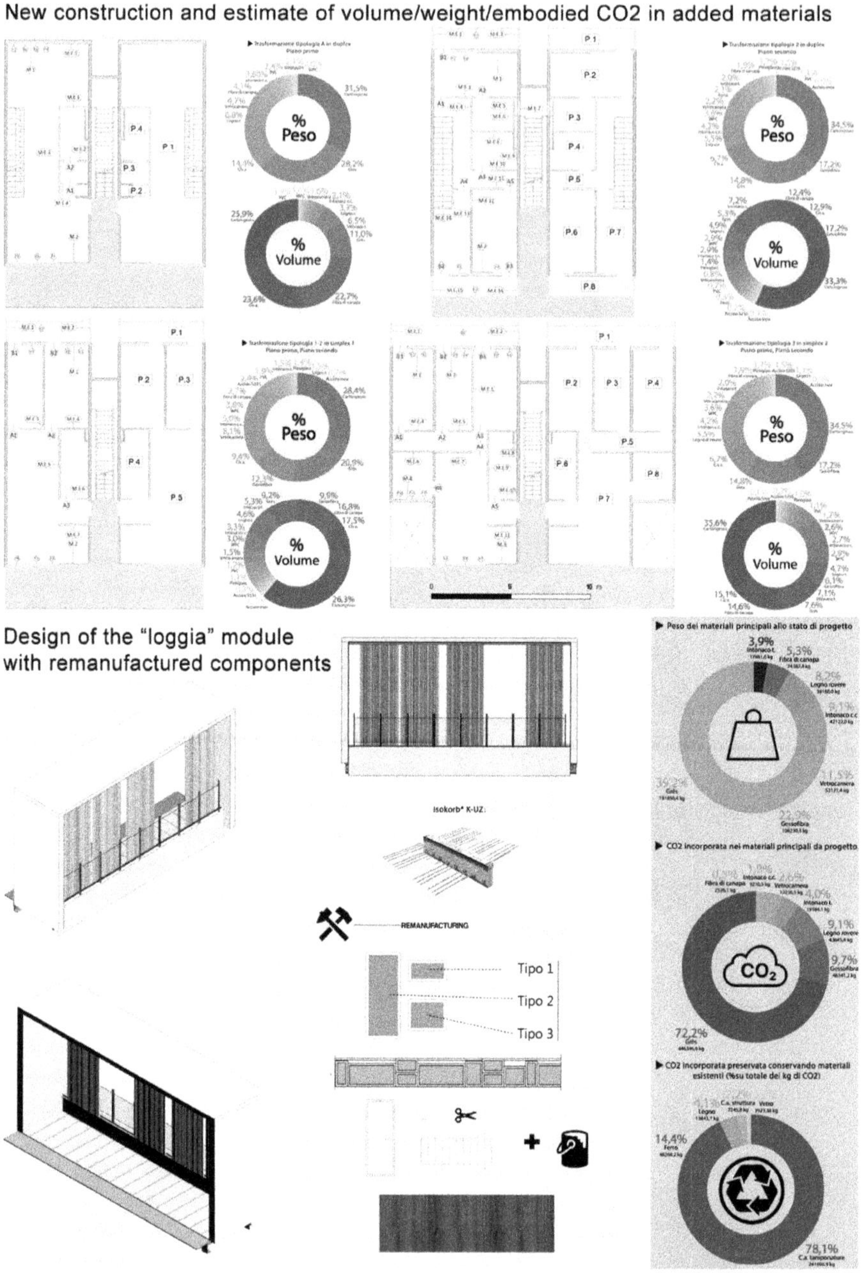

New construction and estimate of volume/weight/embodied CO2 in added materials

Design of the "loggia" module
with remanufactured components

Figure 8.
Design solution for a renovated building in Torrevecchia, Rome, with the quantification of the intervention's materials in weight, volume, and embodied carbon and the calculation of the recovery rate of demolition materials and of the embodied carbon preserved through reuse. Source: Research Studies, S. Baiani, P. Altamura with F. Ianiri, G. Massaroni, N. Taddei (2020).

A comparative assessment was also subsequently conducted to consider the potential effects of resulting demolition waste (in terms of volume/weight). The overall material requirements were also considered under more or less "invasive" intervention scenarios in terms of expanding demolitions/additions. Under these scenarios, various operational choices led to different comparable options based on redefining the housing, introducing/increasing common spaces or living services and identifying components to eliminate or integrate. However, each scenario commonly reflected the guiding technical requirements that interventions be totally reversible, low cost (in terms of environmental, energy, and economic impacts), and material minimizing (in terms of weight and types of materials used).

Estimates were done on materials to be removed from the building in terms of weight and volume, and associated embodied carbon was included in these measurements as well. Estimates were also made in terms of the volume of materials needed to execute each different scenario (these materials were selected based on a set of performance criteria that included maximum decarbonization). This made it possible to come up with a matrix of technical systems, components, and materials, which permitted considering "materials to look for" versus "materials to let go." The Harvest Map was consulted to this end to identify supply "mines." Defining technical systems for each of the options identified (addition, integration, grafting, replacement) has made it also possible to evaluate which existing elements could be recovered and reintroduced over the building life cycle. It also affords systematizing processes of disassembly, micro-demolition, and material or component replacement and recovery. It additionally permits calculating the material/component shares (in terms of percentage by weight and volume), which may come from on or off-site sources. This all made it possible to develop technological solutions while applying a "circular" and "reversible" view of the various elements involved. In doing this, particular attention was paid to the building envelope and the "passive" bioclimatic control devices to be introduced. To this end, verification of energy effectiveness took place as well, alongside with the assessment of the embodied carbon indicator (**Figure 8**).

5. Results discussion and conclusions

By assessing the technical feasibility and environmental potential of adaptive reuse in an urban context, based on the available sources of secondary materials, the experimentation demonstrates the transferability of this strategy in different contexts. This is achieved by proving its significant effectiveness and relevant potential contribution to decarbonization and resources conservation targets. It is possible to identify some specific contributions of the research work, at different levels.

First, the identification of the instrumentation to support the development of a circular design methodology that focuses on the action of recovery and reuse (buildings, components, materials) in order to evaluate how reuse can contribute, at the scale of an urban district—and then scaled up at the city level—to reduce raw materials consumption, waste production, energy, and emission in the production and transportation of building materials and components. In particular, the tools supporting designers and operators (construction and demolition companies) in the estimation, monitoring, and exchange of waste materials in the design and building phases play a crucial role in optimizing the level of material resource efficiency and circularity in the construction sector. Among these: the Site Waste Management Plans (UK) containing an estimate of the waste that will be produced, as well as the accounting of waste actually produced, providing an accurate data collection in real construction/demolition projects, which can effectively integrate statistical surveys; the Reference Practice UNI/PdR 75: 2020 "Selective deconstruction—Methodology for selective deconstruction and waste recovery from a circular economy perspective" (IT) with a view to the digitalization of the management of demolition waste recovery processes, aimed at optimizing their environmental and economic sustainability; the DECORUM platform (IT) supporting the actors of the construction process for enhancing the use of secondary materials; the Démoclès platform (FR), a traceability model for construction waste. Among the tools assessed in the research activities, the Harvest Map was identified as a fundamental reference for the circular project, because it involves sourcing all types of waste materials locally (mines) and enhancing their potential through a new design and construction process.

Second contribution of the research work was the development of a methodological and operational structure, also based on the transfer of international experiences, appropriate to the Italian context, with an experimental approach for the verification of the phases and the evaluation of the results achieved.

Thirdly, through the systematic identification, for each building typology, of elements and technical systems suitable for the realization with reclaimed components, the research validated the compliance of reclaimed elements and materials with specific requirements, with a performance verification procedure.

One potential research perspective opens up, in the definition of an appropriate methodology for the identification and "promotion" of reclaimed components in the urban environment, with a focus on the characterization of virtual and physical spaces (hubs) where materials can be collected and shared with potential users. These local hubs, developed on the basis of the potential demand, which is difficult to correlate with the supply, could constitute an advanced system that could also favor the on-site production of technical components, reducing the considerable impacts caused by transport. The possibility to foresee the potential impact of a greater use of reused materials and components in the building industry favors the reduction of the demand for new materials and opens up new design opportunities in regeneration interventions.

This defines a design vision that focuses on "circularity" in its broadest sense, capable of characterizing the multiple phases of the life cycle of an urban district, through the circular use of materials from regeneration and construction interventions and integrated management of ecological and energy systems, in the broader vision of "reusable cities."

Author details

Serena Baiani* and Paola Altamura
Planning Design Technology of Architecture Department, "Sapienza" University of Rome, Rome, Italy

*Address all correspondence to: serena.baiani@uniroma1.it

IntechOpen

References

[1] Cheryl M. Foreword. In: World Economic Forum. Circular Economy in Cities. Evolving the Model for a Sustainable Urban Future. White Paper 5; 2018

[2] United Nations. Transforming Our World: The 2030 Agenda for Sustainable Development. General Assembly; 21 October 2015

[3] ENEL. Circular Cities—Cities of Tomorrow. 3rd ed. October 2020. Available from: https://www.enel.com/content/dam/enel-com/documenti/media/paper-circular-cities-2020.pdf

[4] European Commission. Communication from the Commission to the European Parliament, the Council, the European Economic and Social Committee and the Committee of the Regions. A New Circular Economy Action Plan for a Cleaner and More Competitive Europe. COM/2020/98 Final. 2020

[5] Sukhdev A, Vol J, Brandt K, Yeoman R. Google, the Ellen MacArthur Foundation. Cities in the Circular Economy: The Role of Digital Technology. 2017. Available from: https://www.isb-global.com/cities-in-the-circular-economy/

[6] World Economic Forum. Circular Economy in Cities. Evolving the Model for a Sustainable Urban Future, White Paper 9. 2018

[7] Circle Economy, TNO and Fabric. Circular Amsterdam. A Vision and Action Agenda for the City and Metropolitan Area. Delft: TNO; 2016. Available from: http://resolver.tudelft.nl/uuid:f7d0eaf1-8625-4439-ae8e-2168bfc20e95

[8] ICESP, GdL5, La transizione verso le città circolari. Vol. 2. Available from: https://www.icesp.it/sites/default/files/

DocsGdL/Rassegna%20GdL5_Volume%202%20-%20La%20transizione%20verso%20le%20citt%C3%A0%20circolari.pdf

[9] OECD. The Circular Economy in Cities and Regions. Paris: OECD Publishing; 2020

[10] Braungart M, McDonough W. Cradle to Cradle: Remaking the Way we Make Things. New York: North Point Press; 2002

[11] ENEA, INEC, ACR+, EEB, ECOPRENEUR. European Circular Economy Stakeholder Platform. Orientation Paper [Internet]. 2020. Available from: https://circulareconomy.europa.eu/platform/sites/default/files/leadership-groupconstruction.pdf

[12] ISPRA. Rapporto Rifiuti Speciali. 2021. Available from: https://www.isprambiente.gov.it/files2021/pubblicazioni/rapporti/rapportorifiutispeciali:ed-2021_n-344_versioneintegrale.pdf

[13] Addis B. Building with Reclaimed Components and Materials. London: Earthscan; 2005

[14] Sassi P. Designing buildings to close the material resource loop. Engineering Sustainability. 2004;**157**:163-171

[15] Hartman H. London 2012 Sustainable Design. Delivering a Games Legacy. Chichester: Wiley; 2012

[16] H2020 Building As Material Banks [Internet]. 2019. Available from: https://www.bamb2020.eu/. [Accessed: 13 February 2022]

[17] Schiller G, Deilmann C. Ermittlung von Ressourcenschonungspotenzialen bei der Verwertung von Bauabfällen und Erarbeitung von Empfehlungen zu deren Nutzung. Dessau-Roßlau:

Umweltbundesamt; 2010. Available from: http://www.uba.de/uba-info-medien/4040.html

[18] Volk R, Stengel J, Schultmann F. Compilation of regional building stock inventories under uncertainty. In: Proceedings of the SB 13 Singapore. Realising Sustainability in the Tropics. Singapore: Research Publishing; 2013. pp. 493-500

[19] Ajayabi A, Chen HM, Zhou K, Hopkinson P, Wang Y, Lam D. REBUILD: Regenerative buildings and construction systems for a circular economy. In: Buildings As Materials Banks. A Pathway for a Circular Future. SBE19 Brussels BAMB-CIRCPATH. Brussels: IOP Conference Series EES 225; 2019

[20] Heinrich MA, Lang W. Capture and control of material flows and stocks in urban residential buildings. In: SBE19 Brussels BAMB-CIRCPATH. Brussels: IOP Conference Series EES 225; 2019

[21] Heinrich M. Erfassung und Steuerung von Stoffströmen im urbanen Wohnungsbau—Am Beispiel der Wohnungswirtschaft in München-Freiham [thesis]. Germany: Technical University of Munich; 2019

[22] Gepts B, Meex E, Nuyts E, Knapen E, Verbeeck G. Existing databases as means to explore the potential of the building stock as material bank. In: SBE19 Brussels BAMB-CIRCPATH. Brussels: IOP Conference Series EES 225; 2019

[23] Adams K, Blackwell M, Holt A. Saving Money, Resources and Carbon through Smartwaste. Watford: IhsBrePress; 2013

[24] Reference Practice UNI/PdR 75: 2020 "Selective Deconstruction—Methodology for Selective Deconstruction and Waste Recovery from a Circular Economy Perspective". February 2020. Available from:

https://www.bauschutt.it/media/9af07049-6542-494a-9f3f-49a743d64595/uni21001058-eit.pdf. [Accessed: 13 February 2022]

[25] Luciano A, Cutaia L, Cioffi F, Sinibaldi C. Demolition and construction recycling unified management: The DECORUM platform for improvement of resource efficiency in the construction sector. Environmental Science and Pollution Research. 2021;**19**:24558-24569

[26] Elcimaï, Girus, Terra, RDC Environment. Démoclès. Étude préalable d'un dispositif de traçabilité des déchets de chantiers du bâtiment. 2019. Available from: https://www.democles.org/uploads/2020/01/democles-rapport-etude-tracabilite-vf.pdf. [Accessed: 13 February 2022]

[27] Jongert J, Peeren C, Van Hinte E. Superuse: Constructing New Architecture by Shortcutting Material Flows. Rotterdam: Oio Publishers; 2007

[28] Superuse Studios. Harvest! Collect! Re-use! Available from: https://www.superuse-studios.com. [Accessed: 13 February 2022]

[29] Baiani S, Altamura P. Waste materials superuse and upcycling in architecture: Design and experimentation. Techne. 2018;**16**:142-151

[30] ARUP. The Circular Economy in the Built Environment. London: ARUP; 2016

[31] Baiani S, Altamura P. Mapping the sources of secondary building materials. First experiences in Rome. In: Baratta A, editor. Dal downcycling all'upcycling verso gli obiettivi di economia circolare. Roma: Timía; 2019. pp. 120-131

[32] Tucci F, Baiani S, Altamura P, Cecafosso V. District circular transition and technological design towards a circular city model. Techne. 2021;**22**:227-239

The Street Edge: Micro-Morphological Analysis of the Street Characteristics of Baghdad, Iraq

Haider Jasim Essa Al-Saaidy

Abstract

In Arabic cities, diversity can be seen in the development of the same underlying order. This assists to manage to qualify well-defined relationships with various levels of movement in the urban setting. The micro-morphological examination is used to emphasise further the spatial pattern at a micro-level within a macro-scale scope. Hence, micro-level studies are essential in evaluating the built environment with regard to private and public domain. Terminologically, the notion of the symbiosis of how the private and public domains interact with each other is needed. Also, there is a need for people to know their rights when using the street edge and the extent to which they (the owner/user) have the authority to modify the public space. The transition of the urban pattern from the traditional order (spontaneous pattern) to the modern model (pre-planned system) not only changes the spatial morphological structure entities but also transformed the association of the private and public domain.

Keywords: street pattern, private control, public control, traditional (spontaneous) order, modern (pre-planned) system, urban symbiotic relationship

1. Introduction

The pattern (Edge–Edge interface) is controlled by morphological parameters that manage the street network regarding the binary element: plots and blocks. The level of superposition between the two domains is evaluated by various indicators based on the specifications of the street pattern. According to Al-Saaidy [1] 'the assets of Baghdad today belong to this historical period of the city with its significant monuments and organic street pattern. Otherwise, the urban areas that settled outside the historical zone were designed according to a modern scheme and a modernist ideology' ([1], p. 6). Moreover, Marshall [2] confirms that 'land use zones and roads, in a modernist urban structure, [are] represented separately as nodes and links, but in a traditional urban street network, the streets themselves are significant spatial entities' ([2], p. 112). The mechanism in operating the street edge differs when it comes to comparison between two patterns: traditional and the modern network. The convergence between the individuals and their adjacent edges, which mostly relates to the street life and social interactions. The pattern

(Edge–Edge interface) is also responsible for defining the boundary between two realms, private and public. The degree of overlap between the two territories is measured by different indices based on the characteristics of the street edge, such as porosity, transparency and permeability (**Figure 1**).

Using a fine-scale analysis by examining the street pattern seems to be a more effective means of understanding the urban characteristics of streets over large-scale classifications. There is a definite pattern of activity about the order process of compound parameters, which increases in an area or within set spatial dimensions. Conversely, large-scale order is influenced by minimum or single settings, and this due to the comprehensive analysis system of streets, which are expected to be unrelated in formulating distinguishing urban characteristics for the city [1, 3–5]. Morphologically, the leading characteristics of Baghdad Street pattern combine variety and difference between the pristine and new model. Both patterns, historical and modern, are managed by two distinct generative orders: spontaneous (bottom-up approach) and pre-planned (top-down procedure) [1].

Recently, the debate is not only between conventional and modern concepts in urban studies but also what can be understood between two domains: private and public in formulating street edge [6]. Micro-level explanations are essential in evaluating the adjacent edges of a street. The notion of a symbiotic relationship between private and public spaces, with the effectiveness of street life, plays an essential role in advancing the quality of social interaction. In this article, knowing the affiliation of the spatial attributes at a micro and macro scale needs an interpretation of how spatial ingredients in an urban environment are placed. There could be a range of expectations when the community uses the private and public realms. These two domains express individual action and group behaviour. The shared beliefs, norms, values, economics, politics and the natural and built environment, these considerations are the predominant aspects of living in a particular community [7, 8].

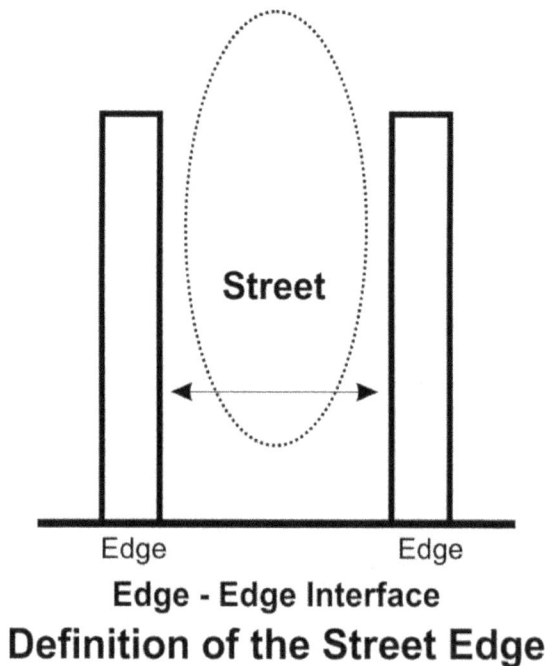

Street

Edge   Edge

Edge - Edge Interface

Definition of the Street Edge

Figure 1.
The fictive image illustrates the interface between two opposite street edges. Source: Drawn by the author.

2. Responding to the urban edge

The manner in which a community, particularly in Iraq, manages public spaces could lead to a spectrum of opportunities that inform individual action and collective behaviour alike. Primarily, human behaviour tends to conform to the predominant dimensions of living in a particular community, for example, by embracing the common beliefs, norms, values, economics, politics and the natural/built environment. In this regard, Bianca [9] states that the physical environment represents, '... every genuine cultural tradition, architecture and urban form' and that this '... can be seen as a natural expression of prevailing spiritual values and beliefs .... it is an outcome of tradition and daily practices which correspond to certain spiritual principles' ([9], p. 22). These factors embed the interrelationship between space, time and culture. Moreover, the tripartite connections among these three social parameters (sociocultural, sociophysical and socioeconomic) are rooted in and formulate both the ecological pattern and different responses to the surrounding environment (**Figure 2**).

Even though there are different urban patterns, people who perceived the public spaces shared the same cultural patterns but not the same behavioural actions. The observation is based on the ethnographic method and quantifies the responses of people who use the street. This technique facilitated the documentation of people's responses and interactions without any interference or effect on the subjects' actions. In this regard, to a large extent, the historical and traditional area in Baghdad grants an opportunity for persons to share the public space and involve in such activities as walking, staying, sitting, standing, watching and chatting where the street edge works to interconnect the accommodation of such activities [10, 11].

In the contemporary neighbourhoods, the lost knowledge of public and semi-public or semi-private spaces can be experienced in various ways. Moreover, in modern areas, the human scale, enclosure and definition, and the authority of its public space are missing (**Figure 3**). As the public space can be distinguished according to the activity pattern of the adjacent context, it is possible to recognise different types of more common street edges, such as residential, commercial and mixed (**Figure 4**).

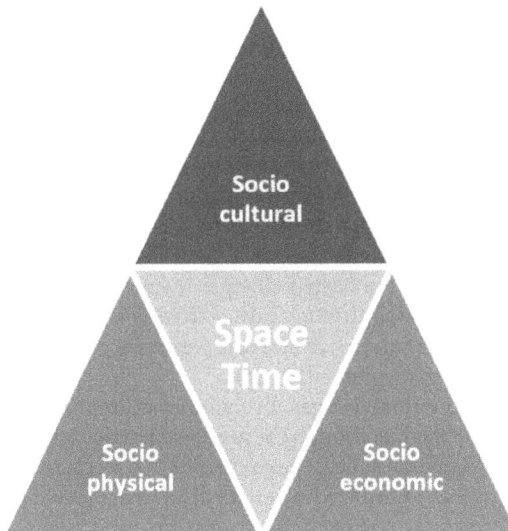

Figure 2.
The tripartite connection among three social parameters (sociocultural, sociophysical and socioeconomic), which are governed by spaceandtime. Source: Drawn by the author.

Figure 3.
In the modern pattern, some public spacesare not within the authority of residents, where there is no explicit declaration about the claim overthis type of territory. Source: Photographed by the author's team, 12/December/2016.

Figure 4.
Residential edges: Multi-storey residential areasversus traditional low-rise neighbourhoods. Source: Photographed by the author's team, 12/December/2016.

These edges are entirely responsible for shaping people's responses, particularly within residential areas where people react spontaneously to the private, semi-private and even semi-public realms (**Figure 4**). Inhabitants in these areas tend to change the characteristics of the semi-public and sometimes public spaces. These changes manifest differently, such as through soft treatments or hard borders when a resident illegally occupies the adjacent realm (**Figure 5**). However, the residential edges are likely to be used by their inhabitants even if these edges face the public space or link directly to the street. Moreover, some proprietors cut off the adjacent part of the street in order to change the primary land use from residential to commercial. Unfortunately, this transformation of purpose, and any misunderstanding of the rights to do so, leads to uncharted changes in land use. Hence, the residential edge is then be used for walking through rather than as a place to stop. According to Alexander [12] and Hillier et al. [13], a street is generally designed for staying in, or movement-to rather than movement-through.

On the commercial and mixed edges, lively interactions between the people and street spaces are experienced. Although these edges, particularly in the historical area, are still lacking in maintenance, they represent an attractive spine for the neighbourhood (**Figure 6**). People who benefit from this type of edge show different responses based on the particular activity of each unit along with the adjacent edge. Those who use public space can be classified according to their two primary activities: walkers and stayers. The aims of these two classes are varied in terms

Figure 5.
Inhabitants have dealt with the public edge by turn it intoa private space. Source: Photographed by author's team, 08–13/December/2016.

Figure 6.
The commercial and mixed edgesin samples a (left) and D (right). Source: Photographed by author's team, 12–21/December/2016.

of their exchange purpose and/or movement-to/movement-through [6, 14]. The expression of public space and its investments differ considerably between the traditional area and the more modern design. Whether in the traditional or modern area, the quantity of the public spaces is generally required, except in the areas offered by the adjoining edges.

Therefore, there is a need to not only examine the traditional part of a city as an isolated pattern but also to understand the comparison with other, new neighbourhoods in terms of the different perspectives afforded, particularly via the urban form and urban life. The traditional urban fabric arose in response to indigenous cultures and traditions; thus, Remali [15] explains that the 'traditional urban form is the result of [the] "selectionism" of an evolutionary process, whereby a built environment gradually become[s] congruent with activity systems, lifestyles, meaning and values by applying rules, which are often unwritten, as in most cultural landscapes' ([15], p. 57). Moreover, there is also a need for individuals to understand their rights when using the street space and the extent to which they (the owner/user) have the authority to alter the public space. Commonly, individuals tend to extend their territoriality, even in temporary activities. This includes peddlers and the owners of adjacent units (shops) who tend to extend the commercial edge by elongating the boundary of their activities. These expansions differ entirely from one individual to another, and from one street to another. One of the main reasons for such territorial extensions is to attract customers by making the adjacent spaces particularly enticing; nevertheless, a critical issue remains concerning the authority for these expansions.

3. Edge: Edge Interface

3.1 Interfacing between street and private-public edges

The main question is 'to what extent individualistic lifestyles can interfere with street life and vice versa' ([16], p. 2). The relationship between private and public would exist within a micro-spatial configuration. Van Nes and López [16] state that the main street network in the urban context is a factor that influences the microscale spatial variables. Spaces that mediate between buildings and streets create social interactions, which help to form human behaviour. These spaces could be part of a buildings' interior that causally link with the public space, such as courtyards and balconies or through spaces in front of buildings, such as sidewalks. They encourage a social encounter and promote street life at different levels, whether in terms of culture, norms and religion or the physical conditions of the built environment [17, 18] (**Figure 7**). According to Jacobs ([19], p. 59), a relationship between the private and public realms requires 'a good city street neighbourhood [that] achieves a marvel of balance between its people's determination to have essential privacy and their simultaneous wishes for differing degree of contact, enjoyment or help from the people around'. According to Marshall [20], the relationship between private and public is neither only determined through physical expression, nor a volumetric enclosure that regulates the public-private border, but rather functions as a social filter.

Marshall ([2], p. 13) states that 'the movement space constituted by streets forms the essential connective tissue of urban public space – from the micro scale

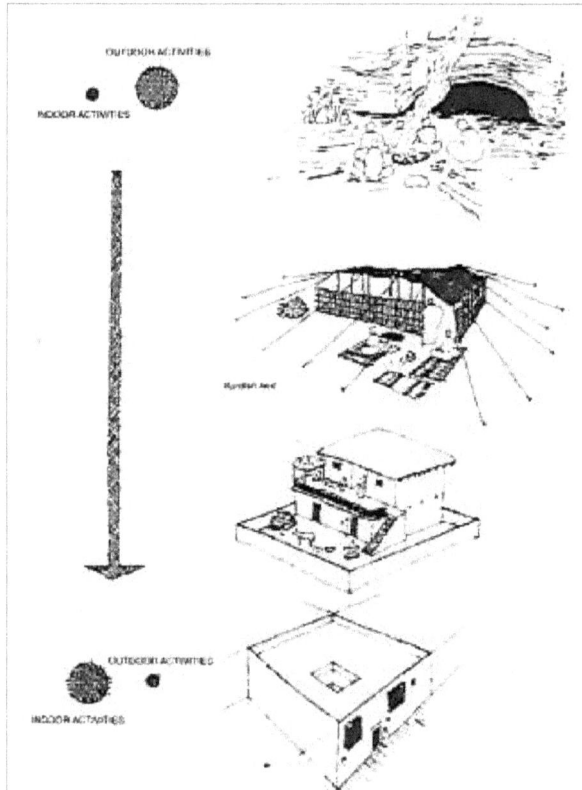

Figure 7.
Transforming outdoor space activities into indoor space activities from ancient to urban settlement. Source: Nooraddin [18].

of circulation within building to the macro scale of whole cities'. Therefore, 'street space forms the basic core of all urban public space – and by extension, all public space – forming a continuous network or continuum by which everything is linked to everything else. This continuum is punctured by plots of private land. The plots of private land surrounded by public streets are like an archipelago of islands set in a sea of public space' ([2], p. 13). Thwaites et al. [21] address different aspects of urban spaces as a betweenness milieu, which mediates between private and public. Also, they sought to highlight the role of the community in making the urban decisions in order to draw at least the local scale or micro scale of their neighbourhoods. This contribution has been defined as the *Transitional Edge*.

According to Thwaites et al. ([21], p. 85), 'a public to private gradient that works in a continuum from private to public and vice versa... [it is] a smooth and complex gradient of subtle changes where a greater range of spaces allows greater diversity of intimacy and social interaction'. Jacobs identifies three main qualities required to successfully encourage people into the street: (1) the situation requires visible demarcation between private and public areas; (2) a particular level of surveillance regarding eyes upon the street and (3) users who exploit the street reasonably, continuously and as effective eyes, in turn, induce others in adjacent buildings into the street to watch not the sidewalk but the pedestrians [19].

Marshall [20] states that there are several subtle complications when understanding privacy; it is not only a single modest linear movement between public and private. Private (exclusive space) means operating the action, giving control of space to reserve a specific area for specific individuals or even a group, contributes to raising the overall supervision and shapes the pattern of difference between public and private. The public (inclusive space) infers to an area where people are able to move, meet, mix and interact [20] (**Figure 8**).

3.1.1 Street edge characteristics

The street is the artery of a city regardless of its classification; for example, the street form (straight, irregular or zigzag), street function (residential, commercial, mixed-use), street dimensions (its length and width), street class (main, secondary, connected street) and street type (open-ended, cul-de-sacs). One of the main aims of the street network is to enable people to access and move to/through the street network towards their destinations. The street is much more than an urban spatial element; it has a crucial space that is to manage the entire movement and people influx. Besides, the street can be 'regarded as a fundamental building- block of urban structure, where, the public street system forms the principal part of the urban transport system' ([2] a, p. 14–15). Hillier [22] states that good spaces are utilised spaces; in this respect, an urban area is utilised by the movement to and/or the through movement. Furthermore, the street proffers routes from everywhere to everywhere else, and its influence on movement is a fundamental source of the multifunctionality that promotes vitality in the city.

Marshall ([2], p. 15) states that 'the challenge is to address the street as an urban place as well as a movement channel, and how to make this connection of the street work – not just as an isolated architectural set piece, but as a contribution to wider urban structure – otherwise, streets are for people'. Thwaites et al. [21] refer to ten themes that characterise the street edge and provide valuable insight into the socio-spatial properties relevant to transitional edges. The ten themes are: 'social activity, social interaction, public-private gradient, hide and reveal, spatial expansion, enclosure, permeability, transparency, territoriality and looseness' ([21], p. 78–79). Hillier et al. [13] denote that the integration of core maps covers the main streets and shopping areas.

Shopping streets tend to become viable when they have a high level of retail that is integrated with the global network and local pedestrian movement.

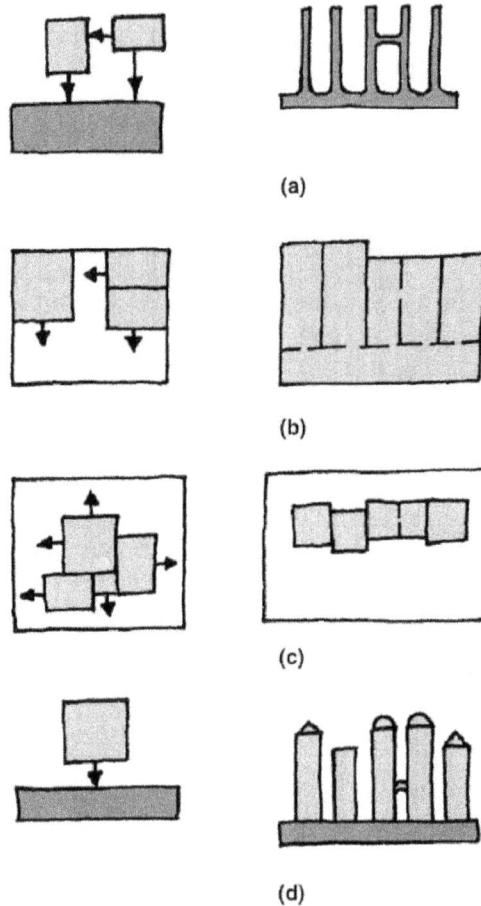

(a)

(b)

(c)

(d)

Figure 8.
Street syntax: (a) all strategic roads connect to form a single network; (b) all private spaces connect to the single public space; (c) all buildings have an interface with the single outside space; (d) all buildings connect to the single ground surface. Source: Marshall [20].

Less integration tends to occur in monofunctional areas, such as residential areas [13, 23]. The proportional place of the street and its integration within the entire network system play a crucial role in shaping the street edge characteristics. The configurational properties of the urban fabric are the primary influence on shaping two types of movement; through-movement and to-movement [13] (**Figure 9**). Movement and multiplier effects are significant prerequisites to promote the quality of street life. The multiplier effect attracts new development, new buildings and uses [22, 24].

Transitional edge is defined by Thwaites et al. [21] as the street edge and its multiple functions. It combines three dimensions: social, participatory and structural components. The *Transitional edge* encourages and diversifies the territorial experience to keep the community sustainably for those who use urban space efficiently. Therefore, this can be regarded as a spatial and sociological line. *Transitional edges* 'are coherent socio-spatial domains and not simply boundaries between the architecture and the external public realm .... offer the potential to achieve a more socially optimal balance of form, place and understanding [therefore] transitional

Origin   **To - movement**   Destination

Tend to be:
Symmetrical system
Nondistributed system

Space   **Through - movement**   Space

In-between Spaces

Tend to be:
Asymmetrical system
Distributed system

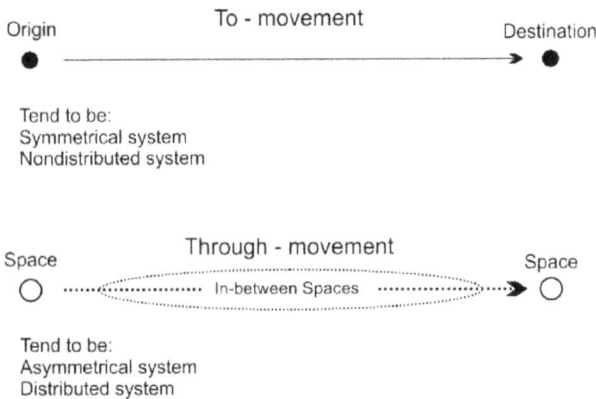

Figure 9.
Two patterns depend on the spatial arrangement: Through-movement and to-movement. The first occurs with the layout as a system of possible paths; however, when the layout can be considered as a system of origins and destinations, then movement is a to-movement. Source: Drawn by the author based on Hillier's concepts.

edges as key components in the socio-spatial order of the urban habitat' ([21], p. 23, 53, 71). Furthermore, Thwaites et al. [21] refer to another idiom of the street edge: the broken and unbroken edge that governs the degree of social interaction. It also conveys an impression of the extent to which people can interact with whatever broken or/and unbroken edges (**Figure 10**).

According to Jacobs ([19], p. 380) *Visual Street Interruption (VSI)* is when, 'a good many city streets (not all) need visual interruptions, cutting off the indefinite distant view and at the same time, visually heightening and celebrating intense street use by giving it a hint of enclosure and entity'. VSI encompasses a set of considerations when 'there is no visual tale of street intensity and detail to tell … and should be in functional terms, not dead ends but corners' therefore, 'visual street interruption is a natural eye-catcher, and its own character has much to do with the impressions made by the entire scene' ([19], p. 382–383). The street edge should be characterised by catching the eye and giving the space a rooted sense of place Buchanan [25]. According to Segall

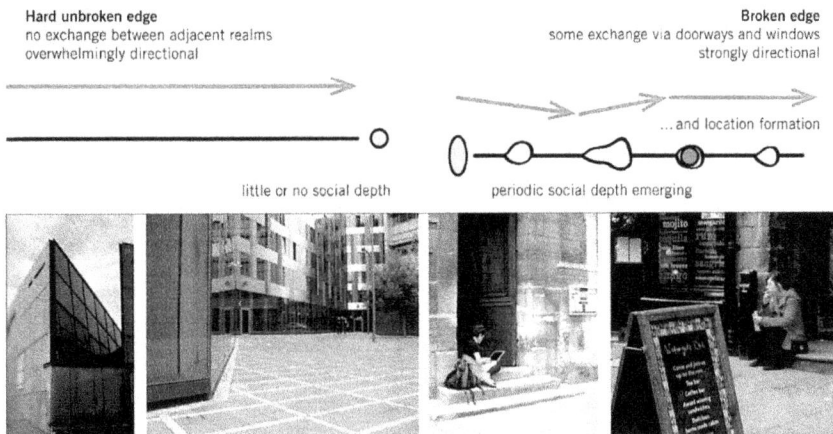

Hard unbroken edge
no exchange between adjacent realms
overwhelmingly directional

Broken edge
some exchange via doorways and windows
strongly directional

… and location formation

little or no social depth   periodic social depth emerging

Figure 10.
Unbroken, an abrupt edge between architecture and the public realm with little hope of encouraging life (left). Abrupt edges are broken by doors and windows which begin to act as catalysts for social activity (right). Source: Thwaites et al. [21].

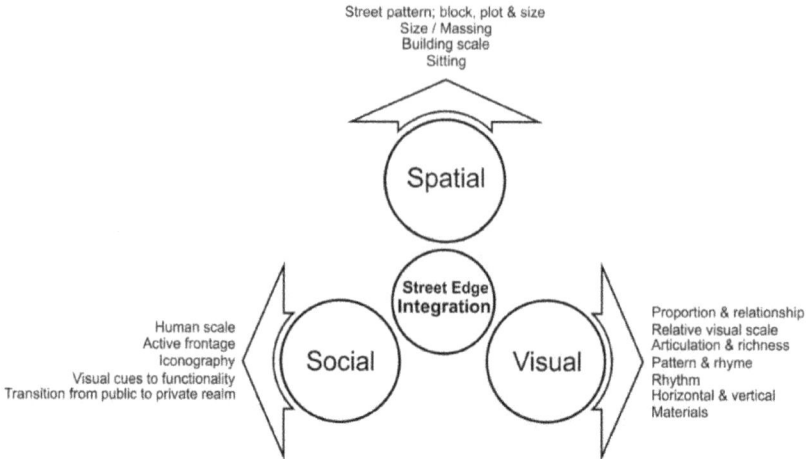

Figure 11.
The criteria of the street edge integration based on Carmona et al. [27]. Source: Drawn by the author.

([26], p. 51, 73), 'the visual experiences most generally available in a particular environment predispose one to identify most readily material similar to the content of those experiences ... the pattern of visual experiences in the lifetime of a person can modify his perceptions of objects in space' (**Figure 11**). Understanding the concept of *Visual Street Interruption* and its role in Baghdad could be a critical issue. Meaning that there is a delicate line between *Visual Street Interruption* and urban chaos in reading the street edge, the second phenomenon one could recognise in some urban areas of Baghdad city.

3.1.2 Private edge characteristics

Alexander [12] offers 253 patterns that are divided into 36 categories. One of these patterns is path shape, which is a crucial component in the built environment and contributes to other patterns in drawing the whole context of a city. Alexander([12], p. 590) advocates that the 'street should be for staying in, and not just for moving through, the way they are today'. Alexander ([12], p. 593) opposes the concept of setbacks, stating that 'buildings' setbacks from the street, originally invented to protect the public welfare by giving every building light and air, have actually helped to destroy the street as a social space ... the setbacks do nothing valuable and almost always destroy the value of the open areas between the buildings' (**Figures 12** and **13**).

Marshall ([20], p. 105-112) states that the '... private plots and buildings... [where]... buildings and cities are different kinds of social container, reflecting their

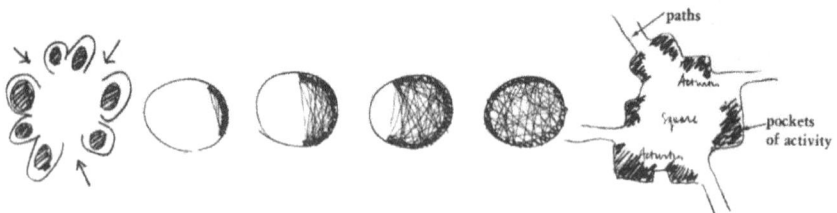

Figure 12.
As the activities grow around the space, it becomes more lively. Source: Alexander [12].

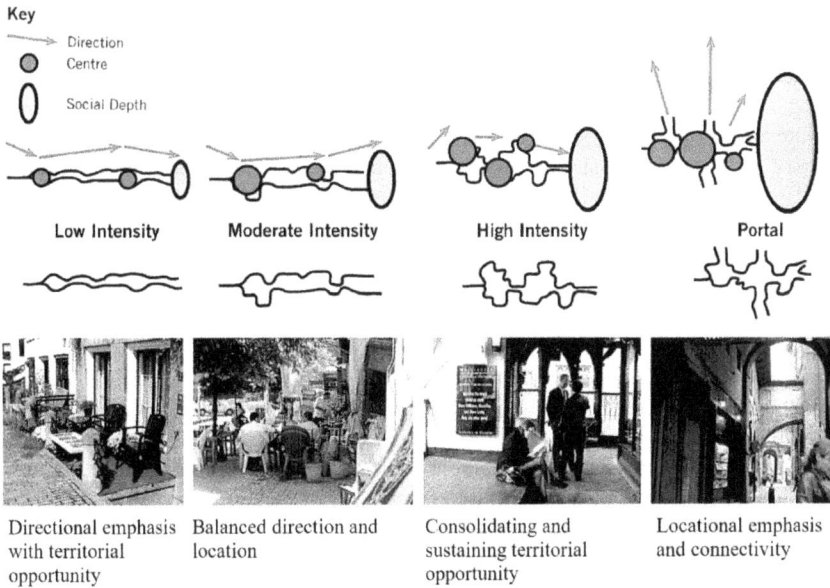

Figure 13.
Four different intensities in four segments. Source: Thwaites et al. [21].

differences in social structure. A city is not a big building, but is articulated into different buildings, mediated by a social fabric of public streets'. Additionally, one side of the street edge is subjected to private demands where the plots and buildings are located; this means that the facades are designed for private benefits [27, 28]. The expression of privacy ranges from soft control, like colour, texture and level, to hard control, such as fence and/or wall.

Furthermore, moveable and invariable can also classify the nature of privacy. Kostof [29] refers to the authority of using private space within the space of the street. He states that the buildings' edges need not be completely subjected to owners' desires, as there are public authority regulations that organise the street edge in such a way to increase the variety of building façades from block to block. The diversity works to impart the beauty of the block and street edge to the city; it incites attraction and surprise in people, whether inhabitants or visitors.

The *Filter Edge* is defined by Marshall [20] as a *social filter* when he states that the city is heterogeneous and involves different kinds of people who move through the social filter system. The systematic circulation of people ranges from loose filters, like streets, to reach fine filters, such as building edges. Therefore, he states that 'a building is an environmental container and filter; the building-plot-street system is both a social container and a social filter.... the importance of public streets as being not void, but as integral to the notion of a city, a kind of mortar binding between social units. Without this system of public spaces, a city would not be a city' ([20], p. 105, 112). Canter ([30], p. 9) argues that 'the environment providing perceptual stimuli [and] also be thought of as a filter ... we are always in the environment to carry out certain activities, and we usually carry out these activities with other individuals ... this is the fact that we actively modify, build and influence our physical surroundings'.

3.1.3 Public edge characteristics

The sense of public space is one of the main concerns and dialogue in generating social interaction and improvements in street life. The public edge embeds a broad spectrum of events, activities and social assemblage. It is a place where people should feel free to express their aspirations and desires. It 'host[s] structured or communal activities—festivals, riots, celebrations, public executions—and because of that, such places will bear the designed evidence of our shared record of accomplishment and our ritual behavior' ([29], p. 124). Accordingly, 'the main public places of a city are its most vital organs [thereby] if a city's streets look interesting, the city looks interesting; if they look dull, the city looks dull' ([19], p. 29). Banerjee ([31], p. 14) suggests that 'the sense of loss associated with the perceived decline of public space assumes that effective public life is linked to a viable public realm. This is because the concept of public life is inseparable from the idea of a public sphere'.

The public edge forms the third domain for social interaction, and investment in the function of the street edge encourages people to collect. The variety in the function of the public edge promotes street life and maximises social interaction [32]. It is necessary for the humanisation of public urban space such that the activities taking place contribute to the continuous surveillance of the space [33]. Oldenburg [34] adopts '*Third Place*' as an expression of other places, apart from settled places and workplaces. The third place should conjoin people in a free and mixed way by presenting exceptional comfort which is important to public life. The third place is a sorting edge that filters interests, and that people admire or 'un-admire' when using such places.

The highest value of the third place lies in its potential to encourage the meeting of people from diverse classes, age groups and with varied interests. It is important for the third place to be accessible, easy to reach and comfortable both for regular frequenters and newcomers. Furthermore, *unplanned, unscheduled, unorganised* and *unstructured* are four characteristics of the third place, which define it as essential, universal and pivotal to informal participation and social interaction [34].

Hall [35] refers to two types of spaces; *Sociofugal* is a space that keeps people apart, and *Sociopetal* is a space that brings people together. Engwicht [14] states that 'a vibrant spontaneous public realm, therefore, allows greater flexibility in our private relationship' ([14], p. 27). However, motorised regulations and their requirements exploit the public realm and drive away from the realm of the spontaneous encounter, which forces people into what Sennett [36] called the '*polarization of intimacy*'. Jacobs [19] states that the spatial social theme within the public space should be employed to capture what she called *self-appointed public characters*.

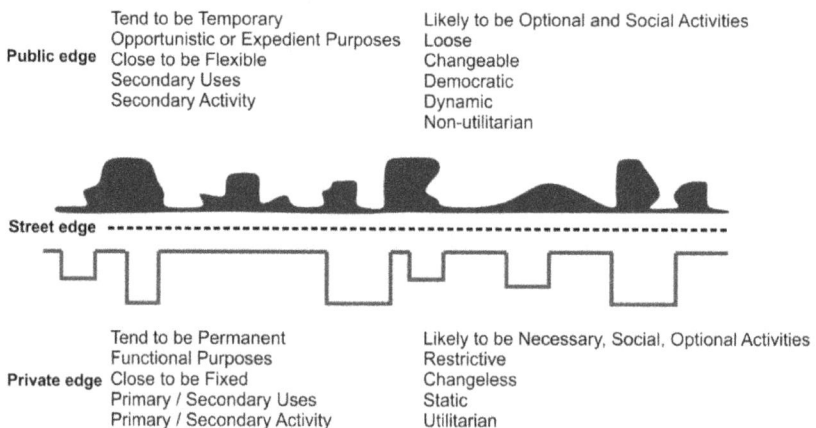

Public edge	Tend to be Temporary	Likely to be Optional and Social Activities
	Opportunistic or Expedient Purposes	Loose
	Close to be Flexible	Changeable
	Secondary Uses	Democratic
	Secondary Activity	Dynamic
		Non-utilitarian

Private edge	Tend to be Permanent	Likely to be Necessary, Social, Optional Activities
	Functional Purposes	Restrictive
	Close to be Fixed	Changeless
	Primary / Secondary Uses	Static
	Primary / Secondary Activity	Utilitarian

Figure 14.
Demonstrating the ability of different types of urban space to shape three street edges. Source: Drawn by the author.

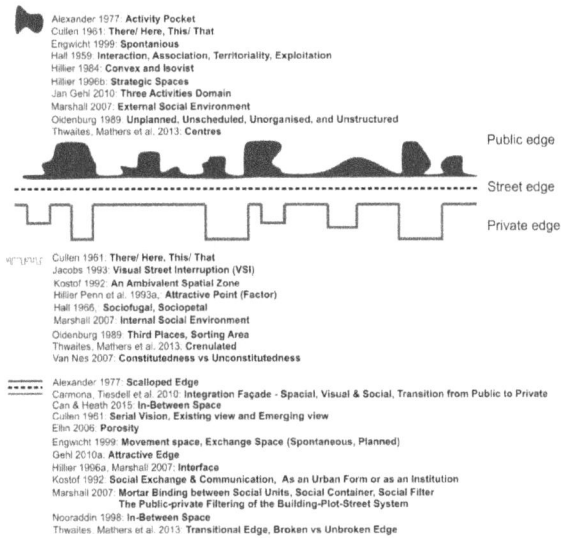

Figure 15.
Illustrating the main characteristics of the street edge and the edge of its private and public domains (based on a considerable number of authors and scholars). Source: Drawn by the author.

A public character is a person who frequently maintains contact with those who also use the same edge. Prosperous public places, according to Carmona et al. [27], are characterised by the frequent attendance of people in self-reinforcing ways. The public space is an optional and available environment, that people can choose whether to visit. Hillier [22] states that the street as a space for movement shapes the primary activity for those who prefer to stay or go.

Self-control can characterise public edge even though space is designed for public benefit. At the same time, it characterises private control by those who regularly use it. Zukin [37] suggests that there is not only social diversity but also the diversity of buildings and that helps to give a city its 'soul'. He mentions that, "the paradox of public space is that private control can make it more attractive, most of time, to a broader public, but state control can make it more repressive, more narrowly ideological, and not representative at all"([37], p. 158). Moreover, Jacobs [19] defined the 'eyes upon the street' where these eyes belong to people who contribute to shaping the property of the street [38–43] (**Figures 14** and **15**).

4. Three edges' characteristics in referring to Baghdad City

According to Hillier [22], in Arabic cities, diversity can be found in the development of the same underlying law. This tends to enable well-defined relationships between different levels of movement in the urban context. The old urban fabric seems quite complex in its street network, particularly within traditional Arabic cities; however, three domains play a crucial role in formulating the character of the street edge in such cities, namely street, private and public. Islamic cities are associated with what is called pre-Islamic regions, which inevitably have their own entities and identity regarding urban patterns, building typologies and construction techniques, besides, the natural and physical environment [44]. The ancient Mesopotamian model of clustered courtyard buildings, which date back to the 2500 B.C., provide evidence of the traditional settlement areas in other surrounding regions. Ur city is an ancient

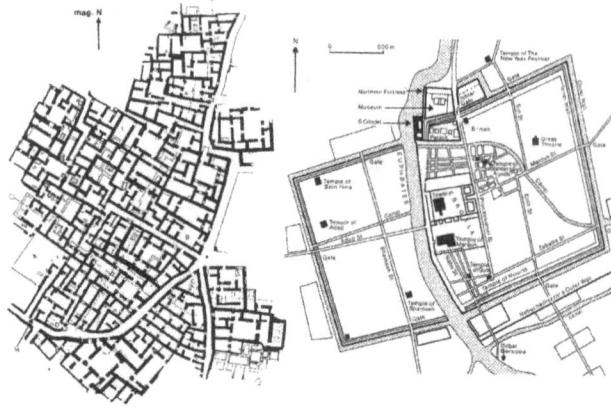

Figure 16.
Plan of a portion of the ancient Sumerian city of URas it was about 1900 B.C. (left). Plan of the city of Babylon at the height of its power, about 600 B.C. the religious features were dominant (right). Source: Lynch [45].

town situated to the south of Mesopotamia where its construction pattern matches the Islamic traditional cities that emerged later (**Figure 16**) [44, 46].

Urf is a systematic generative process (relating to the concept of habit or custom in English). *It* is a hidden order grounded in the consciousness of a community without the need to be listed, where every single member of the society is aware of what *Urf* is. Its principles and influences formed a pattern within traditional Arab and Islamic cities over time. *Urf* is initially based on human behaviours and the acceptance and satisfaction of these among a community, which generates these behaviours; these are compatible and match with Islamic rules. Otherwise, such individuals are rejected by society. Repeating the action means it becomes a habit, 'for every act there must be an impetus or reason... therefore every *Urf* is a habit, but not every habit is *Urf*" ([47], p. 110). *Urf* has become a source of legislation: it is flexible, changeable and dynamic in simulating the reality of life and its conditions. Recently, the term *Urf* could be pretty limited in serving and formulating public and private space use criteria. Unfortunately, this phenomenon is quite evident in a significant number of neighbourhoods of Baghdad. This paper tends to highlight the idea of *Urf,* apart from going deeply in explaining this concept in terms of implication.

Three factors, identified by Hakim [44], affected the nature of Islamic traditional settlements regarding their building patterns and planning. These are: (1) Pre-Islamic urban models and their people, culture and civilisations in territories that converted to Islam, where the norms and customs have continued their influences on the Islamic culture hitherto. (2) A transport pattern was made by the two-primitive means (camel and horse), which affected the street network patterns and the urban fabric of traditional cities between the fourth and sixth centuries A.D. (3) The surrounding natural environment embraced most Islamic regions located between latitudes 10 and 40. Thus, the microclimate was shared with the same analogical conditions.

The emergence of the Arabic/Islamic city was based on three processes. Firstly, it renewed an existing city founded in old colonial areas to meet the prerequisite for a social life among those people at that time. Secondly, they were pre-planned or planned cities, which were designed and pre-planned in accordance with Islamic rules and authorities. Historical resources and archaeologists confirm that the first primary planned city in the Islamic era was the round city of Baghdad, which was

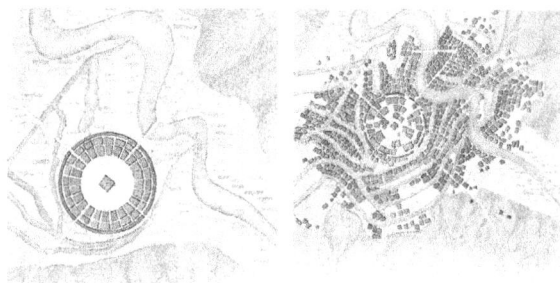

Figure 17.
Baghdad-Iraq, the geometric 8thCentury ground plan, organised around the Caliph's palace, was a casualty of the city's success. By the 9thCentury the sprawling growth of a thriving community had obliterated the original autocratic diagram. Source: Kostof [49].

situated to the east of the Islamic region [48, 49] (**Figure 17**). Thirdly, as a sponta-neous model, it can be identified as 'the most enduring and pervasive, and today most of the older areas of capitals and major towns in Muslim world evolved out of this model' ([44], p. 88).

However, Hakim ([50], p. 84) states that 'an important observation is that when colonialism ended, it left a gap between past and present and also left technology which did not evolve out of the past and has affected architecture considerably and in many ways, colonialism turned into cultural and technological dependency'. Consequently, serious negligence occurred with the introduction of new regulations for city planning. These new demands considerably affected the whole system of the old urban fabric; it was designed on a human scale and their needs aside from the large-scale urban spaces.

4.1 Street Edge's characteristics

The three street space components (street, private and public) integrate with the other components to provide one entity. The essential urban components that con-stitute the main character of the old urban fabric are the clustered courtyard build-ings, street networks and the hanging elements. Two predominant types of street networks are embedded within the old urban fabric. The first is the open-ended street, through which pedestrians publicly flow, and the second one the cul-de-sac, which is governed by inhabitants, is a private zone, and thus not normally permis-sible for other people to enter or to use this type of street [40, 50, 51] (**Figure 18**).

The old part of Baghdad is characterised by a maze of narrow streets continued, designed to meet the needs of pedestrians (**Figure 19**). The traditional pattern forms a more preferable sense of community, which appears serene and shadowed for the most significant part of *Zugag* during the day. Adjacent houses, *Zugag*, are varied in width; in some cases, these are no more than 3 m. While at the top, because of the *Shanashil* (prominent windows as a hanging or high-level protru-sion) the street was almost covered over. The main *Zugags* in the residential quarters of the old part of Baghdad are usually found on mosques and bazaars. This feature can also be observed in Arab cities [51–53] (**Figure 19**).

The hanging element is a '*high-level protrusion*' that can easily be seen during peregrination throughout the old urban fabric; the component was constructed above the street. This element has a unique name in the traditional area of Baghdad city '*Shanshul* (the plural is *Shanashil*)' is an oriel window. It is an upper-floor projection of a courtyard house, varied in size and shape in terms of ornamentation and decoration, and juxtaposed against the mass and shadow of the adjacent street [53, 54]. *Bridging the*

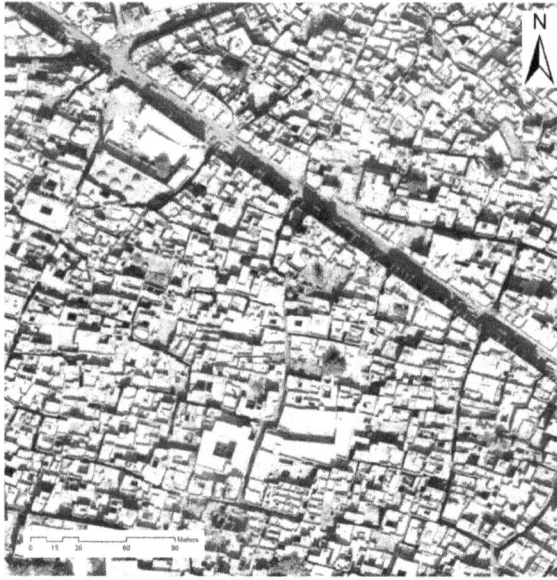

Figure 18.
The contemporary aerial scenery from the traditional part of Baghdad, al-Karkh. It shows the street pattern, cul-de-sacs and open-ended. Source: Prepared by the author based on the georeferencing aerial imagery authorised by R.S.GIS.U [51].

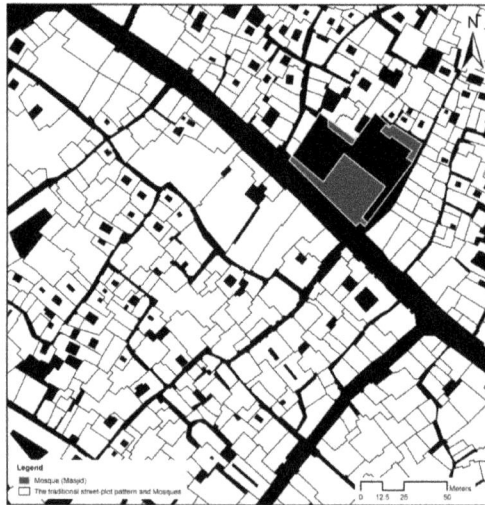

Figure 19.
A magnifier slice of the traditional part of Baghdad city, Al-Rusafa. It displays the traditional street pattern; Zugagsand surrounding neighbourhoods (Mahallahs). Source: Drawn by the author based on the georeferencing aerial imagery authorised by the G.I.S. Department [52] and R.S.GIS.U [51].

street denotes 'bridging the street, and the buttressing arches spanning between walls on either side of the street to provide structural strength and support' ([44], p. 89).

In traditional Islamic cities, a street refers to the central market. The street market on both sides is several repetitive small chambers that are opposite to each other and separated by about 10–20 feet. To enable pedestrian flow, the street is mostly covered by vaults that include skylights, which allow sunlight to pass through and

Figure 20.
Safafeer Suq (market) is one of the oldest traditional market where copper plaques and plates are attached to the shops. This Suq was delegated for copper works, but has since been occupied by the textile merchants, thus minimising the number of artisans who work withthe copper products. Source: Photographed by author's team, 04/December/2016.

protect the customers from undesirable climate conditions. Mostly, each street market is connected by the organic network of the narrow lanes or by other street markets. The other public facilities, such as mosques, baths, hotels or Khans are located close to shopping streets and thus, as an access network are maximally utilised ([44], p. 101, [55]). The traditional *Aswaq* (markets) are still alive in the old part of Baghdad, where each *Suq* is delegated for specific products and purposes, such as the Textile *Suq*, Book *Suq*, Copper *Suq*. The specialisation of functional uses is one of the leading characteristics in the old traditional markets, which are placed close to each other in a harmonic way (**Figure 20**).

4.2 Private edge characteristics

In Islamic cities, privacy is a central factor in determining the use of space; this includes direct visuals, particularly in residential areas. The cooperation between people and other institutions in formulating a generative system worked

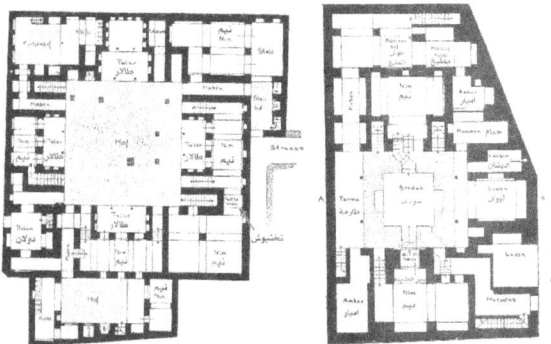

Figure 21.
Two Iraqi traditional courtyard house that illustrates the dogleg (broken) entrance that links the courtyard of the house and the street as a public space. Source: Reuther [58]. All right reserved for Al Warrak Publishing Ltd., London, UK.

to maintain the rhythm and hierarchy between the private and public domains [56, 57]. Furthermore, the Muslim community tends to be more concerned with preserving privacy, not only from physical connections but also in terms of visual contact. The privacy factor significantly affects the morphology of the urban form in Islamic/Arabic cities and gives a distinct shape to the city. For example, the external street edge contains the main dogleg entrance that leads to the courtyard house [58] (**Figure 21**). The dogleg technique gives a high level of privacy for inhabitants where there is no direct access to the private space from the public realm. Despite the fact that entrances are on opposite sides and directly adjacent to the street, no entry directly faces another.

In the residential area of the old part of Baghdad, the lower level of the external wall that is adjacent to the street is almost blind and as solid as a windowless wall to the outside. To attract lighting and ventilation in the courtyard house, all rooms are oriented inwards to the courtyard. Therefore, the external façade lacks apertures except, occasionally, small niches beside the upper level that are designed with *Shanashil*. The *Zugag* exhibits simple façades with a minimum of details at the lower level; instead, rich detail and decoration is placed on the *Shanashil* and main entrance (**Figure 22**) [55].

Moreover, to avoid straight visual connections, people in traditional cities tend to adopt the overlooking technique in setting doors, windows, openings and heights, where 'in Islamic culture, protection from visual intrusion into the private realm of houses was the paramount consideration. Views were appreciated when available, but they took second place to the blocking of visual corridors into the private realm' ([59], p. 29). It allows for inhabitants to observe outdoor activities and pedestrian movement, but those who use the street were not able to see inside properties. This technique used the concept of *Shanshul/Shanashil* as the external element of the *Zugag* (local streets within traditional neighbourhoods) in the traditional area of Baghdad.

Figure 22.
The traditional Baghdad Zugag where the street level tends to have a windowless and solid frontage (left). The main entrance of the courtyard house with the richness in decoration and detail (right). Source: Makiya [55]. All right reserved for Al Warrak Publishing Ltd., London, UK.

Figure 23.
Shanashil; as a serrated row of oriel windows. Source: Makiya [47].

Shanashil were made up of smaller, modular, sash-window units; they are attractive architectural elements employed to promote the external edge of the *Zugag* (street). *Shanshul* includes wooden sliding windows and produces extra shadow for pedestrians against the direct sunlight, particularly in the summer season. Furthermore, the *Shanshul* technique traditionally plays a significant role in social interactions and allows inhabitants to conduct conversations through opposing rooms on the upper floor (**Figure 23**) [54].

The concept of *bridging the street* also has been observed in the traditional area of Baghdad but did not spread widely, like *Shanashil*. Technically, this type of high-level protrusion belongs to one owner or exploits links between two properties that belong to the same inhabitant (**Figure 24**). In the non-residential property, they are employed for public use and have the same characteristics. Traditional shape complements the street pattern and the socio-physical structure of Baghdad, for instance, mosques and hammams. These types of buildings are oriented entirely towards the internal courtyard.

A street provides a distinction between the private and public space in the traditional area of Baghdad; it is very controlled and restricted regarding the degree of permeability, transparency, accessibility and connectivity. The street is almost

Figure 24.
The concept of bridging the street in the traditional area of Baghdad, but it does not spread commonly like Shanashil. Source: Reuther [58]. All right reserved for Al Warrak Publishing Ltd., London, UK.

solid on the ground floor and semi-closed or closed by *Shanashil* on the first floor. The house entrance and *Shanashil* form the only two channels to link between the private streets. In the modern context of Baghdad, within the residential area, the private edge varies from direct adjacency with the street to set backwards. The differences in street pattern, plot layout and block size, and the location of the building within the plot area play a key role in formulating the spatial organisation and provide distinct characteristics for each area of Baghdad (**Figure 25**).

The characteristic of the private edge in the modern pattern of the street network has different criteria and considerations. This leads to different interpretations of the private edge and the extent to which inhabitants have the authority to claim the juxtaposing space located in front of their property. It also influences the boundary of the street width, and to what extent it is for public use. The absence of a clear definition for the private, public and street edges, particularly in commercial streets which broke through the traditional area, has resulted in complicated situations and difficulties in how to manage this critical area of Baghdad (**Figure 26**).

4.3 Public edge characteristics

The public edge formulates the vitality of the street, where it enables people to interact either with the street edge or with other people. Tolerance depends on different criteria and rules, besides the norms of society (values and *Urf*). In the traditional area, the norms and *Urf* can be realised as concealed orders indoctrinated in the consciousness of society without the need for documentation. People realise the system of norms and *Urf* and then accordingly, shape their behaviour. The concept of *Urf* is related to the traditional area of Arabic/Islamic cities, where these types of areas were normally based on a set of treaties accepted by people.

The idea of *Al-Fina* can be addressed as one of the public edge's characteristics in the old traditional part of Arabic/Islamic cities, such as Baghdad. *Al-Fina* is a spatial element that distinguishes the street edge and interior courtyard of a house.

Figure 25.
The modern context of Baghdad within the residential area, the private edge manifests an important transformation which is adjacent with the street edge towards the back. The differences are apparent in street pattern, plot layout and block size, and the location of the building within plot area. Source: Prepared by the author based on the georeferencing aerial imagery authorised by R.S.GIS.U [51].

Figure 26.
The absence of a clear definition of the private, public and street edge, particularly in commercial streets that go through the traditional area, have resulted in complicated situations where sellers deliberately present their products within the street arcades. Al-Rasheed street. Source: Photographed by author's team, 21/ December/2016.

It is located immediately adjacent to the peripheral exterior wall, opposite the street space. It serves daily and temporary uses without a need to own the space [60]. Moreover, the combination of zigzag and a string of narrow and wider areas along one street provides visible evidence of the design of a traditional city in the Islamic/ Arabic world (**Figures 27** and **28**).

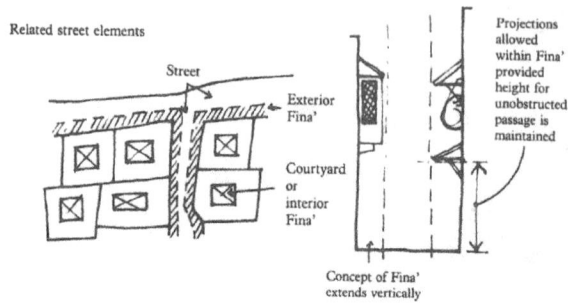

Figure 27.
al-Fina is one of the leading characteristics of traditional Arab/Islamic cities. It refers to different purposes which both private and public domains can benefit from, but it is never be owned by anyone. Source: Hakim [60].

Figure 28.
A street in the traditional part of Baghdad, Al-Karkh. Note the steps of the houses on the left of the picture, and verandahs on the upper level; both are located within the Al-Fina domain, besides having other hanging features. Even the car stop is subject to the same concept, despite the limitations of public edge. Source: Photographed by author's team, 05/December/2016.

Al-fina is completely changed in the modern neighbourhoods, where the built-up area is placed with the frontage set back. However, the area adjacent to the front wall of their properties is still used by people for different purposes, meaning that the authority of *Al-Fina* has been adopted differently. It might be recognised as a type of soft territorial space, 'if territories are relatively small (garden or house versus park or apartment building, for example), and if they can be modified or maintained with modest effort, then it is easier for individuals or small groups to achieve control' ([45], p. 213) (**Figure 29**).

Furthermore, the concept of the '*in-between*' space is used by Nooraddin [18] to denote a transitional milieu that mediates between the street and private space often in Arabic/Islamic traditional cities. In fact, there are no hard barriers between the in-between space and the street. The in-between space is generated by the consumer of street space, 'In between phenomena, how it was organised

Figure 29.
Two examples from modern neighbourhoods in Baghdad exhibits how people pretend to make territorial space at the front wall of their properties. Partially covered area (left). Entirely paved by the inhabitant and using a ramp to access to the house located outside the private edge (right). Source: Photographed by author's team, 12–13/December/2016.

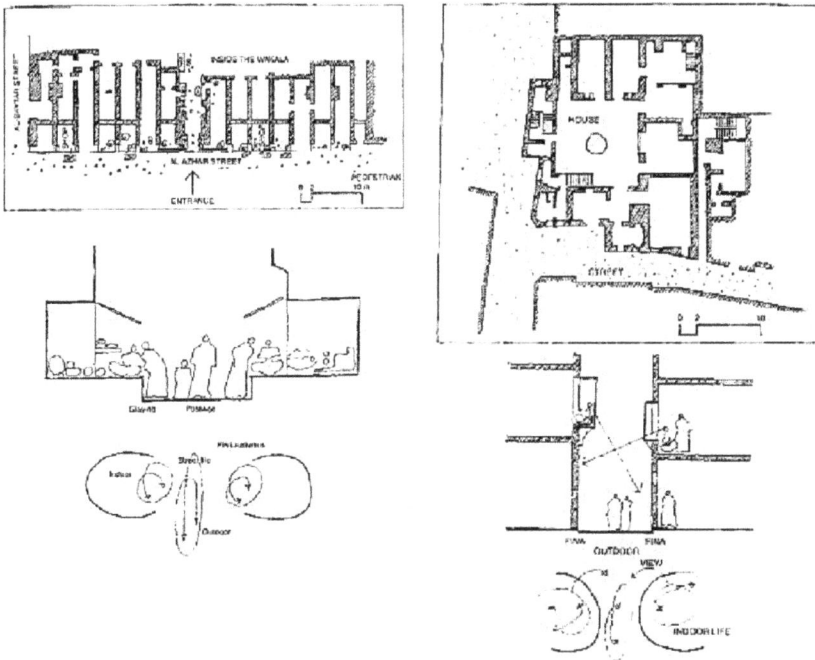

Figure 30.
Left, reconstruction of the in-between spaces of a commercial street in traditional Islamic Arabic cities. Right, reconstruction of the in-between spaces of a residential street in traditional Islamic Arabic cities Nooraddin [18].

in the old Islamic cities and how it contributed to the character of their street environments' ([18], p. 66). The in-between space is mainly located in the front of the private area where it is used as a gathering space and for different activities. This type of space has loose meaning that there is no specific shape to give it a final form. Instead, it is flexible in both investment and appearance. It enables people

Figure 31.
Al-Mutanabbi as one of the more traditional vital streets in the heart of the old part of Baghdad. It is vivid in attracting a massive amount of people from different gender, ages and for various purposes. It is the main resource for books, publishing, stationery, and knowledge exchange. Source: Photographed by author's team, 21/ December/2016–202106/January/2017.

to meet their needs and desires as much as possible regarding comfortable climate, religion, lifestyle, community and cultural aspects. Two types of in-between space are defined as: (1) related to the commercial street, and (2) located on the residential street (**Figures 30** and **31**) [18].

According to Hall [35], the in-between space shapes the microcultural theme, where it attracts the people to share the same territorial area. This notion, to a large extent, is rooted in the old part of cities. Hall [35] distinguishes three types of proximate behaviours that manifest in a space – *Infracultural*: behaviour rooted in the human biological past, *Precultural*: the physiological level in the present, *Microcultural*: based on which most proxemic observations achieve. The *Microcultural* pattern encompasses three aspects (buildings, space and the distances maintained in encounters with others) which define territorial patterns, '... in every sense of the word an extension of the organism, which is marked by visual, vocal, and olfactory signs. Man has created material extensions of territoriality as well as visible and invisible territorial markers ([35], p. 103).

Can and Heath [17] use the term *In-between* to study social interaction and the morphological form of a city. They examine spatial configurations that occur in different street patterns: traditionally and modern. In traditional Islamic and Arabic cities, the *In-between* space reflects a social interaction between neighbours where it offers a niche within the street edge and in turn, improves the street life. The pockets of activity play a significant role in shaping a live space; it mediates activities and the path as an in-between area in order to create an attractive space for people to pause and get involved [12].

Kostof [29] states that public spaces are defined by residual, interstitial spaces located between neighbourhoods' cells, such as bazaars or *Aswaq* and *Maidans* (public squares). The contrast with Western urbanism lies in the fact that it pays more attention to the street system and public spaces. The urbanism process in traditional Arabic/Islamic cities is based on the inside to outside, and is understood as a bottom-up approach, or, in other words, from private tendencies to public propensities. In Arabic/Islamic cities, the sense of public space is defined in soft boundaries rather than hard borders, where the users and visitors share the same norms and values by investing in the public spaces. Kostof ([29], p. 127) states that 'regardless of the private use of these resources, they could never be privately owned. Every member of society had equal claim to public places, be Muslim or

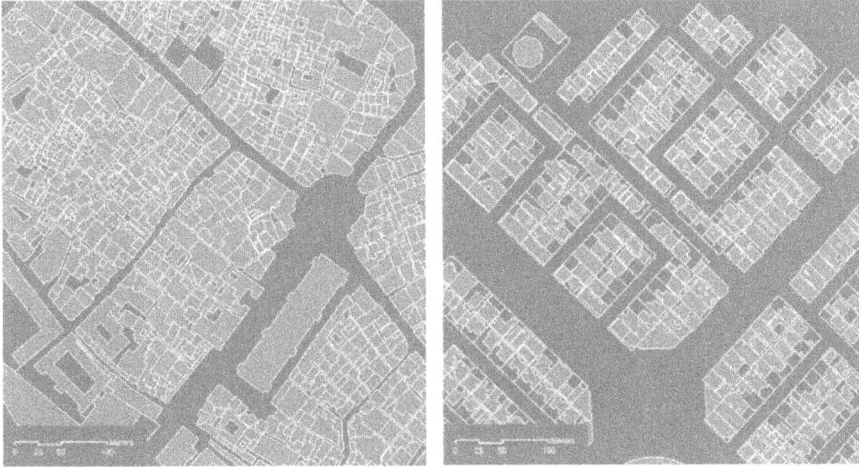

Figure 32.
Two symmetric scale of street patterns in Baghdad (left: Case A) and (right: Case D) show the dramatic transformation in the concept of Suq (Aswaq), from organic theme to loop-grid where the movement is changed from the spontaneous streamlined to a planned change direction. Source: Drawn by the author based on the georeferencing aerial imagery authorised by the G.I.S. Department [52] and R.S.GIS.U [51].

non-Muslim. Whoever comes earliest to a public place has the right to make use of it through that day'. The *Suq* (the Arabic plural is *Aswaq*), in traditional areas of Baghdad, derives from the organic street pattern as there is no dramatic change between the *Suq* space and other networks, but rather movement is spontaneous and streamlined. The transformation is not based on the morphology of the space and its distinct characteristics, but also on the functional pattern of the street. This, in turn, results in a new vista with each movement (**Figure 32**).

A *Suq* is a crucial morphological urban element that is spontaneously subjected to the hierarchy of location. *Aswaq* were organised in different ways, whilst the linear *Suq* functions as a continuum spatial route, where both its sides have opposite shops. As an area, the *Suq* denotes a series of back-to-back rows that are situated to face each other, whilst the units of a *Suq* are located against the perimeter of buildings [61, 62] (**Figure 33**). The pattern of a *Suq* and its spatial configuration came to exist as an assortment of different types of *Aswaq* (Arabic plural). It had various functions in order to serve a significantly sized community within the same scope, where *Aswaq* were located to be proximate and adjacent to each other. The proximate pattern of

Figure 33.
Spatial configuration of Suq pattern in the traditional area of Baghdad; Al-Safafeer Suq (left) and Al-Mutanabbi Street (right). Source: Photographed by author's team, 04/December/2016–21/December/2016.

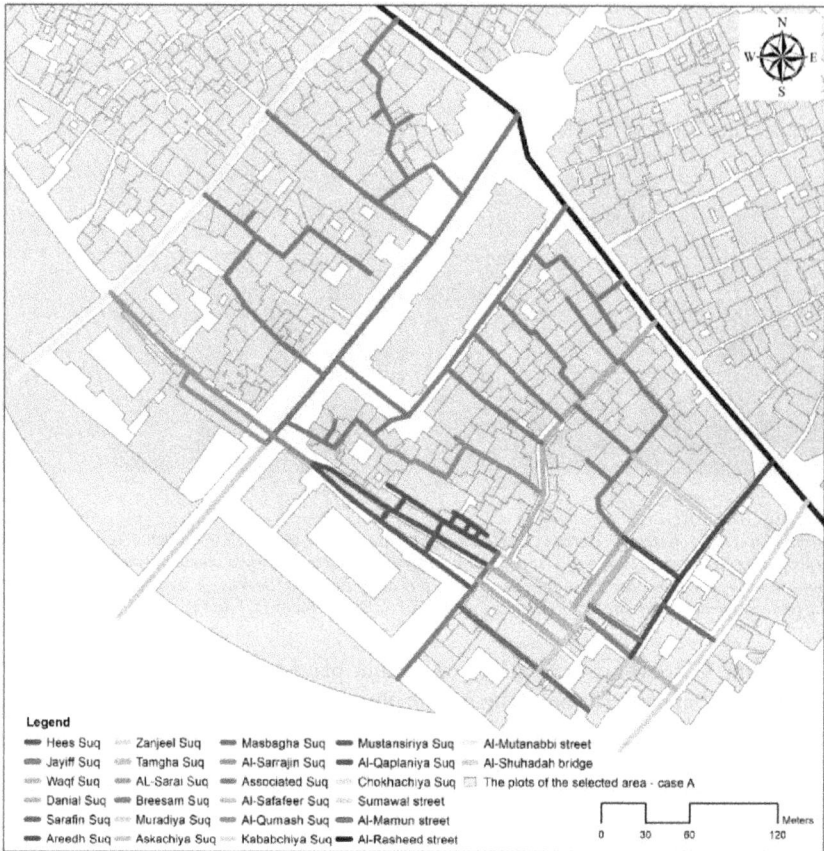

Figure 34.
Sample of the pattern of a Suq with its spatial configuration came into existence as an assortment of different types of Aswaq. This aimed to provide various functions in order to serve a significant amount of the community within the same area. They were located near and/or next to each other. Source: Drawn by the author based on the georeferencing aerial imagery authorised by the G.I.S. Department [52] and R.S.GIS.U [51].

distribution of *Aswaq* provides a distinct morphological form in the traditional area of Baghdad. The proximity enables people to combine shopping and viewing the sights (**Figure 34**).

Functional proximity is often one of the important criteria for the closeness of the *Aswaq*, for example, *Al-Mutanabbi* street is designated for bookshops, publishing and stationery storage; moreover, it is close to *Al-Sarai Suq* to provide stationery and books (**Figure 34**). Along with the value of proximity, the pattern of the street in this part is more complicated and based on the organic network, which developed spontaneously. The proximity governs the location of the *Aswaq* and the other social facilities, such as the *Masjid* (mosques), *Gahwah* (café), *Hammam* (public bath) and *Sahaht* (squares), where people are able to access different activities. This sense of proximity, to a large extent, is lost in some modern neighbourhoods in Baghdad, with their new street patterns that were established according to car-based movements. This resulted in minimal social interaction, and it maximised the distance between settled units and the *Aswaq*, besides other facilities and services (**Figures 32** and **34**) [63, 64].

Proximity, in this regard, is based on human demands, regarding accessibility and connectivity. This considers '*where*' as the settled units in which people live,

Figure 35.
Al-Sahah (Sahaht–as plural), Baghdad, al-Karkh (case B -left). It emerges spontaneously and is embedded in the urban fabric and customarily located where two or more streets come together to form a connective space. Case D (right) is a new modern pattern which exhibits fragmented squares as superfluous. Source: Drawn by the author based on the georeferencing aerial imagery authorised by the G.I.S. Department [52] and R.S.GIS.U [57].

and *'there'* where their needs are located. The traditional part of Baghdad, like other Arabic/Islamic cities, emerged initially from a bottom-up strategy, where the community had the authority to shape and reshape the built environment so that it harmonised with their needs [56, 65]. In the traditional area of Baghdad, Al-Ashab [53] refers to another morphological element that characterises the urban context of the street pattern in this area, namely, *Al-Sahah (Sahaht* – as plural). It emerged spontaneously within the urban fabric and is normally placed where two or more streets come together to shape the space of connection. This space has been used as a meeting place for neighbours. *Sahah* within *Mahallahs* (neighbourhoods) were full of life and attractive as social interaction spots that eventually spread through the traditional urban fabric [53].

Sahah space refers to several *Sahaht* that are varied regarding the size; it ranged from the more private space, such as the courtyards of traditional houses, tertiary *Sahah*, and more public areas like sub-*Mahallah* or secondary *Sahah* and the primary *Sahah*. The hierarchy of accessibility from small to the large *Sahah* was perceived both by those who lived there as well as visitors (**Figure 35**) [53]. In a new urban context, where the squares (*Sahaht*) develop from a planned process, open spaces are meaningless and void from any common function. They, however, enable the unnecessary physical expansion of the city, and the sense of human scale is lost as their geometrical dimensions are not subjected to other surrounding urban elements; thus, the enclosure is also missing. In the modern pattern, squares fall out of the authority of inhabitants, as there is no explicit declaration about the claim for this type of area. Moreover, there is a deficiency in determining the nature of use, even though they are designed for public use (**Figure 35**).

5. Conclusion

Defining the street edge was the primary aim of this paper in order to highlight the different interpretations and meanings of the three fundamental elements that function together to formulate the street. These elements are the street, the private edge and the public edge. The transformation in the urban structure from the traditional pattern to the modern model not only changes the morphological dimension but also influenced the relationship between the private and public realm. The manipulation of private and public relations could be the primary condition to figure out the street life and how people interact with each other. Hence, different variables could be employed to measure the relationship between the private realm and the street space at a micro-level. These factors are experienced by those who use a street when dealing with the street scope within a specific segment. The permeability, inter-visibility, connectivity and accessibility are different between the traditional area and the modern parts.

The notion of the private-public was examined to investigate the street's edges. Each realm was addressed in detail by emphasising the basic morphological process based on the edge's characteristics. The critical interrelationship between two edges: private and public represents the micro-level of a street segment that is used to evaluate the interrelationship and how could affect street life and social interactions along the street edge. Across the traditional pattern and modern model of the neighbourhood, there was a significant disparity between the private and public edge.

People have a set of expectations when they determine their interactions with individual action or collective behaviour in a particular street edge. Classifying the street space into three edges is an essential method in order to understand human behaviour thoroughly and how could people respond to each other and the three edges: street, private and public. Fine-scale is another aspect to address the physical environment at the street scale; also, the micro-level could be one of the strategies to deal with the street parameters in terms of the ability of the different edges in managing human behaviour. Indeed, there is a need to distinguish between two patterns; the first one is based on the bottom-up approach as a spontaneous pattern, and the second is the up-down method as a pre-planned model. Once understand the differentiation of traditional area (spontaneous) and new neighbourhood (pre-planned), it would be there a thoughtful procedure to deal with the private and public edge.

There is a lack of required building legislation and maintenance monitor programs for planning and urban design, including a lack of commitment to restrict initiatives to assure they conform to traditional patterns. Therefore, addressing the central gap means verifying the most critical indicators of the street edge problem, both in traditional and modern patterns, which necessitates the detection of related studies that try to link the urban form with active-ties and human behaviour alike. The traditional region emerged spontaneously apart from the notion of land use or zoning diagram. This characteristic is a crucial point within urban development schemes. The street pattern and paradigm of the buildings in the area are intricate; accordingly, there is a need to develop particular standards and regulations to preserve the identity of old Baghdad and understand the contemporary objectives of the new pattern. In this regard, proffering more further consideration to the centre of Baghdad is required, meaning that the control of this traditional region ought to be studied and systematic to advanced quality of life and to improve urban sustainability.

Terminologically, urban symbiosis could be aligned with sustainability, but this term can cover what is related to human behaviour and street activity. The main aim of symbiosis is to create a high interaction productive relationship between creatures. The notion of symbiosis holds three kinds: Commensalism, Parasitism and Mutualism. To comprehend the fundamental system in the traditional pattern of Baghdad city, understanding the micro-level of activity and fine-scale of the urban fabric is required to form a symbiotic platform. Including the processing of the morphology and the relationship between the plot, block and street network, also the symbiotic relationship between the public and private domains. European experiences in dealing with outdoor activities and how people respond to the street edge have been highlighted thoroughly by different scholars and significant studies. People effectively experience this urban knowledge and urban life at the micro-level of the street edge in participating in the fine characteristics of such edge. In this respect, Jan Gehl's school could be one of the more significant experiences in dealing with street life.

Acknowledgements

I would like to express my gratitude for the thoughtful guidance from my supervisor Professor Sergio Porta, from the Urban Design and Director of UDSU at the Urban Design Studies Unit, Department of Architecture, University of Strathclyde, sergio.porta@strath.ac.uk

Author details

Haider Jasim Essa Al-Saaidy
Department of Architectural Engineering, University of Technology, Iraq

*Address all correspondence to: haider.j.essa@uotechnology.edu.iq

IntechOpen

References

[1] Al-Saaidy HJE. Lessons from Baghdad City Conformation and Essence. In: Almusaed A, Almusaed A, Truong-Hong L, editors. Sustainability in Urban Planning and Design. London, UK: IntechOpen; 2020. pp. 387-417. DOI: 10.5772/intechopen.88599

[2] Marshall S. Streets and Patterns. New York: Spon; 2005

[3] Al-Akkam AJ. Urban characteristics: The classification of commercial street in Baghdad City. Emirates Journal for Engineering Research. 2011;**16**:49-65

[4] Al-Saaidy HJE. Urban Form Elements and Urban Potentiality (Literature Review). Journal of Engineering. 2020;**26**:65-82. DOI: 10.31026/j.eng. 2020.09.05

[5] Al-Saaidy HJE. Urban Morphological Studies (Concepts, Techniques, and Methods). Journal of Engineering. 2020;**26**:100-111. DOI: 10.31026/j.eng. 2020.08.08

[6] Al-Saaidy HJE. Measuring Urban form and Urban Life: Four Case Studies in Baghdad, Iraq," Doctor of Philosophy. United Kingdom: Architecture, University of Strathclyde, Department of Architecture; 2019

[7] Al-Saaidy HJE, Alobaydi D. Studying street centrality and human density in different urban forms in Baghdad, Iraq. Ain Shams Engineering Journal. 2021;**12**:1111-1121. DOI: 10.1016/j.asej. 2020.06.008

[8] Al-Saaidy HJE, Alobaydi D. Measuring geometric properties of urban blocks in Baghdad: A comparative approach. Ain Shams Engineering Journal. 2021;**12**:3285-3295. DOI: 10.1016/j.asej.2021.04.020

[9] Bianca S. Urban form in the Arab World: Past and present. Vol. 46. USA: vdf Hochschulverlag AG; 2000

[10] Chen H, Jia B, Lau B. Sustainable urban form for Chinese compact cities: Challenges of a rapid urbanized economy. 2008;**32**:28-40. DOI: 10.1016/j.habitatint.2007.06.005

[11] Gehl J, Svarre B. How to Study Public Life. USA: Island Press; 2013. DOI: 10.5822/978-1-61091-525-0

[12] Alexander C. A Pattern Language : Towns, Buildings, Construction. New York: Oxford University Press; 1977

[13] Hillier B, Penn A, Hanson J, Grajewski T, Xu J. Natural movement-or, configuration and attraction in urban pedestrian movement. Environment and Planning B-Planning & Design. 1993;**20**:29-66. DOI: 10.1068/b200029

[14] Engwicht D. Street Reclaiming: Creating livable Streets and Vibrant Communities. British Columbia: New Society Publishers; 1999

[15] Remali AM. Capturing the Essence of the Capital City: Urban form and Urban Life in the City Centre of Tripoli, Libya. Department of, Architecture, University of Strathclyde; 2014

[16] Van Nes A, López MJ. Micro scale spatial relationships in urban studies: the relationship between private and public space and its impact on street life. In: Proceedings of the 6th Space Syntax Symposium (6SSS), Istanbul, Turkiye, June 12-15, 2007. Istanbul, Turkiye; 2007

[17] Can I, Heath T. In-between spaces and social interaction: a morphological analysis of Izmir using space syntax. Journal of Housing and the Built Environment. 2016;**31**:31-49. DOI: 10.1007/s10901-015-9442-9

[18] Nooraddin H. Al-fina', in-between spaces as an urban design concept: making public and private places along streets in Islamic cities of the Middle East.

Urban Design International. 1998;**3**:65-77. DOI: 10.1080/135753198350532

[19] Jacobs J. The Death and Life of Great American Cities. New York: Random House; 1961. DOI: 10.4324/978191 2282661

[20] Marshall S. Cities Design and Evolution. New York, NY: Routledge; 2007

[21] Thwaites K, Mathers A, Simkins I. Socially Restorative Urbanism: The Theory, Process and Practice of Experiemics. UK: Routledge; 2013

[22] Hillier B. Cities as movement economies. Urban Design International. 1996;**1**(1):41-60. DOI: 10.1057/udi.1996.5

[23] Penn A, Hillier B, Banister D, Xu J. Configurational modelling of urban movement networks. Environment and Planning B-Planning & Design. 1998;**25**:59-84. DOI: 10.1068/b250059

[24] Hillier B. Space is the Machine: A Configurational Theory of Architecture. Cambridge; New York, NY, USA: Cambridge University Press; 1996a

[25] Buchanan P. Facing up to facades: a report from the front. Architects' Journal, UK. 1988;**188**:21-56

[26] Segall MH, Campbell DT, Herskovits MJ. The Influence of Culture On Visual Perception. USA: Bobbs-merrill; 1966

[27] Carmona M, Tiesdell S, Tim H, Oc T. Public Places, Urban Spaces: The Dimensions of Urban Design. Amsterdam; Boston: Architectural Press/Elsevier; 2010

[28] Carmona M. The existential crisis of traditional shopping streets: the sun model and the place attraction paradigm. *Journal of Urban Design*. 2021:1-35

[29] Kostof S. The City Assembled: The Elements of Urban form Through History. London: Thames and Hudson; 1992

[30] Canter DV. Environmental Interaction: Psychological Approaches to our Physical Surroundings. London, UK: London: Surrey University Press: Distributed by International Textbook Co.; 1975

[31] Banerjee T. The future of public space: Beyond invented streets and reinvented places. Journal of the American Planning Association. 2001;**67**:9-24. DOI: 10.1080/01944360108976352

[32] Carmona M. The "public-isation" of private space–towards a charter of public space rights and responsibilities. Journal of Urbanism: International Research on Placemaking and Urban Sustainability. 2021;**19**:1-32. DOI: 10.1080/17549175.2021.1887324

[33] Moirongo BO. Urban public space patterns: human distribution and the design of sustainable city centres with reference to Nairobi CBD. Urban Design International. 2002;**7**:205-216. DOI: 10.1057/palgrave.udi.9000083

[34] Oldenburg R. The Great Good Place: Café, Coffee Shops, Community Centers, Beauty Parlors, General Stores, Bars, Hangouts, and How They Get You Through The Day. New York, USA: Paragon House Publishers; 1989

[35] Hall ET. The Hidden Dimension. New York, USA: Doubleday; 1966

[36] Sennett R. The Uses of Disorder: Personal Identity & City Life. London: London: Faber; 1996

[37] Zukin S. Naked City: The Death and Life of Authentic Urban Places. Oxford: Oxford University Press; 2010

[38] Hall ET. The Silent Language. New York, USA: Doubleday; 1959

[39] Cullen G. The Concise Townscape. London; Boston: Butterworth Heinemann; 1961

[40] Hillier B, Hanson J. The Social Logic of Space. Cambridge Cambridgeshire; New York: Cambridge University Press; 1984. DOI: 10.1017/cbo9780511597237.004

[41] Jacobs AB. Great Streets. Cambridge, MA: MIT Press; 1993

[42] Ellin N. Integral Urbanism. New York, USA: Taylor & Francis; 2006

[43] Gehl J. Cities for People. Washington: Island Press; 2010a

[44] B. S. Hakim, "Islamic architecture and urbanism " in Encyclopedia of Architecture: Design, Engineering & Construction. vol. 3, T. P. Robert and A. W. Joseph, Eds., ed: New York: Wiley, 1989, pp. 86-103.

[45] Lynch K. Good City Form. Cambridge, MA; London: MIT Press; 1984

[46] Woolley LS. Ur: the First Phases. London: Penguin Books, New York; 1946

[47] Hakim BS. The" Urf" and its role in diversifying the architecture of traditional Islamic cities. Journal of Architectural and Planning Research. 1994;**11**:108-127

[48] Hakim BS. Arab - Islamic urban structure. The Arabian Journal for Science and Engineering. 1982;7:69-79

[49] Kostof S. The City Shaped : Urban Patterns and Meanings Through History. London: Thames & Hudson; 1999

[50] Hakim BS. The Islamic city and its architecture: a review essay. In: Third World Planning Review. Vol. 12. Liverpool University Press; 1990. pp. 75-89. DOI: 10.3828/twpr.12.1.8t5440243m754365

[51] R.S.GIS.U. The Georeferencing Aerial Imagery. Baghdad, The official letter,

No.: 1578, Date: 01/11/2017: Remote Sensing and GIS Unit, Building and Construction Engineering Department, University of Technology; 2017

[52] G.I.S.Department. The Traditional Area of Baghdad based on Polservice, Geokart, Poland, Rectified by Dep. of GIS. Department of Geographic Information System GIS, Mayoralty of Baghdad, ed: The official letter, No.: O.P.U 420, Date: 24/10/2016 issued by University of Technology, Office of the President. 2016

[53] Al-Ashab KH. The Urban Geography of Baghdad. UK: University of Newcastle Upon Tyne; 1974

[54] Fethi I. Urban Conservation in Iraq: The Case for Protecting the Culteral Heritage of Iraq with Special Reference to Baghdad Including a Comprehensive Inventory of its Areas and Buildings of Historic or Architectural Interest. UK: University of Sheffield; 1977

[55] Makiya M. Baghdad. London, UK: Alwarrak; 2005

[56] Hakim B. Generative processes for revitalizing historic towns or heritage districts. Urban Design International. 2007;**12**:87-99. DOI: 10.1057/palgrave. udi.9000194

[57] Hakim BS, Rowe PG. The representation of values in traditional and contemporary Islamic cities. Journal of Architectural Education. 1983;**36**:22-28. DOI: 10.1080/10464883.1983.10758321

[58] Reuther O. Das Wohnhaus in Bagdad und Anderen Stadten Des Iraq (Arabic: The Iraqi House In Baghdad and Other Iraqi Cities). London, UK: Alwarrak (Arabic), Verlag Von Emst Wasmuth - AG Berlin; 2006. p. 1910

[59] Hakim BS. Mediterranean urban and building codes: origins, content, impact, and lessons. Urban Design

International. 2008a;**13**:21-40. DOI: 10.1057/udi.2008.4

[60] Hakim BS. Arabic - Islamic Cities: Building and Planning Principles. UK: Marston Gate; 2008b

[61] Morris AEJ. History of Urban Form: Before the Industrial Revolutions. 3rd ed. Harlow, Essex, England: New York: Longman Scientific & Technical: New York: Wiley; 1994

[62] Oliveira V. Urban Morphology: An Introduction to the Study of the Physical Form of Cities. Cham: Springer International Publishing; 2016. DOI: 10.1007/978-3-319-32083-0

[63] Al-Azzawi SHA. A Descriptive, Analytical and Comparative Study of Traditional Courtyard Houses and Modern Non-Courtyard Houses in Baghdad. UK: University College, University of London; 1984

[64] Al-Hasani MK. Urban space transformation in old city of Baghdad – Integration and management. Megaron. 2012;7:79-90

[65] Mohareb NI. Land use as a sustainability indicator for Arab cities. Proceedings of the Institution of Civil Engineers-Urban Design and Planning. 2010;**163**:105-116. DOI: 10.1680/udap. 2010.163.3.105

Chapter 12

Models and Strategies for the Regeneration of Residential Buildings and Outdoor Public Spaces in Distressed Urban Areas: A Case Study Review

Alessandra Battisti, Livia Calcagni and Alberto Calenzo

Abstract

Given three-quarters of the European population living in urban areas, cities are expected to deliver sustainable growth if they will be able to further thrive and grow, while improving resource use and reducing pollution and poverty, as highlighted also by Sustainable Development Goal 11. In the context of vulnerable and marginal areas within cities, which suffer from multiple deprivations, regeneration processes at the building and district-scale play the most significant role in making cities more inclusive, sustainable and resilient. Reuse and refurbishment strategies, measured building replacement and stratification, redevelopment and enhancement, nature-based solutions and bioclimatic technological devices, are all tools for an integrated regeneration process capable of stimulating the urban metabolism and act as a driving force for the self-regeneration of the city. A comparison of two different building typologies, brought about by a review of existing public housing case studies in the outskirts of Rome, Italy, allowed us to define efficient, sustainable strategies and guidelines, that can be adapted to similar contexts in terms of building typology, social and economic conditions and of relationship to the rest of the city.

Keywords: urban regeneration, distressed urban area, sustainable technologies, public housing, energetic retrofitting

1. Introduction

The 2030 Agenda adopted by the United Nations General Assembly in 2015 and defined by the subscribing members as "a plan of action for people, planet and prosperity" has identified 17 goals in order "to take the bold and transformative steps which are urgently needed to shift the world onto a sustainable and resilient path" [1]. The goals refer to different fields of social and economic development and must be addressed through an integrated approach, aimed at achieving sustainable progress. The United Nations Inter Agency Expert Group on SDGs (UN-IAEG-SDGs) has developed 169 global targets, and 234 indicators that have to be monitored—as a global reference framework—in the period 2015–2030. In particular, Goal 11 deals with the urban sustainability issue and emphasizes how cities play an essential role in achieving the Sustainable Development Goals since half of the world population and three-quarters of the European population live in urban areas. All over the world, cities are responsible for the largest share of energy consumption and carbon emissions, for the growing pressure on the environment and the related public health issues [2].[1] The governance of urban space, therefore, represents a crucial development factor capable of posing worldwide challenges and opportunities. Several aspects must be considered in a systemic, inclusive and integrated way to ensure that cities thrive in a sustainably. It is vital to ensure that the population living, working or passing through the city has access to mobility, quality housing and safe conditions, both in terms of structural stability of public and private buildings and infrastructures, and protection from crime, violence and harassment.

Moreover, the presence of green spaces and public spaces, the protection of the cultural and natural heritage, the redevelopment of run-down areas, the relationship between the city and peri-urban and rural areas are as crucial as the aspects mentioned before. Yet, to be able to proceed in this direction it is essential to work according to an integrated approach that addresses the physical and structural aspects of the city, as well as the intangible ones. These last ones range from social and cultural aspects to those related to work and local economies, within broader processes that activate latent or already existing projects and social energies, which very often require policies from below. This process has already been triggered with the 2007 Leipzig Charter together with the related integrated urban development strategies that at a national, regional and local level focused on the cultural and architectural qualities of cities, conceived as strong tools for social inclusion and economic development useful to positively affect economic prosperity, social balance and the environment, within a coordinated process between spatial, sectorial and temporal aspects of urban areas. This process continued with the Toledo Declaration of 2010, which suggested a transversal, multidimensional and holistic design approach to achieve multiplying, complementary and synergistic effects, solving conflicts and finding the right balance between temporal (short, medium, long term) and spatial (region, metropolitan area, city, neighborhood) needs. These recommendations are reiterated and strengthened in the newborn Renovation Wave strategy, part of the European Green Deal promoted by Brussels which places the redevelopment of the building stock in a relevant position as an essential measure for decarbonization and reduction of emissions and as a tool for boosting the economy and European competitiveness. The new Renovation Wave strategy aims to double the urban regeneration rate, currently at 1%. According to Brussels estimates, a significant share of 35 million renovated and regenerated buildings could be reached by the end of the decade [3]. This situation would lead not only to significant ecological and energy benefits, but also to social ones considering that a recent report on sustainable recovery asserts that building renovation offers the greatest employment leverage: 12–18 local jobs for every million investments. This potential would create by 2030 as many as 160,000 new jobs in the EU construction sector [4].

More specifically, the Renovation Wave strategy will prioritize action in three areas: decarbonization of heating and cooling; tackling energy poverty and energy inefficiency; renovation of public buildings (schools, hospitals and offices). It will do so through several measures that make energy redevelopment operations easier

---

[1] There are various estimations of urban consumption of energy and related emissions. According to the World Energy Outlook (November 2008) http://www.worldenergyoutlook.org/index.asp, much of the world's energy is consumed in cities. Cities today house around half of the world's population but account for two-thirds of global energy use. Because of their larger consumption of fossil fuels, cities emit 76% of the world's energy-related CO2).

and faster.[2] "The green recovery starts from home," said Energy Commissioner Kadri Simson, "with this initiative we will face the numerous obstacles that today make the restructuring complex, expensive and slow, slowing down many necessary interventions" [5].

Furthermore, the recent COVID-19 emergency sets before us a new vision of residential heritage, having highlighted its limits—in particular those of the public residential heritage. Therefore, urban regeneration offers the opportunity to rethink housing models. Today, more than ever, the challenges posed by epidemiological and climate changes bring to light more intangible realities which are more oriented towards generative social action. These realities require the involvement of actors, not only of the construction sector but also of the local community through the implementation of complex and long-lasting social projects, which must be designed to support first of all the most vulnerable groups.

2. Italian intervention policies

In Italy, the evolution of the concept of urban regeneration can be re-read within the relevant legislation, that marks the transition from the concept of recovery to the concept of rehabilitation, within the legislation presented and approved in the period

---

[2] The EU must adopt an encompassing and integrated strategy involving a wide range of sectors and actors based on the following key principles:—'Energy efficiency first'8 as a horizontal guiding principle of European climate and energy governance and beyond, as outlined in the European Green Deal9 and the EU strategy on Energy System Integration10, to make sure we only produce the energy we really need; – Affordability, making energy-performing and sustainable buildings widely available, in particular for medium and lower-income households and vulnerable people and areas; – Decarbonization and integration of renewables11 . Building renovation should speed up the integration of renewables in particular from local sources, and promote broader use of waste heat. It should integrate energy systems at local and regional levels helping to decarbonize transport as well as heating and cooling; – Life-cycle thinking and circularity. Minimizing the footprint of buildings requires resource efficiency and circularity combined with turning parts of the construction sector into a carbon sink, for example, through the promotion of green infrastructure and the use of organic building materials that can store carbon, such as sustainably-sourced wood; – High health and environmental standards. Ensuring high air quality, good water management, disaster prevention and protection against climate-related hazards12, removal of and protection against harmful substances such as asbestos and radon, fire and seismic 8 See Article 2(18) Governance Regulation (EU) 2018/1999: "'energy efficiency first' means taking utmost account in energy planning, and in policy and investment decisions, of alternative cost-efficient energy efficiency measures to make energy demand and energy supply more efficient, in particular by means of cost-effective end-use energy savings, demand response initiatives and more efficient conversion, transmission and distribution of energy, whilst still achieving the objectives of those decisions". 9 The European Green Deal, COM(2019) 640 final. 10 Powering a climate-neutral economy: An EU Strategy for Energy System Integration, COM(2020) 299 final. 11 This refers to energy from renewable sources produced on-site or nearby. 12 Climate resilient buildings mean that the buildings are renovated to be resilient against acute and chronic climate-related hazards relating to temperature, wind, water and solid mass, as appropriate. A complete list of such hazards is included in Table 1 of Annex I of Commission Implementing Regulation (EU) 2020/1208. 4 safety. Furthermore, accessibility should be ensured to achieve equal access for Europe's population, including persons with disabilities and senior citizens— Tackling the twin challenges of the green and digital transitions together. Smart buildings can enable efficient production and use of renewables at the house, district or city level. Combined with smart energy distribution systems, they will enable highly efficient and zero-emission buildings.—Respect for esthetics and architectural quality. 13 Renovation must respect design, craftsmanship, heritage and public space conservation principles.

between the 90's and the early 2000s. Indeed, national legislation moves from integrated intervention programs to urban redevelopment programs (L. 179/1992); from urban recovery programs (L. 493/1993) to district contracts (D.M. n. 1071–1072, del 1° dicembre 1994); from urban regeneration and sustainable development programs (D.M. dell'8 ottobre 1998) to urban rehabilitation programs (L. 166/2002).

It is precisely the building and urban rehabilitation programs (L. 166/2002) that introduce, alongside the concept of transformation of physical space, that of performance, especially linked to the concept of efficiency, taking into consideration also economic and social issues, including physical deterioration. In 2015, all these experiences led to the *Piano nazionale per la riqualificazione sociale e culturale delle aree urbane degradate* (national plan for the social and cultural redevelopment of deteriorated urban areas), where the concept of rehabilitation gained, within the concept of redevelopment, the meaning of quality not only of the physical heritage, but also of the intangible one. Still, in Italy, the concept of urban regeneration has been introduced only in recent decades, addressed by national and local policies (regional legislation) as a matter of territorial governance ascribed to the concurrent jurisdiction of States and Regions. In these policies, a strategic vision from above based on urban planning and programming comes along with a bottom-up regeneration process of common goods which starts directly from the citizens. Within the 2007–2013 programming of structural funds, which conceived intervening on cities as one of the priority actions, the term urban regeneration was reinterpreted as an integrated approach to urban development, capable of overcoming the fragmentation and sectoral nature of interventions in this field. This approach has found further confirmation in the 2014–2020 fund programming. The urban regeneration issue can also be found related to the measures on land consumption and on reuse of built land, where the term is intended, above all, as the recovery of the existing building heritage (DDL. C.2039 "Containment of land consumption and reuse of land built "approved by the Chamber on May 12, 2016). In the Code of public contracts (D.Lgs. 50 del 2016) instead, we find the concept of urban regeneration combined with horizontal subsidiarity interventions (art. 189) and administrative bartering (art. 190), where social partnership contracts are introduced based on projects proposed by an individual or associated citizens.

In recent years, several regional regulations in the field of urban planning and construction have been introduced within the framework of urban regeneration with a strategic vision of territorial planning, implemented through complex plans and programs. Many regional regulations reveal that the regulatory concept of urban regeneration differs more and more from that of building recovery and urban planning, and is gradually including complex actions for the urban, environmental and social rehabilitation of degraded urban areas. Examples are the regional law of Emilia-Romagna (L.R. n.24, 21 dicembre 2017) on the protection and use of the territory, the regional law of Tuscany (L.R. n.65, 10 novembre 2014), which lays down rules for the government of the territory, the regional law of Lazio, (L.R. n.7, 18 luglio 2017) containing regulations for urban regeneration and building recovery, just to name a few.

The law L. n. 158 dell'8 ottobre 2017—containing measures for the support and enhancement of small municipalities, as well as regulations for the redevelopment and recovery of their historic centers—complies with the goals mentioned above, where the concept of regeneration takes on a connotation of territorial and environmental protection and where small municipalities are recognized as a resource due to their role as a territory presidium, especially about their role in contrasting hydrogeological instability and in preserving and protecting common goods.

Urban regeneration, which is gaining an important space in regional legislation, still struggles to find a precise definition in the national one, where it is addressed as an emergency measure by the D.L. 18 Aprile 2019, n. 32—" *Disposizioni urgenti per il rilancio del settore dei contratti pubblici, per l'accelerazione degli interventi infrastrutturali, di rigenerazione urbana e di ricostruzione a seguito di eventi sismici*" (urgent regulations for the relaunch of the public contract sector, for the acceleration of infrastructural interventions, urban regeneration and reconstruction following seismic events). The decree fosters the reduction of land consumption in favor of regenerating the existing building heritage by encouraging the redevelopment of degraded urban areas. Eventually, the unified text of the D.D.L. on urban regeneration, which is now under discussion, provides a state fund of 500 million euros per year until 2040 to co-finance regional tenders for the urban regeneration plans presented by the municipalities, thanks to an alliance between state, regions, municipalities and private individuals.[3]

3. Sustainable urban regeneration

In this context marked by a European policy strongly focused on energy saving and consumption reduction, the existing building stock and its redevelopment play an important role, especially the energy requalification of public housing (ERP) [6]. By public housing we refer to the residential real estate built, directly or indirectly, by the State, to be assigned, at particularly good economic conditions, to citizens with low incomes or who find themselves in poor economic conditions. The law regulating public housing identified three areas of intervention onto which allocate the available economic resources: subsidized housing of exclusive public ownership, assisted housing in property and/or with controlled rent and housing with agreements on surface or property rights. The fact that the European public housing heritage is plentiful and assorted, is a clear expression of the cultural and economic differences of our continent. In Europe, a significant share of the housing stock was built in response to the demand for housing following the Second World War. To date, more than 220 million buildings, representing 85% of the European building stock, were built before 2001. The majority of them are not energy efficient, as a result of old technologies and bad insulation—and account for around 40% of total European energy consumption and 36% of greenhouse gas emissions [3]. The physical (technical-functional), social and economic conditions of degradation that characterize the public building heritage demand the identification of immediate intervention strategies. The aim of the research is to show how certain strategies, in particular bioclimatic, modular and low-cost ones applied to the small building scale, can become the main tools for rehabilitating relevant parts of the contemporary city. For this reason, the research work described in this essay aims to give back urban quality to the suburban fabric which hosts public residential buildings through an architectural, energetic, bioclimatic and environmental requalification. This operation provides an attempt to read the peripheral palimpsest, through punctual and diversified interventions involving the description of the physical space, of the biophysical one and the understanding of bioclimatic phenomena, which cannot be separated from the understanding of the social space. Throughout

---

[3] Moreover, the text of the law provides also for the creation of a "database on reuse" of vacant and abandoned properties, for the right for Municipalities and Regions to raise taxes on unused or unfinished real estate units for over 5 years and for the possibility to resort to two-level design competitions. Finally, it provides for wide use of tax incentives (such as the superbonus, the eco-bonus or sismabonus) and for the establishment of a control room for urban regeneration meant to coordinate the interventions on different levels and to implement the national program goals, planned to be adopted within 4 months from the entry into force of the DDL.

the research, the role of the architect has been reconsidered as a social role that requires the ability to listen, interpret and explore the peripheral space, with the intent of setting shared and experimental assumptions to which reference can be made to overcome with the method and analytical scientificity the contradictions and conflicts of extended urban peripheries. Every overall transformation program, each detailed project is declined in this sense by associating several reinterpretations of the space and considerations on the possible scenarios of transformation, to develop, through a critical synthesis, an innovative conception of urban peripheral environment. This environment is meant to dialog with the consolidated city and to connect the redevelopment process to the cultural, social and technological transformations that affect society and urban form and that emphasize the need for new ideas, innovative models and relevant examples.

Overall, the analyses and evaluations of this research work start from the identification of the problems within the analytical phase and then explore the feasibility of intervention scenarios that can help to achieve greater urban quality. At the heart of the work, there are three research survey directions concerning:

1. The ability to measure oneself with the characters of the places and the territory;

2. An environmentally sustainable and energetically and ecologically efficient transformation-development;

3. The reconstruction-enhancement of the public space.

Given the first research direction, the operational strategic lines address:

1. The broadening of the analysis framework concerning the anthropic, biophysical and bioclimatic factors to support the development of meta-project scenarios;

2. The optimization the control system of environmental components in the development phase of the intervention program-project;

3. The maximization of instrumental skills to verify the quality and eco-efficiency levels of the architectural project and urban reality;

4. The development of verification steps concerning the different degrees of quality and environmental sustainability to be assigned to the different phases of the project;

5. The optimization of systems for monitoring and controlling the behavior of outdoor, intermediate and confined spaces.

Regarding the implementation of the second research axis, the following operational strategic lines have been identified:

1. to optimize the specific conditions of mobility-transport connecting the periphery with the consolidated city;

2. to enhance-protect the local landscape-vegetation assets;

3. to enhance the soil conditions in terms of site orography, lithology and stratigraphy;

4. to optimize local hydrogeological conditions, considering rainwater runoff;

5. to introduce morphological-typological solutions aimed at integrating residential functions and extra-residential activities;

6. to organize and manage the material, energy and information flows at the district level;

7. to introduce innovative building solutions aimed at energy and ecological efficiency.

Finally, the third research axis aims at:

1. controlling and enhancing greenery in its role as a filter between environmental factors and functionality of outdoor spaces;

2. using outdoor spaces for meeting and social occasions within the general reorganization of paths and mobility;

3. enhancing the relationship between mobility, vegetation, bioclimatic and urban furniture in the design of in-between outdoor spaces;

4. maximizing psycho-perceptive comfort conditions related to the morphology of the intermediate outdoor spaces;

5. maximizing energy and ecological efficiency of the public space.

4. Sustainable technologies, hypotheses and solutions for two case studies in Rome, Italy

The case studies illustrated in the essay focus on the analysis and regeneration of two types of ERP multi-storey buildings, *in linea* (linear) and *a torre* (tower) blocks located in two different peripheral areas of the city of Rome.

The two-building typologies are taken into consideration using their similarities from an administrative and legal point of view and of their differences on a typological and environmental level and about their relationship with the context (**Figure 1**). The case studies analyzed are chosen to be representative of the respective building typology which recur in the ERP Roman context. The choice to study buildings located in peripheral areas is linked to the desire of investigating distressed urban areas [7], areas where it is even more urgent and necessary to intervene and where regeneration processes have a reorganization potential that transcends the architectural level and bring along positive effects on the socio-economic conditions of the inhabitants.

Specifically for the tower high rise building typology, an ERP condominium was identified—part of a plot of four twin condominiums—in the north-western suburbs of Rome, XIII–XIV municipality, while for the linear multi-storey building typology an ERP condominium was identified—which is repeated 18 times with different orientations according to *Piano di zona 02v*—in the north-eastern outskirts of Rome, IV municipality. Both buildings were designed in the late 70s and built in the 80s in distressed urban areas of the city that differ in density (150 inhabitants/ha in Torrevecchia and 98 inhabitants/ha in San Basilio), in the degree of marginality compared to the adjacent context and the rest of the city as well as in the plano-volumetric system.

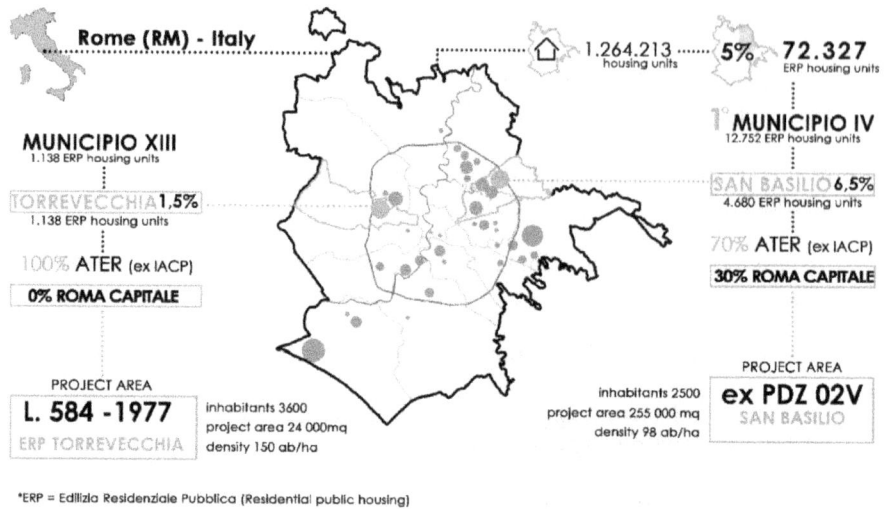

Figure 1.
Territorial framework: two ERP case studies in Rome.

The methodology adopted in this research allows, once the different hypotheses have been identified, to evaluate both the technical (energy savings that would result) and economic feasibility, as well as to verify the overall compliance with national and local regulations.

The methodological approach followed for both case studies provides for an analysis of the context and the current situation of the buildings starting from a territorial and urban framework with a specific focus on mobility, facilities and greenery. Subsequently, demographic, socio-economic and housing demand surveys were carried out. Ultimately, an environmental and micro-climatic analysis both of the entire context and the building under examination was carried out and finally, the architectural and technological components underwent a thorough examination. The aim is to detect critical issues at the building and housing level and to subsequently define the typological and functional program and the overall intervention strategies in line with technological and environmental requirements. The definition of the general morphological-functional characteristics of the intervention about the interaction model between microclimatic factors and the context led to the preliminary design. This initial design stage was developed according to studies and technical validations set at a meta-design level, followed by a functional study and the reinterpretation of housing schemes and supplementary facilities, according to the social demand. At last, a summary report on the identified demand/performance system was drafted: clarification of the environmental technological requirement system and its related design choices. The different strategies and design solutions underwent a definitive design elaboration of the building system and its subsystems and components in line with the environmental context and their interrelations with the transformations induced by the intervention.

5. Tower high rise building typology in Torrevecchia, Rome

The Torrevecchia district (**Figure 2**), built with funding from law L. n. 584 of 1977, is an area of approximately 24 ha with 1074 accommodations for 3600 inhabitants located in the north-west area of Rome in the Primavalle district, XIII–XIV

Figure 2.
Aerial view of Torrevecchia district (source: Google Earth).

municipality. It is owned and managed by the autonomous institute of popular housing (IACP), a specific Italian institution to promote, build and managing public housing to assign to citizens on low income rented at controlled rates. The architects P. Barucci, L. Passarelli and M. Vittorini were in charge of its design and execution. Until the 1960s Torrevecchia was part of the *agro romano,* and during the 1970s the area was affected by strong urbanization. The district is defined to the south by via di Boccea, to the west by via di Casal del Marmo, to the east by via Mattia Battistini and to the north by via Trionfale.

The plano-volumetric scheme is developed around a central square defined by four 15-storey high tower buildings (76 apartments) on which a group of offices and a bar overlook. Long 4/5 storey buildings (192 apartments) branch off from the central square with three levels of housing and a ground floor meant for shops, which were never realized.

Thanks to a progressive series of shifts, these volumes tend to spread out towards the extremity of the area thus creating in-between them two green spaces large enough to host respectively a small public park and a sports field. Car parking spots are located on the external side of the linear storey buildings, thus remaining outside of the central green areas defined by the building volumes.

Overall, the architectural solutions adopted in the different buildings are rather simple and they all respond to the constraints set by the standards imposed by law and by the economic means: prefabricated concrete panels and ribbon windows with metal frames.

5.1 Analysis

The IACP complex is commonly known as the "Bronx" due to its architectural aspect (poor architectural-spatial, environmental and energy quality of the buildings) and the socio-economic conditions of the area. The complex is strongly marked by economic precariousness, by the absence of public spaces and areas for meeting and socializing, by the absence of life and services at ground zero, and eventually by lack of maintenance and building degradation.

According to 2011 ISTAT census data (the Italian National Institute of Statistics, which is the main producer of official statistics in Italy), one out of four residents appear to be unemployed and 9 out of 10 people have reached a level of education below the middle school. The social hardship index, which provides a measure of the possible social-occupational drawbacks, is among the highest in Rome.

An analysis of mobility and facilities (**Figure 3**) reveals a lack of good quality common spaces, a poor and inefficient transport system that makes nearby facilities not accessible. The Battistini metro station (line A), which connects the district

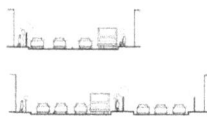

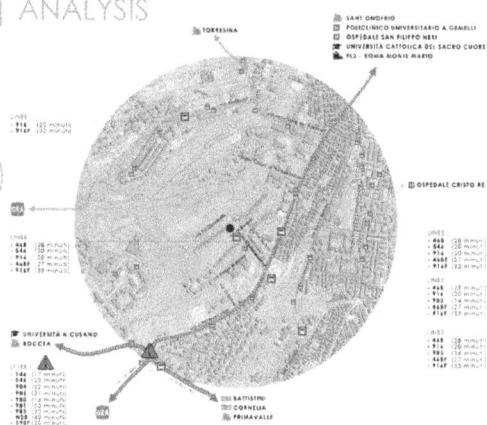

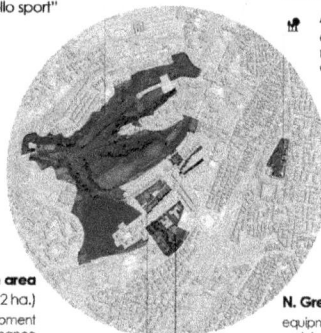

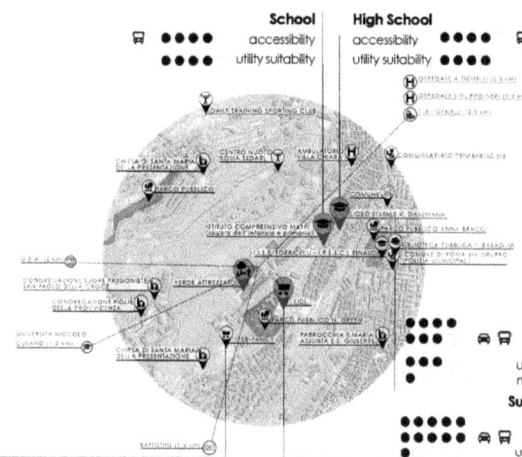

Figure 3.
Torrevecchia district: overall analysis.

with the city centre, is 2 km far from the complex and can be reached in approximately 25 min on foot and in 15 min by public transport. Of the 12 bus lines covering the area, only 4 connect Torrevecchia with the city centre.

Data regarding facilities shows that compulsory schools are insufficient to serve the catchment area. Conversely, there are four technical/professional schools within a 2 km radius of the area under examination. The entire area suffers from a serious lack of facilities which is revealed by the presence of one pharmacy for every 24,500 inhabitants, as opposed to what is required by the Italian law L.362/91 (Art. 1), which is one every 12,500 inhabitants.

Torrevecchia still preserves evident signs of the past rural vocation. The valley between the IACP complex and the Quartaccio district is the least urbanized part of the neighborhood. Despite several green expanses, still not affected by irregular urbanization, the neighborhood is in fact devoid of proper public parks. The only two parks, the Insugherata Nature Reserve and the Pineto Urban Regional Park, are more than 2 km away from the centre of Torrevecchia. Out of more than 65 ha of green areas, only 3% are organized in equipped areas.

Air quality is significantly and positively affected by the abundance of proximity greenery (parks and reserves), although uncultivated and derelict. The average concentration of nitrogen dioxide NO_2 is about 32.13 $\mu g/m^3$, far lower than the annual limit value for human health protection, established by D.Lgs. 155/2010 which provides for a maximum limit of 40 $\mu g/m^3$. As for fine particles, according to what is reported by a PM2.5 map obtained from a dispersion model, the average annual concentration is 18.31 $\mu g/m^3$, which is far lower than the limit value of 25 $\mu g/m^3$ (D. Lgs. 155/2010 in force since 2015).

As regards the microclimatic conditions (**Figure 4**), in the summer season, the area is affected by winds coming from south west (speed of 16 km/h) which considerably contribute to cooling the south-west area that is most of the time subject to direct radiation during the day, given the poor vegetation and the low building

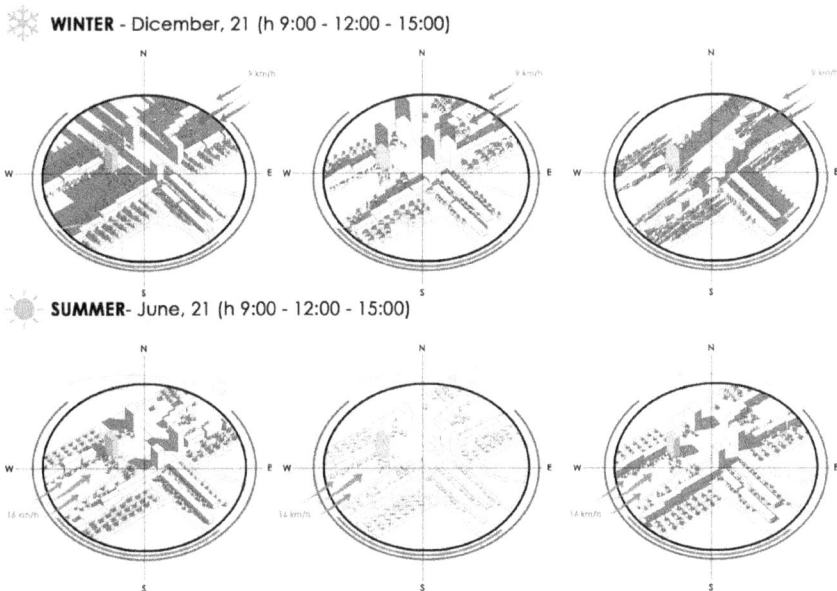

Figure 4.
Solar analysis: the tower building in Torrevecchia district.

heights. The square to the north-east receives just as much radiation, but the layout of the surrounding buildings prevents it from adequate ventilation. In the winter season, the cold wind coming from north east (speed of 9 km/h) sharpens the perception of comfort in the north-east square, which remains cold and in the shade all day long taking into account average winter temperatures of about 7°C. Throughout the day, the south-west square catches the sunlight in its highest part since the shadows cast by the storey buildings are reduced compared to the width of the square.

The tower buildings, especially on the higher floors, enjoy summer ventilation from the south-west but suffer from winter ventilation on the north-east façades, which correspond exactly to the façades with a higher percentage of openings.

The tower buildings, 43.2 m high and with a floor area of approximately 4140 m^2, house 76 accommodations (48 units of 60 m^2 and 28 of 45 m^2) for 248 estimated occupants. A comparison between current users and the availability of floor area, results in 16.7 m^2 per person. According to the national legislation, the building should thus house no more than 207 inhabitants. The 60 m^2 apartments feature a double exposure while the 45 m^2 apartments have a single-exposure. The 60 m^2 apartments are designed to accommodate 4 people but both bedrooms (12.4 m^2 and 12.8 m^2 big) are smaller than 14 m^2, thus do not meet the current minimum standards. It is estimated that the prevalent (50%) family unit typology in Torrevecchia is composed of 1 or 2 members, while families with 4 members account for only 12% of all families.

5.2 Solutions and strategy

Starting from the overall critical evaluation a project concerning both adjacent outdoor and indoor spaces has been developed with particular emphasis on bioclimatic solutions (**Figure 5**). In the first place, it is essential to redesign the building's connection to the ground, its consequent relationship with the street as well as the intermediate in-between public spaces, to provide meeting and relational opportunities. A necessary prerequisite is an involvement of the inhabitants from the very start, ranging from the preliminary design to the future management and care of common spaces. To integrate the facility system, currently rather inadequate, and enhancing the relationship between building and streets, an elevated square is proposed. The new plaza consists of a solid volume with internal excavated patios and courtyards. Vocational training laboratories, also meant to work as local facilities are located on the ground floors of the towers and in the hypogeal areas beneath the elevated square. A system of ramps allows connecting the different levels on which the public outdoor spaces are distributed. Therefore, the square becomes an open space arranged on two levels on which some food services, laboratories and craft shops open. The internal patios enable to host in the hypogeal spaces several other facilities and services, providing them with adequate light and ventilation and with quality green outdoor space of relevance. The derelict south-west green square is transformed into plots of collective urban gardens managed and used by inhabitants.

When it comes to the tower building, the redesign of the ground connection through an excavation on the short façades (north-west and south-west) strongly contributes to improving thermo-hygrometric and visual comfort parameters of the former basement which is thus freed on two sides from direct contact with the damp ground. The ground floor and the basement are merged to obtain double-height rooms suitable for hosting a kindergarten and fab-labs with an external space of relevance gained thanks to the excavation.

To maximize the feasibility of the project, the intention is to limit as much as possible the need for the inhabitants to temporarily leave their homes. Therefore, priority actions address the need to provide all accommodations with adequate living space in terms of square meters to satisfy standards imposed by national law.

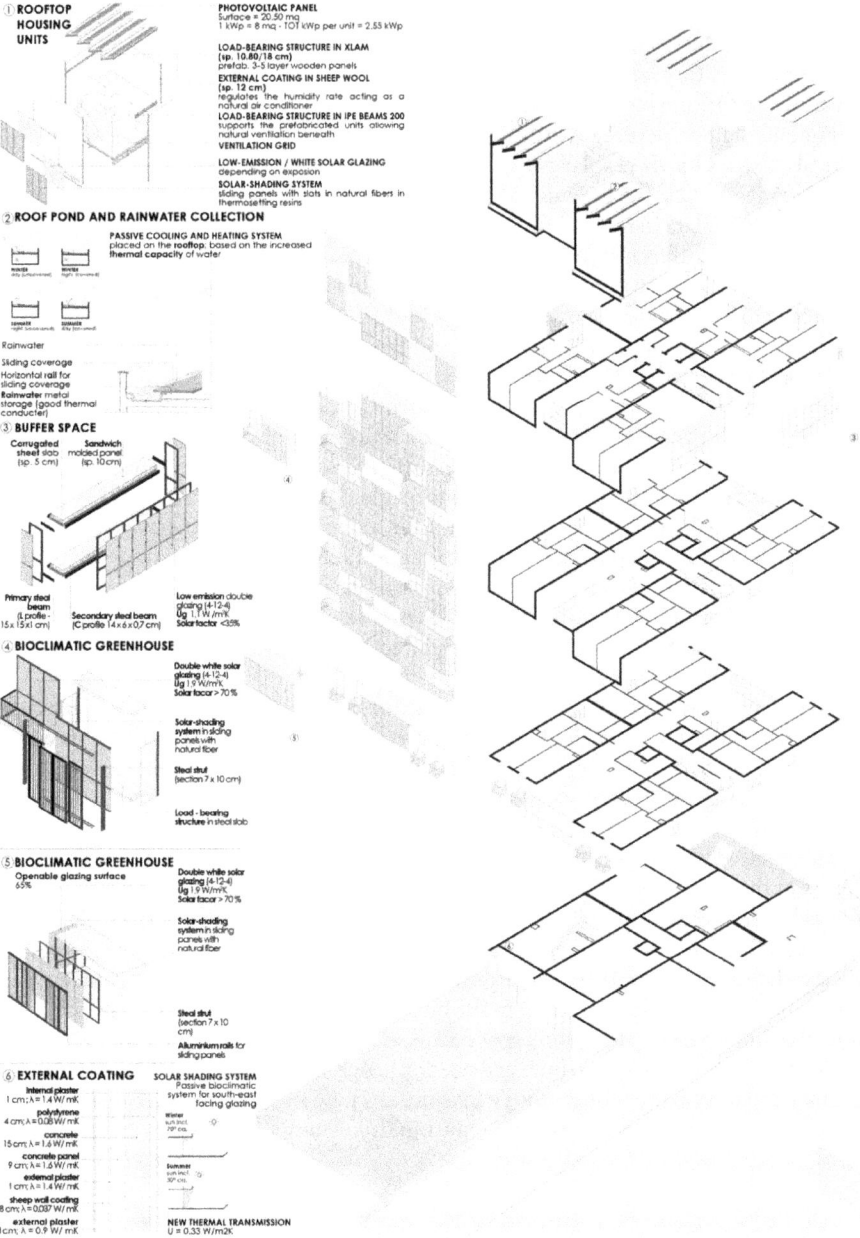

Figure 5.
Exploded axo: overall project and focus on bioclimatic solutions.

The addition of plug-and-play modules on the north-east and south-west façades, diversified on a technical level according to the exposure, can solve simultaneously problems related to poor lighting, poor insulation and consequent thermal comfort and under-sizing. The buffer-space modules added to the north-east façade and the bioclimatic greenhouses added to the south-west façade make up new indoor or outdoor living spaces, diversified internally about the type of environment to which they are added and to the needs of the occupants. It's a system of prefabricated units, ranging from 2 to 8 m^2 big, which add a total of about 10 m^2 to each dwelling, which corresponds to 17% additional surface area in the case of the two-room apartment and 14.5% for the four-room apartment.

The bioclimatic greenhouses leaning on the south-east, south-west and south elevations constitute heat accumulation spaces to introduce preheated air into the apartments. The structure is made up of modular steel elements with transparent vertical closures in white solar glass, with a solar factor higher than 70% and with an openable glass surface of 65% out of the overall transparent closure. Sliding panels with adjustable slats in natural fibers and thermosetting resins work like shields. The modules jut out differently according to the functional and structural needs, ranging from 120 to 240 cm. The buffer-spaces, attached to the north-east elevation, highly contribute to improving the overall energy performance, through the reduction of heat loss and consequent thermal gains in winter and through the dissipation of heat in summer. The overhang is 90 cm and the structure is similar to that of the greenhouses but with low-emissivity glass and a solar factor lower than 35%.

New dwellings can be added on the roof, taking advantage of the incentive offered by the regional law L.R. 7/2017 on urban regeneration which allows adding 20% of the original building volume or the original floor surface in case of energy efficiency interventions on residential buildings. The new volumes have a dry load-bearing structure in X-Lam panels and are placed on a load-bearing structure in IPE steel beams to detach and slightly lift the housing module from the existing roof and thus ensure natural ventilation. Each dwelling is equipped with photovoltaic panels (20.50 m^2) and solar collectors integrated into the roof to assure self-sufficiency in terms of energy.

The original flat roof is replaced with an extensive green roof covered by a pitched canopy which, besides ensuring shading, is also designed to collect rainwater through the central impluvium for irrigating the green roof. The green-blue roof combines different technologies allowing an increase in the storing capacity and a control system of the water flow to release. The green roof helps to cool and humidify the surrounding air, positively affecting the microclimate with slight effects also for the squares located at the street level. In doing so, the storey just beneath the roof slab gains in thermal insulation, therefore less indoor overheating results in less consumption for air conditioning, affecting the overall energy balance. In addition, the vegetated surface effectively protects the waterproofing membrane from UV rays, hail, heat and cold, contributing in the long term to the building envelope maintenance. At the same time, the roof becomes a common space available for all the building users.

Where plug-and-play modules are not applied, an 8 cm sheep wool insulation is laid on the external current envelope to achieve a new transmittance of 0.33 W/m^2 K to ensure an overall optimization of the building energy performance.

6. Linear multi-storey building typology in San Basilio, Rome

San Basilio (**Figure 6**) is located in the IV municipality, in the north east of Rome, in the urban area 5E, and borders the Grande Raccordo Anulare, an orbital motorway that encircles Rome, to the east and Casal de' Pazzi and Tor Cervara to the west. The municipality is delimited by the via Nomentana to the north, by the municipality of Guidonia Montecelio to the east, by via Tiburtina, the Aniene river, the A24 motorway and the Rome-Pescara railway to the south, and the Rome-Florence railway to the west.

Between 1981 and 1988, the *Piano di zona 02v*—San Basilio social housing urban plan—part of the 1981 supplementary variant of the general urban plan, provided for the construction, based on a project by Antonio Salvi, of 18 linear buildings

Figure 6.
Aerial view of San Basilio district (source: Google Earth).

of 6–7 floors over an area of 25.5 ha destined to settle 2500 inhabitants. The intervention is not well integrated with the existing fabric and is characterized by an orthogonal system of roads that shapes and defines the various plots. The buildings too are arranged in an orthogonal way and form green courtyards that open up towards the roads and a series of inter-closed courtyards at the points where the building heads come close to each other. San Basilio hosts 6.5% of the ERP accommodations in Rome and about 36% of those in the IV Municipality, gaining first place in terms of ERP accommodations in the city.

6.1 Analysis

According to the 2011 ISTAT census, it is clear that about 27% of the district inhabitants are between 45 and 60 years old and over half of the families are composed of a single member, 27% of 2 members and only the remaining 20% are families of 3, 4 or 5 components. One-third of the inhabitants have not more than a middle school diploma and only 10% have a university degree, half of the figure for graduates in Rome (20%). The number of unemployed is about 2% higher than the Roman average and about half of the population lives in rented apartments. The neighborhood is also known for the strong presence of petty crime and drug dealing.

Different multi-thematic analyses were carried out concerning mobility, facilities and green systems (**Figure 7**). The mobility analysis highlights how the area under examination is a sort of enclave to the district that stretches to the west, as it is connected to the urban fabric of the old San Basilio district and the rest of the city only by two access roads. The closest metro station is almost 3 km away. A station was supposed to be built in the old San Basilio area but the project for the extension of the metro Line B has never been realized. As for local public transport, although the area is served by several bus lines, these are not sufficient to ensure a direct and rapid connection with the city centre.

With regard to the facilities, since there are no public or private ones within a radius of 250 m from the area under examination, and only a mechanic within a radius of 500 m, one is forced to travel almost 1 km to reach a supermarket, pharmacy or the nearest primary school. 57% of the district facilities are ascribable to retail trade, while public space meant for squares does not exceed 2%.

The several parks and green spaces in the area are unequipped and poorly maintained. In general, the outdoor spaces lack even the most basic elements of street

SAN BASILIO | ANALYSIS

MOBILITY SYSTEM

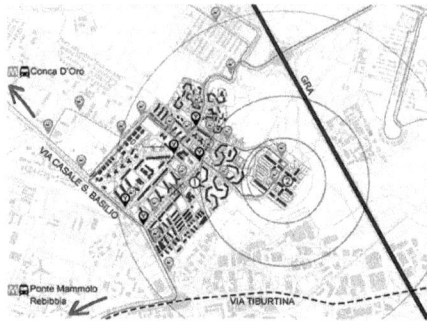

GREEN SPACES SYSTEM

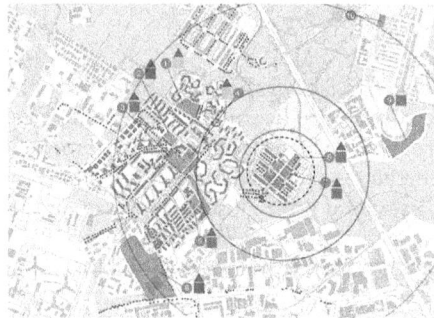

FACILITIES SYSTEM

Figure 7.
San Basilio district: overall analysis.

furniture. This leads inhabitants to use these spaces only to go from one place to another and not for social purposes. Moreover, inadequate public lighting increases the feeling of insecurity among the inhabitants.

The microclimatic analysis (**Figure 8**) shows that the area under examination is affected by cold winter winds coming from east, north-east (about 9 km/h) and by hot summer winds coming mainly from south-west with a speed of approximately 16 km/h. In the summer season, both the north-east square and the southern square adjacent to the building suffer significant overheating phenomena throughout the whole day, yet moderated by ventilation. In winter, the northern square lays most

SAN BASILIO **SOLAR** ANALYSIS

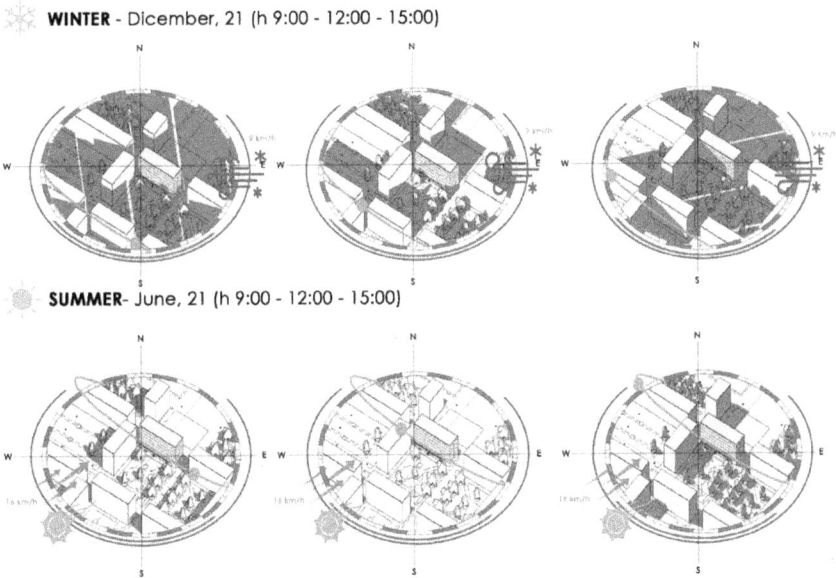

❄ **WINTER** - Dicember, 21 (h 9:00 - 12:00 - 15:00)

☀ **SUMMER**- June, 21 (h 9:00 - 12:00 - 15:00)

Figure 8.
Solar analysis on the linear building in San Basilio district.

of the time in the shade and is constantly exposed to cold winter winds due to the absence of adjacent buildings.

The building under consideration (6G) is a linear multistorey building typology of about 42 × 13 m consisting of 7 floors above ground and a basement floor. The building currently houses about 135 people for a total of 45 apartments of 50, 60, 80 and 100 m², accounting respectively for 40%, 30%, 25% and 5% of the total housing units. The structure is in reinforced concrete bearing walls that define a succession of different sized spans parallel to the short side. The façade is characterized by prefabricated concrete blocks with a minimum insulating layer of glass wool. The joints have not been carefully designed and the discontinuity of the insulating layer causes several thermal bridges. The building is equipped with two staircases and the access is via a gallery located on the ground floor at a height of +1.00 m, above the cellar floor, accessible from the condominium staircases. Currently, the ground floor houses, in the eastern part, two special housing units for people with physical disabilities and two rooms initially designed to be a condominium space and a laundry for common use. At the moment, the western part is occupied by storage spaces but it was meant to be—according to the original project—a *pilotis* floor with a walkway at a height of 0 m from which to access the gallery via the staircase. The upper floors house the apartments and each staircase serve from 3 to 4 units, two-thirds of which have single exposure. The living spaces do not face under-sizing problems since all the indoor spaces meet the minimum surface and height standards for public housing and besides, aero-illuminating ratios are verified in all rooms. The staircases lead up to the roof level.

6.2 Solutions and strategy

The analysis of the current condition reveals the need to rethink the building in all its aspects (**Figure 9**) starting from its connection to the ground not only in terms of spaces and functions but also about the square in front. The special apartments

SAN BASILIO | CLIMATE DESIGN TECHNOLOGIES

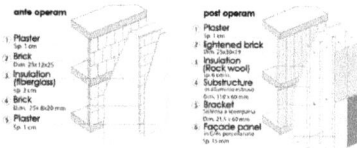

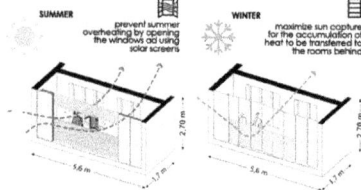

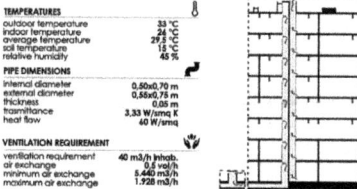

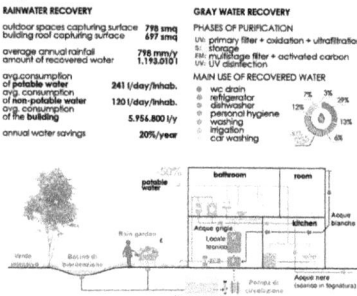

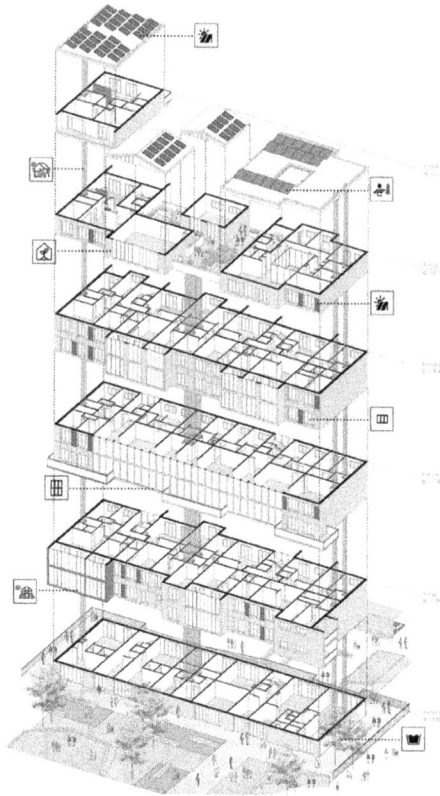

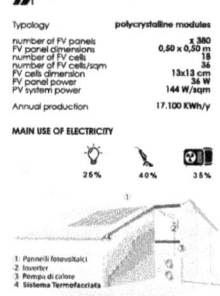

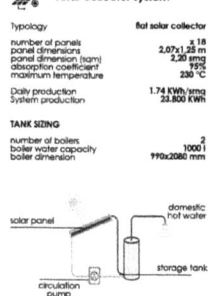

Figure 9.
Exploded axo: overall project and focus on bioclimatic solutions.

on the ground floor are to be relocated, in terms of living space, on the roof level and the currently unused common spaces are to be converted into local and building facilities chosen according to what emerged from the previous social analysis. In rethinking the relationship with the outdoor spaces, the current gallery at a height of +1.00 m is redesigned in terms of accessibility, use and relationship with the square, providing for its expansion in some points to create terraces to serve the new activities and rest areas to encourage meeting and socialization occasions. A co-working, refreshment and internet point could be located in the western part of the ground floor (internal height of 3.70 m) at the same level as the square and near the main road. These spaces are thought of as facilities for the entire local community. The remaining part of the same floor (with an internal height of 2.70 m) could host a series of flexible spaces, including a medical-assistance clinic with an adjoining small outpatient clinic where professionals can offer different services on shifts, multi-purpose spaces for courses and activities and a bicycle repair workshop.

Overall, the existing envelope does not meet current national standards for energy performance for buildings. In this regard, a new insulation layer must replace the previous thin and inadequate one together with the application of a ventilated façade in specific parts. Old windows are replaced by new ones in recycled aluminum with thermal break and double glazing with argon gas inside, with a global transmittance of 1.56 $W/m^2 K$ compared to the maximum 1.80 $W/m^2 K$ required by law.

The current design of the housing units of the standard storey does not make the most out of the living space. A new distribution of indoor spaces in favor of living areas located south and an implementation of new spaces to increase liveability, are required. In this regard, steel plug-and-play modules added to the façade, besides providing additional volume, can also be configured as bioclimatic devices by hosting greenhouses or buffer spaces depending on the orientation or, in some cases, as balconies or galleries. To make the housing units more compliant with the family units—according to socio-demographic data—one standard storey could be turned into a cohousing for about 40 members. This housing typology, mainly designed for relatively young users, single or couples, spreads over one entire floor and is accessible from both staircases. The sleeping area is essentially located in the east and west wings and the north. The south front instead hosts several shared spaces in sequence, such as a kitchen equipped with a dining area, a common living room, a mini cinema/games room and a common laundry room, all joined by a single glazed connection placed in adherence to the south façade. To ensure adequate ventilation inside the single-exposure apartments (about 60% of the total), a geothermal cooling/heating system operated by wind towers is inserted. In the light of the microclimatic analysis, the tower collection heads should be placed where airspeed accelerations occur both in summer and in winter. The air is trapped and then directed through underground ducts—where thanks to geothermal energy it is pre-heated/cooled depending on the season—to the apartments to be, in a second step, introduced into each room through a distribution system installed in the false ceilings positioned over the service and distribution spaces.

With regards to the roof, the availability of such a large free surface allows the implementation of different passive and active strategies as well as technological devices. Special housing units (once located on the ground floor) and common spaces are relocated on the roof. These new accommodations, larger and suitable for families of 4–5 members, have been designed for a different target audience, to encourage different people to approach the neighborhood and thus promote social *mixitè*.

Moreover, a common laundry room, a greenhouse for food production and a common outdoor kitchen/dining area are integrated as new volumes on the roof. The roof is also equipped with a system for the collection and reuse of rainwater and gray water. The uncovered surfaces are redesigned to better capture rainwater and convey it to specific collection points. From here, the water is filtered and purified and then used for cleaning and irrigation purposes for outdoor spaces, toilet drains and washing machines. With regard to gray water—before being stored in the collection point—it undergoes a different purification process and is later reinserted into the general circuit. This system can bring about significant clear water savings accounting for about 20%. The building is also equipped with a photovoltaic system for electricity production. The system is composed of polycrystalline modules of the size of 50 × 50 cm with a nominal power of 35-W peak each. The energy produced, equal to about 17,000 kWh per year, will feed, not only the lighting system—replaced with LED elements—but also the heat pumps for the underfloor heating system and part of the domestic consumption. A small portion of the roof is also meant for a solar thermal system for domestic hot water production, consisting of 18 panels of about 2 m^2, for a total of 36 m^2, able to cover 50% of the annual housing needs.

7. Results and conclusions

Currently, most of the interventions on ERP have an emergency nature: direct operations aimed at solving specific problems in the short term. This logic, devoid of investment, does not allow to respond to broader issues, without halting the heritage deterioration. Today it is even more necessary to outline intervention strategies capable of coping with the technical-functional aging of buildings. The ERP heritage of the city of Rome can become the key element for a qualitative regeneration of the city. According to socio-demographic data, it is possible to point out similar contexts, characterized by strong social unrest, petty crime and a high unemployment rate (23% in Torrevecchia and 16% in San Basilio). About the district population, that of San Basilio is generally younger and made up of larger families compared to that of Torrevecchia. Yet, in both cases, the most recurrent family unit is composed of 1 or 2 members. Not only the different characteristics of the plano-volumetric system but also those related to the socio-economic, environmental and microclimatic context play their part in the choice of which strategies and solutions implemented.

The study carried out allowed to investigate and compare the limits and the potentialities deriving from the building typology and its plano-volumetric system:

- Ground floor. As a direct consequence of the building typology, the linear multistorey building undoubtedly presents a greater availability of space in terms of surface area. This feature allows the introduction of several different facilities on the ground floor, by simply replacing the existing housing units, poorly lit and ventilated and with serious privacy concerns. The new facilities are chosen according to the socio-demographic analysis output and the structural-dimensional characters of the building and address different target users at different scales (inhabitants of the building, of the neighborhood, of the district). In addition, the larger perimeter of the ground floor in this building typology allows greater freedom in rethinking the relationship with the relevant outdoor spaces, enhancing the integration between facilities and pertinent outdoor spaces. In the tower building typology, given the limited availability of surface, the possibility of integrating different facilities is consequently quite limited. The reduced ground floor surface area requires to use of additional storeys to host facilities running into the limits resulting from the reduced internal heights of the upper floors (originally intended for housing). About the useful heights, in the specific case of San Basilio, two public exercises with a district-scale catchment area can be introduced in the area with a useful height of 3.70 m. In the case of Torrevecchia, to overcome these design limitations, the creation of a new elevated public square, in-between the tower buildings, revealed to be the best solution for hosting laboratories for professional training, retail shops and craft labs whilst guaranteeing quality outdoor space at the same time.

- Standard floor. The linear multistorey building, as opposed to the tower typology, tends to house a greater number of single-exposure apartments, which consequently face serious problems related to indoor ventilation. This aspect is mainly linked to the need for a greater surface area for connection purposes to serve all the apartments. Unlike linear typologies, in tower buildings, the connective surface is reduced to the least and corresponds to the stair and elevator block. On the other hand, the linear typology allows greater freedom in redefining the standard floor plan by changing the dimensions of the existing housing units to adapt them to the user's needs. About the apartments, in Torrevecchia their dimensions were deemed suitable for the users and more than two-thirds of them have double exposure. In the case of San Basilio, this proportion is roughly the opposite, with % single exposure apartments accounting for 60%. As mentioned above the single exposure led to the introduction of wind towers to improve ventilation in indoor environments. Some of the apartment rooms in Torrevecchia, such as the double bedrooms, do not meet the minimum standards required by law. Therefore, the addition of new plug-and-play modules on the façades allows for to increase in the limited current surface area and meets the standards.

- Roofing. Taking into consideration the linear typology, likewise the ground floor, the roof level has a greater surface. It offers, in the first place, the chance to implement several technological, active and ecological devices, to gain significant overall energy-water savings:

    ○ A photovoltaic system for electricity production: the larger roof surface allows to install a more powerful photovoltaic systems capable of satisfying a higher share of electricity consumption. In the linear typology, the cost is maximized, since it does not require any integrated systems on the façade, and since the system is more efficient thanks to the possibility of positioning the panels according to the best exposure.

    ○ Rainwater collection system: the larger collecting surface allows to accumulate a greater quantity of water, achieving far higher water-saving percentages.

Furthermore, greater space availability also results in the possibility of adding new accommodations and several common spaces such as a laundry room, multipurpose rooms and a common kitchen in order to provide the inhabitants new spaces in which to spend time and do activities together.

Although, the building typology is quite significant in defining the different possible intervention strategies, these must necessarily be contextualized according to the specific study/project area and its *genius loci*, in other words, the socio-cultural, architectural, economical habits and characters of the place.

Conflict of interest

The authors declare no conflict of interest.

Author details

Alessandra Battisti*, Livia Calcagni and Alberto Calenzo
Department of Planning, Design, and Technology of Architecture,
Sapienza University of Rome, Rome, Italy

*Address all correspondence to: alessandra.battisti@uniroma1.it

IntechOpen

References

[1] UN Desa. Transforming Our World: The 2030 Agenda for Sustainable Development A/RES/70/1. New York: United Nations; 2015

[2] European Union. European Commission. Directorate-General for Regional Policy. Cities of tomorrow: Challenges, visions, ways forward. Brussels: Publications Office of the European Union; 2011

[3] European Commission. Report from the Commission to the European Parliament and the Council. 2020 Assessment of the Progress made by Member States towards the Implementation of the Energy Efficiency Directive 2012/27/EU and towards the Deployment of Nearly Zeroenergy Buildings and Cost-optimal Minimum Energy Performance Requirements in the EU in Accordance with the Energy Performance of Buildings Directive 2010/31/EU. Brussels: European Commission; 2020

[4] IEA. Sustainable Recovery. World Energy Outlook Special Report in collaboration with the International Monetary Fund. Paris: IEA Publications; 2020

[5] EU SHREC | Interreg Europe: Renovation Wave Strategy Targets European Buildings. Retrieved from SHREC | Interreg Europe. [Internet]. 2020. Available from: https://www.interregeurope.eu/shrec/news/news-article/10397/renovation-wave-strategy-targets-european-buildings/#:~:text=Commissioner%20for%20Energy%2C%20Kadri%20Simson,green%20recovery%20starts%20at%20home.&text=The%20Strategy%20will%20prioritize%20action [Accessed: April 10, 2021]

[6] Corrado V, Ballarini I, De Luca G, Primo E. Riqualificazione energetica degli edifici pubblici esistenti: direzione nZEB. Studio dell'edificio di riferimento uso uffici della PA nella zona climatica Nord Italia (zona E: 2100< GG≤ 3000). Report RdS/PAR2017. Roma: ENEA; 2018

[7] OECD. Integrating Distressed Urban Areas. Paris: OECD Publishing; 1998. DOI: 10.1787/9789264162884-en

Chapter 13

Urbogeosystemic Approach to Agglomeration Study within the Urban Remote Sensing Frameworks

Sergiy Kostrikov and Denis Seryogin

Abstract

The spatial arrangement of human activity within urban areas is normally provided by areal management, and its effective provision is a complicated problem. The current urban development causes a number of problems and urgent challenges, which can be met and resolved exclusively on the basis of innovative scientific and technological advances. The main research objective of this chapter is to represent the authors' theoretic concept of the urban geographical system combined with the original Urban Remote Sensing approach based on the advanced technique of airborne LiDAR (Light Detection And Ranging) data processing. The authors attempted to prove that the presented concept could contribute to an understanding of the urban agglomeration as an urbanized spatial entity. The chapter explains in what way the urbanistic environment is a quasi-rasterized 3D model of actual city space, and the urbogeosystem (UGS) is a quasi-vector 3D model of the hierarchical formalized aggregate of UGS elementary functional units–buildings, both can efficiently simulate and visualize an urbanized area. Web-based geoinformation software for LiDAR data processing with the objectives of urban studies has been introduced together with its key functionalities. The population estimation use case has been examined in detail within the presented approach frameworks.

Keywords: urbanistic environment, urbogeosystem, urban remote sensing, LiDAR, automated feature extraction, web-based software, population estimation use case

1. Introduction

The continuing significant growth of population all over the world, but, first of all, in developing countries, forces scientists to seek new advances and solutions in Demography and Urban Studies domains. These two subject areas primarily mean increasing involvement of the innovative approaches and techniques related to geo-information technology (GIS) and to the urban remote sensing (URS) field [1–3]. Since the continuing growth of the total world population takes place together with the phenomenon of urbanization, the relevant information systems intended for the survey of these two connected processes have to possess some bidirectional modeling and analyzing characteristics, which would overlap both demographic and urbanized issues.

We have just mentioned the significance of remote sensing data processing and GIS-modeling tools to the mentioned extent. This role can hardly be overvalued, taking into account that many from contemporary cities and their affiliated areas have become to act for several recent decades as more and more complicated *urban systems* with drastic dynamic changes within the relevant geographical space and with systemic specific impact on involved people movement and behavior [4–7].

If *an urban agglomeration* can be considered as a highly developed spatial entity of urbanized areas [8], then the approach of *the urban geographical system – urbogeosystem* (UGS) [3] should be applied for examining a number of relationships among the constituents of this system, which may definitely demonstrate its core feature – *the complexity*. Since the complexity is a key description of the contemporary urbanization process too, a whole issue of the spatial urban regularities may require to be evaluated by taking into account not only spatial but purely geographic issues. Both the mentioned rapid urbanization growth, and its attendant alterations in old, and in new cities do not allow to examine any other alternative to an acceptance of a city phenomenon as this just mentioned entity – *an urbogeosystem*, which operates within a certain extent of the geographic space.

It is also necessary to emphasize that the key characteristics of contemporary urban development, which has its effect in forming agglomerations, have caused a number of challenges that require innovative technologies in urban studies. These challenges and responses to them can be summarized in the following way [9]:

- With rapid development and alterations in urbanization, the studies of urban systems become more and more sophisticated;

- First of all - in developing countries, the number of cities has been substantially increased and the urban territories have been enlarged with a rapid speed in several years only;

- Fast-growing regions with a huge variety of extensive urban constructions become more and more numerous;

- A necessity for accurate terrain models for urban planning and landscape architecture as well as relevant sophisticated spatial data processing becomes quite necessary;

- A need for an effective automated survey of buildings to determine quantity and quality characteristics of changes that take place over some period of time;

- Provision of precise environmental monitoring over the key cities in the regions with an intention to obtain extensive data of the URS category: optical and infrared imageries, LiDAR (Light Detection and Ranging) point clouds, and radar imageries.

Although the urban areas cover only 2% of the globe surface in recent years, they include more than half of the world population, and consume more than three-quarters of the total generated energy. The latter produces more than 80% of the greenhouse impact [10]. It is understandable then, why a problem of optimized growth of urban settlements has been a major problem for residents, urban developers, and city authorities for many centuries already. The category of "urbanism" itself appeared more than a century ago [11], while the first statement that an urban agglomeration might represent the core definition in the theory of urbanism occurred with the introduction of the "megalopolis" entity in the middle of the twentieth century [12]. The author of this latest reference stated that routine urban areas gradually would transfer into mentioned megalopolises by joining and changing nearest semi-urban areas and rural neighborhoods. Monitoring this settlement growth became more and more complicated phenomenon, that was why some further research focused on the necessity of the urban system approach together with various sophisticated mapping techniques, which we have mentioned already at the beginning of this introduction [4, 6, 13].

Data of various remote sensing approaches, different GIS platforms, and modules provide the application of a variety of modeling techniques for resolving fundamental riddles related, for example, to spatial dimensions of the agglomeration growth. These techniques may belong to different scientific domains, e.g., fractals and theory of chaos [14], unsupervised classification [15], the algorithm of cellular automata [16], fuzzy logic [17], automated feature extraction [18], analytic hierarchy procedures [19], urban change detection [20], and several other ones. Even being quite diverse, all mentioned methodical solutions can effectively contribute to both estimations of the urban agglomeration expansion to the neighboring rural environment, and to the description of a relevant urban system according to key features of its internal and external relationships and impact.

The main **research objective** of this chapter is to introduce the authors' theoretic concept of the urban geographical system, and this concept is combined with the original URS approach to simulation of *the urbanistic environment* as a model of *a real city domain*. This urban remote sensing approach is based on the advanced technique of airborne LiDAR data processing. A use-case of population estimation on the base of building geometries and topology of urban space both modeled within the urbogeosystemic approach is described in detail in the finalized section of the chapter.

2. The concept of the urban geographical system

Earlier research completed in the fifties-seventies of the past century normally defined an urban system as not more significant entity, than a straightforward set of cities (or smaller settlements combined in a united urban territory) with some relations among these separate units. Nonetheless, there were two seminal books in the second half of the seventies, which represented some *regular structure* in the systems of cities [4, 21]. Probably, these publications were that trigger, which initialized actual urbo-systemic research somewhat later. The authors insisted, that they merely summarized within an applied perspective some concepts and methods, that had been developed as earlier as in the fifties [21, 22]. Although, all these publications, from our point of view, represented only few relevant research samples, which could be reliably determined as some phenomena of the pure emergent features of either a system of the city (separate districts within one urban area as a systemic entity) or a system of several different cities.

Introducing once a definition of an urban geographical system [3, 9], we attempted to extend and develop some basic ideas of the urban system delineation represented by various scientists in former publications [4, 6, 23, 24].

Empty city spaces between buildings and other infrastructural objects within urban territories are much more complicated according to their daily dynamics than they were even 10 years before. It means the schedule of these spaces filling during a day with residents, both static, and moving objects has altered drastically. By choosing the appropriate GIS-modeling technique we can simulate the mentioned dynamics and record it in a certain formalized mode within the frameworks of the model of the urbanistic environment mentioned above.

The urbanistic environment (UE) is a quasi-rasterized model of a continual nature of actual city space and this space key features, which can be visualized as a space limited by various surfaces and can be represented directly by these surfaces. Thus, it can be reasonable to suppose, that the UE also possesses a continuality of the object it represents. The UE continual nature can be contrasted with *the discrete nature of an urbogeosystem – the hierarchical formalized aggregate of elementary functional constituents of its natural analog, which may demonstrate some emergent (systemic) properties. The UGS can be visualized by various 2D vector graphical primitives on a plain (points, lines, polygons), and by quasi-vector 3D primitives in the three-dimensional space.* All emphasized 2D/3D primitives combine a particular *formalized view of the urban space.*

Taking into account modeling characteristics of UE and UGS, quasi-rasterized and quasi-vector ones, correspondingly, and referring to the essence of real objects both models represent – physical environment of a real city (modeled by UE) and sets of separate features in it (simulated by UGS), a research and developing procedural consequence *Initial/derivative data* = > *UE*= > *UGS* can be easily placed within the frameworks of raster-vector transformations. The latter is a subject of routine GIS functionality. Applying this functionality is the only understandable procedure, which can contribute to answering the question: if a given city does rather belong either to urban systems or to *urban sprawl* [25].

The first outlining of the urbogeosystem was suggested in our earlier paper and it laid in a completely ontological aspect. According to it, an urbogeosystem is "...The UGS is an urban system located within a definite extent of the geographic space; it is an unsustainable social-environmental system which is also a united entity of various architectural features and dramatically changed natural ecosystems..." [3, p. 110].

Those literature sources, that introduce various descriptions of *the urban system structure* [4, 6, 21–24, 26], imply each separate systemic component in a set of cities as *a point feature*, while interconnections and relations between each pair of these single objects – as *a linear feature*. Then a certain group of cities within the boundaries of a definite region, are located in a certain *areal feature*. Instead of "a city" as a separate unit, we can accept "a city ward (district)", then obtain a set of such units within a particular urban territory. In this way, we can enter a completely another research scale, but in both larger, and smaller scales points, lines, and areal fragments (regions or parcels) are key components of an urban geographical system. The geographical scalability can be applied then, while a single object (a city or a ward) is *a point* in one scale, but on another, larger scale it becomes *an area*. In a similar way, we can apply scalability to *the lines* and obtain the linear features of different magnitude [7, 9].

Let us assume that initially, a set of N cities indicates some $N*N$-matrix, in which "point cities" interact in different terms of human, industrial, trade, transportation, and information traffic, composing a picture of *an external urbogeosystem*. On the first step of scalability, a matrix would also define a number of linear features, which mirror spatial linkages in *an external urbogeosystem* in the mentioned terms. On the second scalability step, not the same, but similar matrix depicts N districts of one city only and all interconnection pairs among them, which exist in an *internal urbogeosystem*. In the simplest definition, it is a set of districts in one city, as we already mentioned.

On the basic fundamentals of the UGS approach introduced above, we elaborated and proposed *the algorithmic sequence of the UGS research with GIS tools* [3]. It consists of several algorithmic blocks that sustainably combine a thematic geographical model, urban remote sensing technique, and both basic and customized GIS functionalities. The key algorithmic blocks in this scheme are as follows (**Figure 1**):

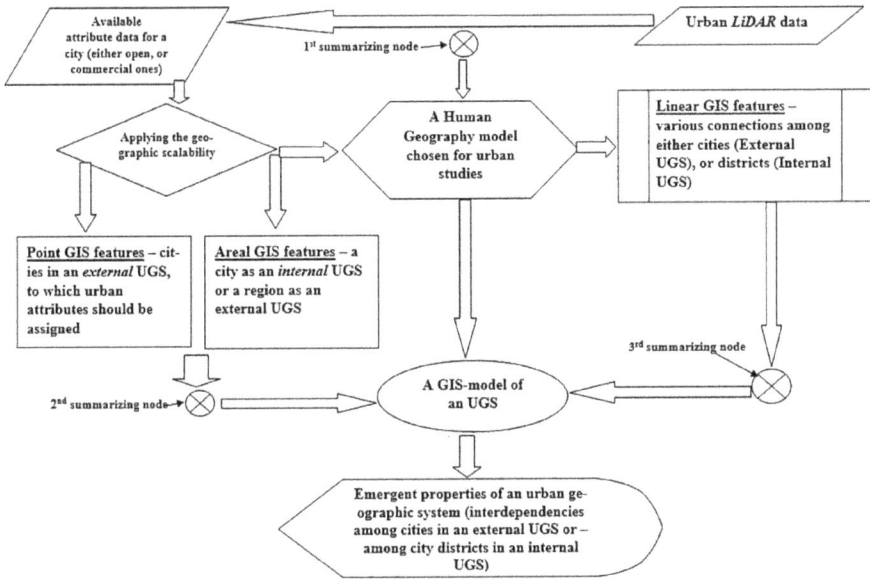

Figure 1.
The algorithmic flowchart of the UGS study with GIS tools [3, p. 111].

- Gathering with LiDAR and initial processing of the urban remote sensing data;

- Adding supplementary attribute data obtained from other than urban remote sensing sources;

- Choosing a thematic model for simulation of the city residents' behavior (a model from Human Geography or Geography of Population domains);

- Applying the geographic scalability and further delineation of point, linear, and areal features as the content of both an external, and an internal UGS;

- Composing an ultimate GIS-model of an urbogeosystem, which consists of two components: a quasi-rasterized model of the urbanistic environment, and, in fact, a quasi-vector model of an urbogeosystem;

- Adding available attribute data (semantic data, as a rule) to a model of the urbogeosystem and generating derivative attributes for this model (geometric attributes and metadata, as a rule);

- Finalized results of applying UGS-approach as delineation of various emergent properties for a given urban territory. These results can be employed for various thematic use-cases in different municipal and other applications.

We have already published several papers in the urbogeosystem approach, examining various aspects of this concept: its basic fundamentals [3], its applicability to the Smart City concept implementation [9], its possible involvement in the multifunctional approach to the 3D city modeling [27], some from UGS-basics were applied to the structural analysis of agglomerations in Kharkiv region (Ukraine) [28]. The latest research accepted a well-known definition of agglomeration as a large city (as an agglomeration core) with all its nearest townships and

#	Agglomeration hierarchical level	A settlement as an agglomeration center	Topical SGS/corresponding urbogeosystem
1	Mega-level	A large city (nearly or over a million of residents)	Interregional, regional SGS/*internal UGS of the first rank*
2	Macro-level	A city as a center of *an oblast* – a larger territorial administrative unit in Ukraine	Regional SGS/*internal UGS of the second rank*
3	Upper meso-level	A town as a center of *a rajon* – a smaller territorial administrative unit in Ukraine	Under-regional SGS/*internal UGS of the third rank*
4	Lower meso-level	A town, a township as a center of a united territorial community	A united territorial community as an SGS/*internal UGS of the fourth rank*
5	Micro-level	A township, a large village	Local SGS/*internal UGS of the fifth rank*

Table 1.
The corresponding agglomerations and urbogeosystems hierarchy for the Ukrainian population settlements (an updated table from [28, p. 4950]).

the suburbs. We attempted to define that all these settlements are characterized by various interrelations. Thus, a new entity of aggregated functioning appears, which is common for this big urban territory, and for small towns and villages around it. This urbanized compact entity of settlements was accepted as a spatial systemic formation with all relevant features of the urban geographical system. Therefore, it can be reasonable to apply to an agglomeration study that algorithmic flowchart presented in the illustration above (**Figure 1**). Socio-geographical survey over the East of Ukraine, in particular – within Kharkiv region, proved that agglomerations as spatial patterns of different hierarchical levels can be delineated, not only as social geographical systems (SGS), but also as both external and internal urgogeosystems, and they are significantly present in the territorial arrangement of this region. Taking into account the general concept and the surveyed results, we suggested the hierarchy of the delineated agglomerations with respect to the necessary update of the territorial division of Ukraine (**Table 1** is updated from [28]). Thus, a regional system of settlements has been proven to be not only a mosaic of all five agglomeration levels, which may overlap each other in the spatial extent but also – *the spatial hierarchy of urbogeosystems*. Consequently, the local agglomerations are the urbogeosystems of the fifth, lowest rank. In other words, they are basic units, elementary ones in the common hierarchy for both urbogeosystems, and for agglomerations. It follows from **Table 1**, that various hierarchical levels of the settlement spatial structure can be distinguished – from microlevel to mega-level, and these levels correspond to a particular social geographical system, and to a particular urbogeosystem.

Concluding the second section of this chapter, which has introduced the UGS approach with this example of agglomeration research, it is necessary to address the following issue. This approach can be directly provided for examining agglomerations according to its main features introduced in this chapter section:

- gathering, combining, and processing urban remote sensing data, in particular – LiDAR point clouds;

- choosing an applicable Human Geography and Demography models;

- refining derivative digital information and converting it into the GIS-primitives;

- defining geospatial aspects of all interrelated contents and conditions of actual urban environment, and consequently generating in three steps *quasi-rasterized model of urbanistic environment = > quasi-vector model of urbogeosystem = > model of agglomeration clusters.*

Further in this text, we examine some steps of this consequence more in detail, while taking it for granted, that a strong spatial aspect of the urban research necessarily implies the GIS/URS processing procedures, tools, and operations efficient involvement in this research, what we attempt to outline as various issues in the text below.

3. Urban remote sensing with LiDAR for digital cities

3.1 Automated feature extraction

Automated reconstruction of the sets of various buildings is yet a serious challenge on the way to 3D digital city modeling. Other significant tasks can be affiliated with it, for example, outlining the Smart City concept implementation [9]. Exactly for the two latest decades, LiDAR data and its processing results have become real alternative data sources to optical and multispectral imageries with respect to generating a three-dimensional representation of urban territories [2, 29, 30]. Being able to collect straightforwardly dense and accurate 3D point clouds over both urban, and rural features, the technology of the LiDAR survey provides a reliable and beneficial data source to this end. Almost all LIDAR devices are either Airborne types (ALS, aircraft-based) or Terrestrial (Mobile, MLS) (vehicle-based), as well as drone-platform ones.

The key processing and simulated procedure intended for building digital cities, while the latter is a basic fundamental for urbogeosystem delineation, is the *automated feature extraction* (AFE) from point clouds generated by LiDAR [31]. Normally the automated feature extraction is based on both optical satellite images of high resolution, and on LiDAR datasets generated by airborne, terrestrial, and drone platforms on regional surveys [32]. The latter ones are usually provided by strips and then combined as three-dimensional point clouds [2]. AFE output is the key tool that makes digital urban models. Various approaches, methods, and solutions that *detect*, *extract*, and *generate* building models with any selected alternative technique, all compose a highly significant research domain [33].

This latter statement can be accepted by default, because a whole approach mandatory means 3D automatic, but desirably - *smart mapping* of the multi-scalable urban environment, that is of the extreme complexity. Moreover, as it has been mentioned already if exactly LiDAR data become in recent decades an efficient alternative to imageries obtained by traditional satellite remote sensing, then this data source should become a subject for various approaches and algorithms, as previously traditional URS was. These approaches and algorithms should differ for various procedural stages, and suggest robust solutions separately for 1) building detection, 2) extraction and 3) building reconstruction steps [34].

The automated building/other infrastructural feature extraction procedures can be fulfilled by three sub-procedures, as was already stated above, i.e., building detection, building extraction/segmentation, and building reconstruction [34–36]. All three sub-procedures mentioned may not be clearly distinguishable. To complete a single stage of automated extraction of buildings may not yet be satisfactory enough for practical applications due to the great complexity of actual urban architecture, which we always face while modeling the urbanistic environment on the first step of the urbogeosystem delineation. Different additional sophisticated algorithmic solutions should be involved, for example, those ones, which assist in distinguishing between building constructions and urban vegetation, while processing an airborne point cloud [37].

Traditionally being within the frameworks of our original multifunctional approach to LiDAR point cloud processing [27, 31, 38, 39] we have to consider only those methods, which use exclusively LiDAR data, so that to utilize the building geometric and topological properties only, and not any other urban landscape characteristic except *urban topography*. In this way we have to pass through the mentioned above trinity of steps: building detection, segmentation, and reconstruction ones, while topography is generated upon the first step from these three while discriminating so-called "ground" and "non-ground" points when processing LiDAR datasets.

It is commonly accepted understanding that the model, which includes not only the ground as the topography, but other features – the discrete ones, is not a digital elevation model (DEM), but *a DSM – a Digital Surface Model*. According to existing references before the sustainable usage of LiDAR point cloud for topographic modeling, the digital surface model was normally calculated using various imageries, hybrids (imageries + point clouds), and feature pyramids [40]. The final DSM surface is refined then on the base of local adaptive regularization techniques provision. While the urban topography has been generated already, the building detection step is grounded on the fact that buildings, as a rule, should be higher than the neighboring topographic surface. This is normally estimated using various mathematical morphology techniques through the DSM [41].

In our original approach to LiDAR point cloud processing with the intention to separate "ground" and "non-ground" point as a mandatory premise for further non-ground features detection, segmentation, and reconstruction, we have provided the following steps, which can be introduced in the following summarized way proceeding from several relevant references [27, 31, 39]. The initial unique step assumes the delineation of both DEM and a DSM from the airborne point cloud raw data, in which point density should be preferably within a range of *10–80 points per square meter*. The proposed method of DEM generation accomplishes a classification of the original data as "ground" points versus "non-ground" points by robust estimating procedure, which has been described in detail in one of our latest papers [39]. In all consequent algorithmic steps of modeling UE, the *heavyweight models* generated by triangulation and interpolation, and *lightweight models* generated by clustering and segmentation are used, but not the original data points. DEM is subtracted from the DSM. The output results of building detection and segmentation, i.e., the delineation of individual building footprints can be provided with a connected component analysis. A set of the selected feature candidate regions can be arranged. Then a planar surface segmentation can be executed, is based on the analysis of the DSM vector variations. The output result of this step is crucial for finding planar parcels of buildings. These parcels are expanded then by applying a bunch of the region growing algorithms. The neighborhood connections of these parcels are determined, and a simplified model resembling the roof structure in a certain building is generated. A Voronoi diagram can be created for extraction of neighboring joints and connections of numerous facets that compound roofs and walls in the heavyweight models, while planar segmentation and customized topological rules are used for segmenting and combining lightweight models of simplified buildings with gable roofs.

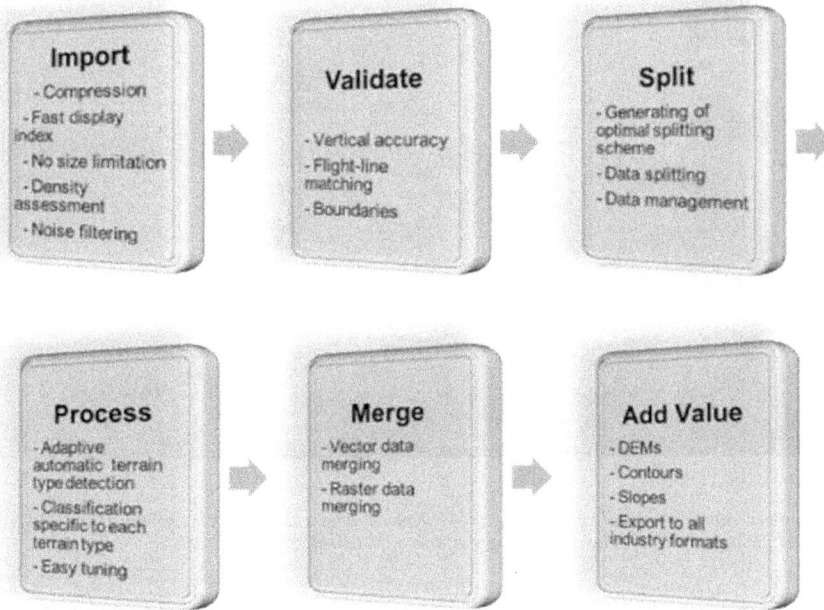

Figure 2.
Some key constituents of the AFE-pipeline are intended for the generation of both urban topographies, and building models from LiDAR point clouds.

A summarized AFE-pipeline relevant to LiDAR data processing, which contains some of the basic fundamentals presented in this subsection, is visualized on the following flow-chart composed by this chapter authors (**Figure 2**):

The flowchart presented not only depicts the main components of the automated feature extraction pipeline but also is some kind of a presentation due to the digital city content creation. The latter with the introduced UGS approach consists of two phases, as we already explained:

1. modeling the quasi-rasterized UE, and

2. simulating the quasi-vector UGS.

Both phases contain in one way, or in another all six blocks of this flowchart. Nonetheless, the *first phase* (directly affiliated with modeling the UE) does definitely include the *urban data mining* complex (*Import*, *Validate*, and *Add Value* blocks), while *the second phase* implies the implementation of pre-processing, processing, and simulating solutions (*Split*, *Process*, and *Merge* blocks) for the presentation of the three-dimensional geometry of each separate building and sustainable topology for the sets of buildings in a digital city. The output results of the flowchart, which is in **Figure 2**, may be provided in several formats, e.g.,. gLTF,. KLM,. DAE,. B3DM, etc. Nonetheless, a core inner format is. OBJ. Simulated features of a digital city are produced with the representation of their borders. A whole *3D urban scene* can be depicted as a set of building constructions with the continuality of their bounding walls, vertices, edges, and supplementary outhouses, and this continuality can be described by certain parameters of *urban geometry*. In this way, the urbanistic environment is simulated. Due to the mentioned continuality, a scene can also demonstrate the topological interdependencies of buildings

Figure 3.
The urbanistic environment and a fragment of the UGS modeled for a district of Washington, D.C. (USA) and visualized in the interface of a cloud processing platform: EOS LiDAR tool – ELiT cloud.

among themselves and with non-housing urban features and various infrastructural objects. Urban features are visualized according to the *CityGML* LODs (Level of Detail) standards [42].

Thus, a partial fragment of an internal urbogeosystem can be modeled and visualized in the 3D scene with spatial, geometric, and semantic characteristics, which can be exposed for each selected feature, or for a number of them. A number of LOD 1 (a simplified box-model of a building) models that correspond to the CityGML 2.0 concept are visualized for the Washington, D.C. urban area in the interface of a web-GIS software, in which elaboration participated both of this chapter authors. This interface sample relates to the cloud processing platform of this software (**Figure 3**).

Our models of urban objects exposed on the illustration above possess all necessary characteristics of 3D digital city models. While many other three-dimensional objects seem to be predominantly used for display, it is reasonable to emphasize that these simulated features presented in a 3D Scene can be increasingly employed in a number of domains within a large range of tasks beyond the direct visualization. Such perspectives can be opened if we accept simulated sets of building models as the aggregations of elementary functional features of an urbogeosystem. The reasonability of such an assumption has been proved by the authors in some previous publications [3, 9, 39].

3.2 Web-based geoinformation software for the urbogeosystem approach implementation

We have already mentioned that both authors of this chapter participated in research and development (the first author – as ahead of this R&D) of the web-based and cloud-based versions of the geoinformation software focused on LiDAR data processing for urban studies, what took place in the EOS Data Analytics Company (https://eos.com/eos-lidar/). Common fundamentals of the Automated Feature Extraction determine our core algorithmic structure named as the *High Polyhedral Modeling* (HPM) and elaborated within the frameworks of the integrated *BE (Building Extraction)* /*BEF (Building Extraction with Footprints)* /*CD (Change*

Detection) /DEM-G (Digital Elevation Model Generation) functional pipeline of ALS/ MLS data processing [27]. HPM produces building models with numerous facets.

According to the whole HPM workflow, two following problematic issues may occur with great probability: #1 – to provide more precise classification of both "vegetation" points, and "building" points is crucially necessary; # 2 – to elaborate a definite method in what we have to define the building topological and geometric properties in those cases when point cloud data are incomplete. All possible solutions for both issues should be preliminary outlined, and we took it into account while developing our basic original algorithm of LiDAR data processing and proposing some supplementary technique that has to be accomplished in parallel with core algorithm operation.

Within frameworks of our conceptual R&D approach buildings are accepted as the key man-made features in the modeled urbanistic environment. According to the HPM output results it consists of numerous continuous surface segments (polyhedrons) that compose *the trinity content of the city space*: urbanized topography, building surfaces, and empty urban spaces between buildings that are separated by two previous issues.

There are two platform versions on which *EOS LiDAR Tool*, *ELIT* software, can operate: a cloud processing version, as *ELiT Cloud*, that applies to AWS instance service power (**Figure 3**), and a typical client–server, web-based application as *ELiT Server*. The urbanistic environment of Toronto-City as a model reconstructed by the HPM pipeline may look like follows in the *ELiT Server* interface (**Figure 4**).

Corresponding functional tools of both *ELiT*-software platforms, which are set within the HPM frameworks are *BE, BEF, CD*, and *DEM-G* tools. The *BE/BEF* tools extract *original building footprints* from point clouds while modeling [39, 43].

In addition to the High Polyhedral Modeling, we have developed the alternative AFE-technique, such as is the Low Polyhedral Modeling (LPM) approach, which is based on procedures of planar segmentation and clustering of LiDAR point clouds rather, than on their classification (in the case of HPM). The LPM technique is primarily intended to extract low-rise buildings of either rural areas, or city suburbs

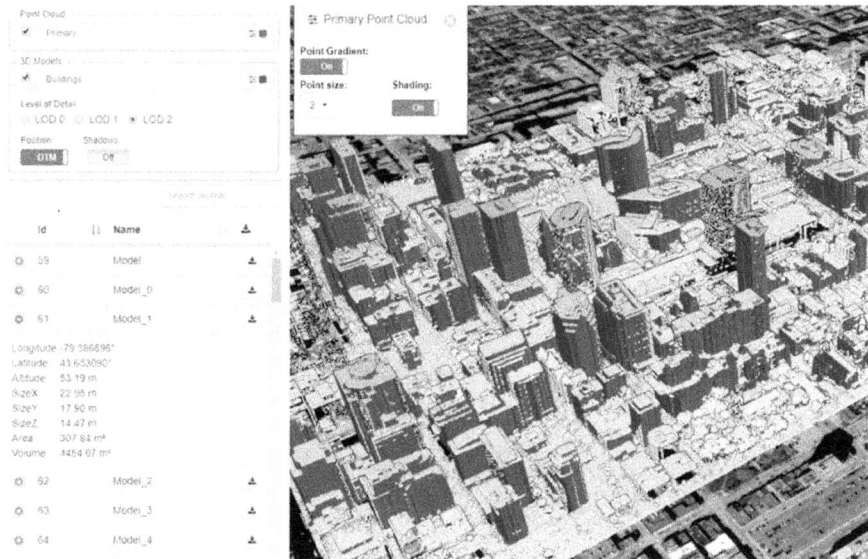

Figure 4.
The UE of Toronto-City (Canada) is modeled in the ELiT server interface.

Figure 5.
Lightweight models of elementary functional constituents of the urbogeosystem of Lubliniec-City (Poland) in the ELiT cloud processing platform interface.

as light-weighted models, which consist of only a few facets. The relevant functional tool of the LPM approach is the *BERA (Building Extraction Rural Area)* tool. The *BERA* instrument employs *the third-party building footprints* while modeling [39, 43]. If the HPM-technique with its *heavyweight models* is more preferable for simulation of the quasi-vectorized UE (**Figure 4**), then the LPM-method – for creation of the 3D quasi-vector *lightweight models* of buildings and other features as elementary functional constituents of a certain urban geographical system (**Figure 5**).

If not taking into account such research entities as UE and UGS, but evaluating only building modeling itself, then it can be emphasized, as we have already mentioned above, that the BERA functionality is an application for detection, extraction, and modeling of low-rise housing located in city suburbs and urban areas. The HPM approach is recognized to be more efficient for simulating high-rise buildings of downtowns.

Contrary to HPM, with which the BE tool is affiliated, this alternative AFE technique, LPM, and the *BERA* tool, as it has been already underlined, are strongly based on planar segmentation, clustering, and reconstruction of polyhedral building models. In comparison with the HPM pipeline, both planar segmentation and clustering substantially decrease the number of polyhedrons as constituents of a building model extracted. Thus, we attempted to provide an efficient update and an applied realization [43] of the advanced theoretical approach known as segmentation and reconstruction of polyhedral building roofs [18].

Both software, a client–server application, and a cloud-processing platform can be run from a web browser installed on a user's workstation. According to its architectural scheme, the *ELiT* software performs transmitting procedures between the Processing Core, that is on a server, and a Client, while providing such operational sets as Data Management (uploading, downloading, etc.), Task Management, and interactions between the Core and a database. Finally, a Client provides a user graphical interface and the building model/topographic surface visualization. A Java Script based library - *Cesium 3D Tiles* is employed for this display, https://cesium.com/

4. Population estimation and mapping on the base of elementary functional units of an urban geographical system

Urbanistic environment and urbogeosystem modeling on the base of automated building feature extraction, further mapping of extracted and reconstructed features, and finalizing 3D digital city generation for both urban, and rural areas can be highly essential for many industrial applications. It would be reasonable to define four next key categories of the urbogeosystemic approach in its applied perspective. Each of these categories may directly relate to an agglomeration study: 1) common urban planning and design, urban environment visualization, promotion and learning of urban information, 2) specific urban planning, 3) those usages that are not directly related either to planning, or to visualization, for example, city population estimation as an operational procedure "on fly", that can be completed on any date between census, 4) commercial sector and marketing, including infrastructure, facility services related to specific urban information visualization, and urban data mining.

The range of those industrial applications that are pertinent, for example, only to *BE* and *BERA* building extraction functionalities may be lengthy enough: urban and municipal planning, augmented reality for gaming industries; environmental planning and monitoring, insurance policy and procedures, optimization of transmitter placement for telecommunication, locational based services, navigation, housing simulations, urban microclimate investigations, and shadow estimation. In all these use cases a building model is the primary object of interest, while exactly the sets of these models examined within the frameworks of the urbogeosystemic approach can, in our opinion, act as those elementary functional constituents of the actual city environment, which compose its *adaptive renewal cycle* with all four basic functions: *exploiting*, *conserving*, *releasing*, and *recognizing* [44]. These functions can be efficiently defined with the UGS approach, if we consider urban (agglomeration) growth in the context of this cycle, while also applying to spatial morphology, as those authors to whom we have just referred to, suggested once.

The point of view introduced in the above paragraph can be accepted as a forcible argument for choosing exactly *a set-of modeled buildings-level* for an urban population estimation use-case as a dominant one in a perspective of that agglomeration research, to which the UGS approach could mostly contribute. If an urban agglomeration is "...the future spatial organization of cities" [8], then any proven method of robust estimation of the population on the base of the urban spatial morphology are expected to be valuable enough.

Taking into account the routine public scarcity of real population values in various city district configurations of a real city, any more or less reliable procedures for evaluating numbers of residents between two censuses, which temporal gap may be up to ten and even more years, can hardly be overvalued [39]. Therefore, even an approximate estimation within a certain selected AOI may be highly necessary for optimizing routine municipal management. It has been evidently proved by the latest events in urban areas due to the modern pandemic phenomenon.

If we accept both separate buildings, and the sets of them as elementary functional urbogeosystemic units within a certain geographical extent of a city, then it is evident that not only different linkages caused by people movement between these sets combined in modeled city districts should be taken into account for calculating a number of residents in a certain area-of-interest (AOI), but also – building geometries themselves. The latter parameters can be the most precisely reconstructed just by LiDAR data processing, which proves the applicability of our approach to agglomeration research in overall extent. The UGS approach to population estimation has been supplemented by some existing methods of GIS /urban remote sensing application within this use case. This GIS/URS application is mainly concerned with the urban block- and census track-level of a number of residents calculating [45–48].

In one of our former publications, we have already presented "the step-by-step building space-metric method (BSMM) of population estimation" [39]. This method presents the series of procedures for any AOI, block, and district population estimations based on the building geometric and city space topological parameters derived from airborne LiDAR data processing. As it has been stated in the second section of this chapter, **Figure 1** summarized our whole research workflow, in which the BSMM was accomplished within three following consequent blocks: 1) *A Human Geography model…* = > 2) *A GIS-model of an UGS* = > 3) *Emergent properties of an internal UGS (interdependencies among city districts in an internal urbogeosystem)*. The blocks *Urban LiDAR data* and *Available attribute data for a city* were completed even before this BSMM block-trinity 1)-3), and their output was transferred through the first and the second summarizing nodes to *A GIS-model of a UGS* block (**Figure 1**). *Point-*, *Linear-*, and *Areal GIS feature* blocks are locked to the second block of the mentioned trinity. A whole introduced configuration of blocks is based on building a model produced by two blocks: 1) *A Human Geography model…* = > 2) *A GIS-model of a UGS*. This model is used for the calculation of interactions due to people movement among city districts and *census tracks* in the internal UGS. Geoprocessing aspect of the methodology introduced in this paragraph consists in adding population data to the metadata of. OBJ files presenting building models, and then visualizing in a *Cesium Scene* by the gradient color method.

A study area and data sources are related to the city of Boston, Massachusetts state, USA, and overlapped most of this urban territory. While completing the *ELiT* Geoportal web resource [39], we applied to airborne LiDAR data of open access as to one of the USGS projects available from: ftp://rockyftp.cr.usgs.gov/vdelivery/Datasets/Staged/Elevation/LPC/Projects/USGS_LPC_MA_Sndy_CMPG_2013_LAS_2015/laz/. The relevant census data were available from the U.S. Census Bureau's Web site (http://data.census.gov/), and from the Bureau of Geographic Information (MassGIS) site – the regional data of the 2010 U.S. Census [49]. A seamless, Massachusetts statewide digital map of land use has been taken from [50]. We assumed that it would be possible to obtain the territorial distribution of the population from UGS elementary functional units – the building models and *their affiliated volumes*. The BFT- parameter – Building Function Type (first of all – *residential, non-residential*) has been used as a key semantic attribute. Because of the lack of reliable semantic data and a certain vagueness of a particular building belonging to a certain land-use class, we had to apply to the original technique of automated definition of building type by its topology and geometry [39]. In total, the following stages complete the whole URS/GIS-tools pipeline of population estimation within the urbogeosystemic approach with BSMM:

1. *The preliminary data preparation stage* for population estimation on the basis of the UGS approach with LiDAR data processing was like follows. LiDAR point clouds as *.LAZ files were downloaded from a few USGS projects through the web reference mentioned above. Building footprints were downloaded from the Open Street Maps (OSM) resource https://developer.here.com/products/data-layers?cid=. All footprints were combined in a united.SHP file by the *Save as* =>. SHP tool, which can be applied for any vector layer in the *QGIS 3.10* GIS platform. This combined file might contain information about 1) building population counts, and about 2) classes of buildings (a class of residential ones and a few classes of non-residential buildings – commercial, industrial, educational). This information can be available from the OSM footprints, but footprints with it overlapped not more, than 5% of their total number only.

Thus we have to apply to alternative information sources from [49, 50], as we have already mentioned above. Thus, after completing the preliminary stage of the population estimation use case we obtain the following:

- three polygonal layers as *.*SHP* files with the U.S. Census 2010 data [49] on the following three levels (both continuous, and random census blocks are used as samples): 1) census parcels; 2) sets of blocks; 3) separate blocks and sub-blocks; for each layer population data are stored in *"POP100_RE"* field;

- a statewide polygonal layer with boundaries of the land use classes [50].

2. *Modeling the urbanistic environment with a number of quasi-vector models* in it. The *BERA* tool has generated more than 350,000 City GML LOD1 models within a selected contour of the urbanized Boston territory. Thus, those sets of. *OBJ* files that can be associated with census parcels, blocks, and groups of blocks, have been obtained. Further, possessing already generated *.*OBJ* files, we have to add census information to them.

3. *Enriching.OBJ files of UE-quasi-vector features with census information*:

- A point layer with geographic coordinates of each *.*OBJ* has been created. It contains the centroids of building footprints. The *BERA* tool has also generated for each *.*OBJ* a *.*JSON* file of the same name (an ordeal model number or *OSM_WAY_ID* of its footprint), and this *.*JSON* contains various metadata for a model, e.g. a computed volume of a building. We have added to the metadata dictionary the key *population* and a quantitative value for it.

- Using a customized Python script, we have processed all *.*JSON*-files in the *BERA* output folder and stored resulted data in a *.*CSV* file with a header as: *name, latitude, longitude, volume*.

- Importing a.*CSV* file to *QGIS 3.10* (menu *Layer= > Add Layer= > Add delimited text layer*), where all points presenting models have been localized, while names and volumes have become their attributes.

- Layers of land use and census tracks have been added to a *QGIS* project. Thus, for example, all models can be added to a *2D* map so that to define spatial belonging to a certain class (**Figure 6**):

In the same way as on the visual above, a layer of building models has been placed on the census parcels.

- The layers of point models, land use, and census parcels have been reprojected into *EPSG 26919* (a projection of.*LAS* files relevant to the territory of Boston) with the *QGIS* tool *Geoalgorithms= > Vector general tools= > Reproject layer*.

- The *Land use class* parameter has been recorded in a point layer of models by the tool *SAGA= > Add polygon attributes to points* (field *LU05_DESK*). According to the rule, *Polygon contains point* a point layer has accepted the information about a land-use class for any model as a point.

- Then a record of a population value of each census parcel or block should be provided as a semantic attribute for each model point, which falls into this

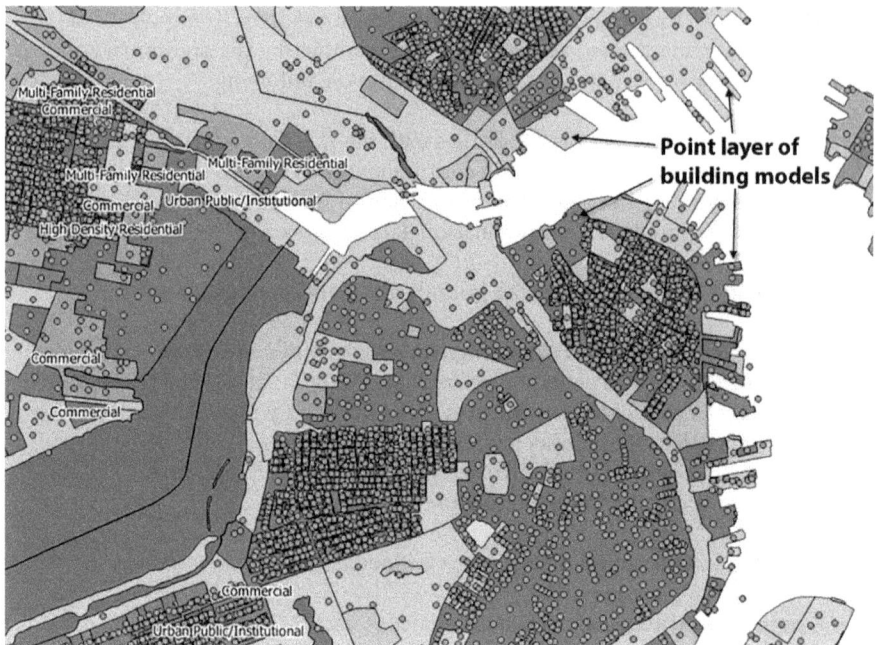

Figure 6.
Visualized in the QGIS-interface the points of building models (footprint centroids) located through different classes of land use in a fragment of the urbanized territory of Boston. The complete land use legend is available from [50].

area. A procedure is completed with the *SAGA = > Add polygon attributes to points* tool. A new file in the point layer attribute table is titled in the same way with the attribute table of the census parcel layer - *"POP100_RE"*.

- A summarized volume of residential buildings for each census parcel should be further provided. Firstly, it has been necessary to use the tool *Select features using an expression* with a query *"LU05_DESK" like '%Resident%'*. Secondly, by the tool *SAGA = > Points statistics for polygons* total volumes of residential buildings have been calculated for each census parcel. Thus, a layer of census parcels has been obtained with a supplementary field – *SUM_volume* (a total residential buildings volume for each census parcel).

- Just as in 3.6 and 3.7 items the polygonal layer information has been recorded in a point layer: a total volume of residential buildings has been recorded in each building centroid (the field *SUM_volume*) that falls in this census parcel.

- The finalized correcting coefficients have been introduced for the sets of buildings located in various census parcels (field *COEF*). These parameters have attempted to take into account the major trends of people movements. It may actually reflect the population spatial distribution dynamics in an internal urbogeosystem, that took place after the latest census, and it was extrapolated from changes that actually occurred between two former censuses. Input for such evaluation can be based both on the information available from [49] and on some supplementary data sources.

- A new float-field has been added to a point layer table – *bldng_popul*. It has computed a ratio through all other fields of this point layer table:

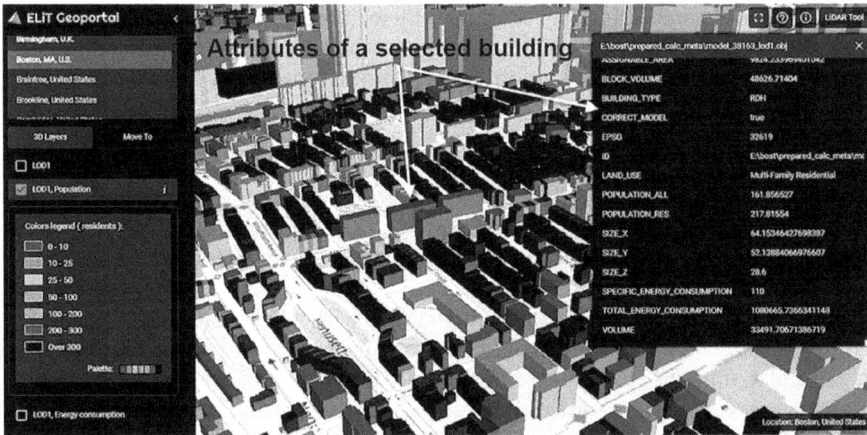

Figure 7.
Resulted from the URS/GIS pipeline stages 1–4 visualization of the building population distribution in the urbanistic environment of Boston-City presented in a 3D scene of the ELiT cloud processing platform interface.

$bldng\_popul = (volume*POP100\_RE*COEF/SUM\_volume$, where *volume* – a building volume, $POP100\_RE$ – a population value for a given census track, in which this building falls in; SUM_volume – a total volume of building in a census parcel.

- *The finalized attribute table* should contain the following fields: model *name* (obtained on step 3.1.); model *volume* (step 3.1); $LU05\_DESK$ (land-use class obtained on step 3.6); $POP100\_RE$ (census parcel population – step 3.7); SUM_volume (a total volume of residential buildings in a census parcel - step 3.8); $COEF$ (correcting coefficients due to probable people movement – step 3.10); $bldng\_popul$ (estimated for a period between census a number of residents in each building – 3.12).

4. *Combined visualization in Cesium 3D Scene of the ELiT software interface* of those results obtained upon the second and third stages of the URS/GIS pipeline – an attribute table from 3.12 has been visualized as a 3D scene. In this way, the urbanistic environment and a viewed fragment of the UGS of Boston-City are presented with the building population distribution evaluated on the base of the urban architectural morphology (**Figure 7**).

While implementing a population estimation use case, it is reasonable to take into account, that some computed extreme numbers of residents can be caused by the errors in the input land use data. For example, a large residential building has been prescribed to the commercial or to any other non-residential class of land-use, while being actually in one census parcel with another, much smaller residential building, and there are only two these buildings in a given parcel. The small building, being prescribed to the residential class properly, has accepted a whole number of residents in a parcel, and a number of residents is drastically exaggerated then.

5. Conclusions

This chapter has introduced the original conceptual research approach concerning the urban geographical system, which is based on urban remote sensing with LiDAR data processing. The authors have made an attempt to prove that the presented methodology and techniques might contribute to the scientific understanding of the urban agglomeration as a highly developed spatial aggregation of urbanized areas. The urbanistic environment as a quasi-rasterized 3D model of actual city space, and the urbogeosystem as a quasi-vector 3D model of the hierarchical formalized aggregate of UGS elementary functional units – buildings, both can efficiently simulate, visualize, and represent an urban agglomeration according to its all representative criteria. The algorithmic flowchart of the UGS study within the suggested approach has been provided, and further research introduction has been affiliated with flowchart blocks.

The URS/GIS pipeline of making a digital city with LiDAR data processing has been examined mainly within an automated feature extraction perspective. In particular, it has been illustrated by the AFE-flowchart of some key processing constituents related generation of both urban topography, and building models from LiDAR point clouds. The possible scheme of digital city creation might consist of two consequent steps: 1) modeling the quasi-rasterized UE, and 2) simulating the quasi-vector UGS.

Web-based geoinformation software for LiDAR data processing due to the objectives of urban studies, in general, and agglomeration research, in particular, should demonstrate its optimal architectural solution as both a client–server application, and as a cloud-processing platform. The latter applies to AWS resources. HPM-technique provided by this software is preferable for the urbanistic environment modeling, while its LPM-method – for model generation of elementary functional units of the UGS – buildings. Each one from the row of software tools – *BE*, *BERA*, *CD*, and *DEM-G* can contribute in a particular perspective to agglomeration research.

Mentioning several thematic applications, which can potentially be resolved within the frameworks of the presented approach, we selected and examined in detail the building population estimation use case as the most relevant one to agglomeration research. A number of building residents, as a rule, are not widely available due to security and privacy reason. Thus, the suggested technique can significantly assist not only in an AOI-population estimation between census but also, e.g., in predicting the agglomeration growth in both short-term and long-term perspectives.

Appendices

AFE	Automated Feature Extraction
ALS	Airborne Laser Scanning
AOI	Area of Interest
AWS	Amazon Web Services
BE	Building Extraction
BEF	Building Extraction with Footprints
BFT	Building Function Type
BERA	Building Extraction Rural Area
BSMM	Building Space-Metric Method
CD	Change Detection
DEM	Digital Elevation Model
DEM-G	Digital Elevation Model Generation
DSM	Digital Surface Model
ELiT	EOS LiDAR Tool
HPM	High Polyhedral Modeling

LiDAR	Light Detection and Ranging
LOD	Level of Details
LPM	Low Polyhedral Modeling
MLS	Mobile (Terrestrial) Laser Scanning
SGS	Social Geographical System
UE	Urbanistic Environment
UGS	Urban Geographical System
URS	Urban Remote Sensing

Author details

Sergiy Kostrikov and Denis Seryogin*
V.N. Karazin Kharkiv National University, Kharkiv, Ukraine

*Address all correspondence to: sergiy.kostrikov@karazin.ua

IntechOpen

References

[1] Weng A, Quattrochi D, Gamba P, editors. Urban Remote Sensing. 2nd ed. Boca Raton, USA: CRC Press; 2018. p. 387

[2] Dong P, Chen Q, editors. LiDAR Remote Sensing and Applications. Boca Raton, USA: CRC Press; 2018. p. 246

[3] Kostrikov S, Niemets L, Sehida K, Niemets K, Morar C. Geoinformation approach to the urban geographic system research (cases studies of Kharkiv region). Visnyk of V. N. Karazin Kharkiv National University, series "Geology. Geography. Ecology". 2018;49:107-124. DOI: 10.26565/2410-7360-2018-49-09

[4] Bourne L, Simmons J. Systems of Cities: Readings on Structure, Growth, and Policy. Oxford: Oxford University Press; 1978. p. 565

[5] Nijkamp P, Perrels A. Sustainable Cities in Europe. London-NY: Routledge; 2009. p. 152

[6] Du G. Using GIS for analysis of urban systems. Geo Journal. 2001;52:213-221

[7] Kostrikov S, Sehida K. GIS-modelling of regional commuting (a case study of Kharkiv region). Actual Problems in Economics. 2016;186:399-410

[8] Fang C, Yu D. Urban agglomeration: An evolving concept of an emerging phenomenon. Landscape and Urban Planning. 2017;162:126-136

[9] Kostrikov S. Urban Remote Sensing with LiDAR for the Smart City concept implementation. Visnyk of V. N. Karazin Kharkiv National University, series "Geology. Geography. Ecology". 2019;50:101-124. DOI: 10.26565/2410-7360-2019-50-08

[10] UNEP. Visions for Change, Recommendations for Effective Policies on Sustainable Lifestyles. The Global Survey on Sustainable Lifestyle. United Nations Environment Programme; Sweden: Regeringskansliet, Ministry of the Environment; 2019. p. 84

[11] Henard E. The Cities of the Future. Royal Institute of British Architects. Town Planning Conference. London: Transactions; 10-15 October 1911. p. 345-367. Available from: http://urbanplanning.library.cornell.edu/DOCS/henard.htm Accessed: March 22, 2021

[12] Gottmann J. Megalopolis, or the urbanization of the North-eastern seaboard. Economic Geography. 1957;33:189-200

[13] Powell R, Roberts D, Dennison P, Hess L. Sub-pixel mapping of urban land cover using multiple end member spectral mixture analysis: Manaus, Brazil. Remote Sensing of Environment. 2008;106:253-267

[14] Triantakonstantis D. Urban growth prediction modelling using fractals and theory of chaos. Open Journal of Civil Engineering. 2012;2:81-86. DOI: 10.4236/ojce.2012.22013

[15] Esch T, Marconcini M, Felbier A, Roth A, Heldens W, Huber M, et al. Urban footprint processor; fully automated processing chain generating settlement masks from global data of the Tan DEM-X mission. IEEE Geoscience and Remote Sensing Letters. 2013;10:1617-1621. DOI: 10.1109/LGRS.2013.2272 953

[16] Li X, Liu X, Le Y. A systematic sensitivity analysis of constrained cellular automata model for urban growth simulation based on different transition rules. International Journal of Geographical Information Science. 2014;2:1317-1335. DOI: 10.1080/13658816.2014.883079

[17] Liu Y. Modelling sustainable urban growth in a rapidly urbanising region using a fuzzy-constrained cellular automata approach. International Journal of Geographical Information Science. 2011;**26**:151-167. DOI: 10.1080/13658816.2011.577434

[18] Sampath A, Shan J. Segmentation and reconstruction of polyhedral building roofs from aerial LIDAR point clouds. IEEE Transactions of Geoscience & Remote Sensing. 2010;**3**:1554-1567. DOI: 10.1109/TGRS.2009.2030180

[19] Park S, Jeon S, Choi C. Mapping urban growth probability in South Korea: Comparison of frequency ratio, analytic hierarchy process, and logistic regression models and use of the environmental conservation value assessment. Landscape and Ecological Engineering. 2012;**8**:17-31. DOI: 10.1007/s11355-010-0137-9

[20] Dong L, Shan J. A comprehensive review of earthquake-induced building damage detection with remote sensing techniques. ISPRS Journal of Photogrammetry and Remote Sensing. 2013;**84**:85-99. DOI: 10.1016/j.isprsjprs.2013.06.011

[21] Helly W. Urban Systems Models. London-NY: Academic Press; 1975. 196 p. DOI: 10.1016/C2013-0-10844-6

[22] Batty M, Hutchinson B. Systems Analysis in Urban Policy Making and Planning. Series II: System Science. New York: Plenum Press; 1983. p. 619

[23] Marshall J. The Structure of Urban Systems. Toronto: University of Toronto Press; 2019. p. 389. DOI: 10.3138/9781487577544-014

[24] Bretagnolle A, Daudé E, Pumain D. From theory to modelling: Urban systems as complex systems. Cybergeo: European Journal of Geography. 2006;**335**:1-26. DOI: 10.4000/cybergeo.2420

[25] Tsai Y-H. Quantifying urban form: Compactness versus "sprawl". Urban Studies. 2005;**42**:141-161. DOI: 10.1080/0042098042000309748

[26] Cabral P, Augusto G, Tewolde M, Araya Y. Entropy in urban systems. Entropy. 2013;**15**:5223-5236. DOI: 10.3390/e15125223

[27] Kostrikov S, Pudlo R, Kostrikova A, Bubnov D. Studying of urban features by the multifunctional approach to LiDAR data processing. In: New Methodologies for urban investigation through remote sensing. Vann, France: Proceedings of the Joint Urban Remote Sensing Event JURSE 2019, UBS; 2019. IEEE Xplore Digital Library, 2019. DOI: 10.1109/JURSE.2019.8809063

[28] Niemets K, Kostrikov S, Niemets L, Sehida K, Kravchenko K. The structural analysis of agglomerations as the ontological basis of territorial planning (a case study of Kharkiv region, Ukraine). In: Proceedings of the 35th International Business Information Management Association Conference (IBIMA); 1-2 April 2020; Seville. Madrid: IBIMA; 2020. pp. 4949-4954

[29] Tarsha-Kurdi F, Landes T, Grussenmeyer P, Koehl M. Model-driven and data-driven approaches using LIDAR data: Analysis and comparison. International Archives of Photogrammetry and Remote Sensing. 2007;**36**:87-92

[30] Wang C, Ji M, Wang J, Wen W, Li T, Sun Y. An improved DBSCAN method for LiDAR data segmentation with automatic eps estimation. Sensors. 2019;**19**:172-187. DOI: 10.3390/s19010172

[31] Kostrikov S, Bubnov D, Pudlo R. Urban environment 3D studies by automated feature extraction from LiDAR point clouds. Visnyk of V. N. Karazin Kharkiv National University, series "Geology. Geography. Ecology".

2020;**52**:156-182. DOI: 10.26565/
2410-7360-2020-52-12

[32] Wehr A. LiDAR systems and
calibration. In: Shan J, Toth K, editors.
Topographic laser ranging and scanning.
Principles and Processing. 2nd ed. Boca
Raton: CRC Press; 2018. pp. 218-272.
DOI: 10.1201/9781420051438-4

[33] Biljecki F, Stoter J, Ledoux H,
Zlatanova S, Coltekin A. Applications of
3D city models: State of the art review.
ISPRS International Journal of Geo-
Information. 2015;**4**:2842-2889.
DOI: 10.3390/ijgi4042842

[34] Rottensteiner F, Sohn G, Gerke M,
Wegner J, Breitkopf U, Jung J. Results of
the ISPRS benchmark on urban object
detection and 3D building reconstruction.
ISPRS Journal of Photogrammetry and
Remote Sensing. 2014;**93**:256-271.
DOI: 10.1016/j.isprsjprs.2013.10.004

[35] Awrangjeb M, Ravanbakhsh M,
Fraser C. Automatic detection of
residential buildings using LiDAR data
and multispectral imagery. ISPRS
Journal of Photogrammetry and Remote
Sensing. 2010;**65**:457-467.
DOI: 10.1016/j.isprsjprs.2010.06.001

[36] Xiao Y, Wang C, Li J, Zhang W,
Xi X, Wang C, et al. Building
segmentation and modeling from
airborne LiDAR data. International
Journal of Digital Earth. 2014;**8**:694-
709. DOI: 10.1080/17538947.2014.914252

[37] Liu K, Ma H, Ma H, Cai Z, Zhang L.
Building Extraction from Airborne
LiDAR Data Based on Min-Cut and
Improved Post-Processing. Remote
Sensing. 2020;**12**:2849. DOI: 10.3390/
rs12172849

[38] Kostrikov S, Pudlo R, Kostrikova A.
Three key EOS LiDAR Tool
functionalities for Urban Studies. In:
Remote Sensing Enabling Prosperity,
Proceedings of a meeting held 15-19
October 2018. Kuala Lumpur, Malaysia:

39th Asian Conference on Remote
Sensing (ACRS 2018), AARS – Curran
Associates, Inc.; 2018 3 p. 1676-1685

[39] Kostrikov S, Pudlo R, Bubnov D,
Vasiliev V. *ELiT*, multifunctional
web-software for feature extraction
from 3D LiDAR point clouds. ISPRS
International Journal of Geo-
Information. 2020;**9**(11):650-885.
DOI: 10.3390/ijgi9110650

[40] Haala N, Rothermel M. Dense
multistereo matching for high quality
digital elevation models. Journal of
Photogrammetry, Remote Sensing and
Geoinformation Processing. 2012;**4**:331-
343. DOI: 10.1127/1432-8364/2012/0121

[41] Cheng L, Zhao W, Han P, Zhang W,
Shan J, Liu Y, et al. Building region
derivation from LiDAR data using a
reversed iterative mathematic
morphological algorithm. Optics
Communications. 2013;**286**:244-250.
DOI: 10.1016/j.optcom.2012.08.028

[42] Gröger G, Plümer L. CityGML –
Interoperable semantic 3D city models.
ISPRS Journal of Photogrammetry and
Remote Sensing. 2012;**71**:12-33. DOI:
10.1016/j.isprsjprs.2012.04.004

[43] Kostrikov S, Pudlo R, Bubnov D,
Vasiliev V, Fedyay Y. Automated
Extraction of Heavyweight and
Lightweight Models of Urban Features
from LiDAR Point Clouds by Specialized
Web-Software. Advances in Science,
Technology and Engineering Systems
Journal. 2020;**5**(6):72-95.
DOI: 10.25046/aj050604

[44] Marcus L, Colding J. Toward an
integrated theory of spatial morphology
and resilient urban systems. Ecology
and Society. 2014;**19**(4):55-67.
DOI: 10.5751/ES-06939-190455

[45] Wu S-S, Wang L. Incorporating
GIS building data and census
housing statistics for sub-block-level
population estimation. The Professional

Geographer. 2008;**60**(1):121-135.
DOI: 10.1080/00330120701724251

[46] Lwin K, Murayama Y. A GIS
approach to estimation of building
population for micro-spatial analysis.
Transactions in GIS. 2009;**13**(4):401-
414. DOI: 10.1111/j.1467-9671.
2009.01171.x

[47] Dong P, Ramesh S, Nepali A.
Evaluation of small-area population
estimation using LiDAR, Landsat TM
and parcel data. International Journal of
Remote Sensing. 2010;**31**(21):5571-5586.
DOI: 10.1080/01431161.2010.496804

[48] Lu Z, Im J, Quackenbush L. A
volumetric approach to population
estimation using Lidar remote sensing.
Photogrammetric Engineering &
Remote Sensing. 2011;77(11):1145-1156.
DOI: 10.14358/PERS.77.11.1145

[49] MassGIS Data: Datalayers from the
2010 U.S. Census. Bureau of Geographic
Information: Commonwealth of
Massachusetts, 2012. Available from:
https://docs.digital.mass.gov/dataset/
massgis-data-datalayers-2010-us-census
[Accessed: July 5, 2020]

[50] MassGIS Data: Land Use (2005).
Bureau of Geographic Information:
Commonwealth of Massachusetts, 2009.
Available from: https://docs.digital.
mass.gov/dataset/massgis-data-land-
use-2005 [Accessed: July 19, 2020]

www.ingramcontent.com/pod-product-compliance
Lightning Source LLC
Chambersburg PA
CBHW062015210326
41458CB00075B/5529